Natural Computing Series

T0181415

Sanghamitra Bandyopadhyay · Sankar K. Pal

Classification and Learning Using Genetic Algorithms

Applications in Bioinformatics and Web Intelligence

With 87 Figures and 43 Tables

 Springer

Authors

Sanghamitra Bandyopadhyay
Indian Statistical Institute
Machine Intelligence Unit
203 Barrackpore Trunk Road
Kolkata 700108
India

sanghami@isical.ac.in

Sankar K. Pal
Indian Statistical Institute
Machine Intelligence Unit
203 Barrackpore Trunk Road
Kolkata 700108
India

sankar@isical.ac.in

Series Editors

G. Rozenberg (Managing Editor)
rozenber@liacs.nl

Th. Bäck, J.N. Kok, H.P. Spaink
Leiden Center for Natural Computing
Leiden University
Niels Bohrweg 1
2333 CA Leiden, The Netherlands

A.E. Eiben
Vrije Universiteit Amsterdam
The Netherlands

ACM Computing Classification: F.2.2, G.1.6, H.3, I.2.8, I.5, J.3

ISSN 1619-7127
ISBN 978-3-642-08054-8   e-ISBN 978-3-540-49607-6

Springer is a part of Springer Science+Business Media

springer.com

© Springer-Verlag Berlin Heidelberg 2007
Softcover reprint of the hardcover 1st edition 2007

Cover Design: KünkelLopka, Heidelberg

Utsav and Ujjwal, and my parents Satyendra Nath Banerjee and
Bandana Banerjee
Sanghamitra Bandyopadhyay

The workers of the Indian Statistical Institute
in its Platinum Jubilee Year
Sankar K. Pal

Preface

Genetic algorithms (GAs) are randomized search and optimization techniques guided by the principles of evolution and natural genetics; they have a large amount of implicit parallelism. GAs perform multimodal search in complex landscapes and provide near-optimal solutions for objective or fitness function of an optimization problem. They have applications in fields as diverse as pattern recognition, image processing, neural networks, machine learning, jobshop scheduling and VLSI design, to mention just a few.

Traditionally, GAs were designed to solve problems with an objective to optimize only a single criterion. However, many real-life problems involve multiple conflicting measures of performance, or objectives, which need simultaneous optimization. Optimal performance according to one objective, if such an optimum exists, often implies unacceptably low performance in one or more of the other objective dimensions, creating the need for a compromise to be reached. A suitable set of solutions, called the Pareto-optimal set, to such problems is one where none of the solutions can be further improved on any one objective without degrading it in another. In recent times, there has been a spurt of activities in exploiting the significantly powerful search capability of GAs for multiobjective optimization, leading to the development of several algorithms belonging to the class of multiobjective genetic algorithms (MOGAs).

Machine recognition of patterns can be viewed as a two-fold task, comprising learning the invariant properties of a set of samples characterizing a class, and deciding that a new sample is a possible member of the class by noting that it has properties common to those of the set of samples. Many tasks involved in the process of learning a pattern need appropriate parameter selection and efficient search in complex and large spaces in order to attain optimal solutions. This makes the process not only computationally intensive, but also leads to the possibility of losing the exact solution. Therefore, the application of GAs for solving certain problems of pattern recognition that require optimization of computation requirements, and robust, fast and close approximate solution, seems appropriate and natural. Additionally, the existence of the proof of convergence of GAs to the global optimal solution as the number of iterations goes to infinity, and the ability to estimate the ϵ-*optimal* stopping time, further strengthens the theoretical basis of its use in search problems. Finally, the development of new models, e.g., that of chromosome differentiation, and sophisticated genetic representation and operators, like variable-length real-coded representation with corresponding operators, makes the application of GAs more effective and efficient.

Over the past few decades, major advances in the field of molecular biology, coupled with advances in genomic technologies, have led to an explosive growth in the biological information generated by the scientific community. Bioinformatics, viewed as the use of computational methods to handle biological data, has evolved as a major research direction in response to this deluge of information. The main purpose is to utilize computerized databases to store, organize and index the data, and for specialized tools to view and analyze the data. It is an interdisciplinary field involving biology, computer science, mathematics and statistics to analyze biological sequence data, genome content and arrangement, and to predict the function and structure of macromolecules. The ultimate goal of the field is to enable the discovery of new biological insights as well as to create a global perspective from which unifying principles in biology can be derived. Data analysis tools used earlier in bioinformatics were mainly based on statistical techniques like regression and estimation. Recently GAs have been gaining the attention of researchers for solving certain bioinformatics problems with the need to handle large data sets in biology in a robust and computationally efficient manner. Moreover, GAs can process, in parallel, populations billions times larger than is usual for conventional computation. The usual expectation is that larger populations can sustain larger ranges of genetic variation and thus can generate high-fitness individuals in fewer generations. Laboratory operations on DNA inherently involve errors. These are more tolerable in executing GAs (where in fact they may be regarded as contributing to genetic diversity, a desirable property) than in executing deterministic algorithms. All these factors make the application of GAs in bioinformatics more natural and appropriate.

Mining the Web is another recently emerging research direction due to the immense popularity, usefulness and ubiquity of the Web. It may be considered to be a repository of a mind-boggling amount of heterogeneous data that needs to be searched and mined using minimum human intervention. Research in Web mining is at the crossroads of research in information retrieval, text mining, databases, artificial intelligence, machine learning and natural language processing in particular. In this regard, GAs are seen to be useful for prediction and description, efficient search, and adaptive and evolutionary optimization of complex objective functions in dynamic environments.

Several research articles integrating genetic learning with pattern recognition tasks, and their applications to bioinformatics and Web intelligence have come out in the recent past. These are available in different journals, conference proceedings and edited volumes. This scattered information causes inconvenience to readers, students and researchers.

The present volume is aimed at providing a treatise in a unified framework, with both theoretical and experimental results, describing the basic principles of GAs and MOGAs, and demonstrating the various ways in which genetic learning can be used for designing pattern recognition systems in both supervised and unsupervised modes. Their applications to bioinformatics and Web mining are also described. The task of classification, an integral part of pattern recognition, can be viewed as a problem of generating appropriate class boundaries that can successfully distinguish the various classes in the feature space. In real-life problems, the boundaries between different classes are usually complex and nonlinear. It is known that any nonlinear surface can be approximated by a number of simpler lower-order surfaces. Hence the problem of classification can be viewed as searching for a number of simpler surfaces (e.g., hyperplanes) that can appropriately model the class boundaries while providing the minimum number of misclassified data points. It is shown how GAs can be employed for approximating the class boundaries of a data set such that the recognition rate of the resulting classifier is sometimes comparable to, and often better than, several widely used classification methods including neural networks, for a wide variety of data sets. Theoretical analysis of the classifier is provided. The effectiveness of incorporating the concept of chromosome differentiation in GAs, keeping analogy with the sexual differentiation commonly observed in nature, in designing the classifier is studied. Since the classification problem can be naturally modelled as one of multiobjective optimization, the effectiveness of MOGAs is also demonstrated in this regard.

Clustering is another important and widely used exploratory data analysis tool, where the objective is to partition the data into groups such that some similarity/dissimilarity metric is optimized. Since the problem can be posed as one of optimization, the application of GAs to it has attracted the attention of researchers. Some such approaches of the application of GAs to clustering, both crisp and fuzzy, are described.

Different features of the aforesaid methodologies are demonstrated on several real-life problems, e.g., speech recognition, cancer detection and remotely sensed image analysis. Comparative studies, in certain cases, are also provided.

Several tasks in bioinformatics, e.g., sequence alignment, superfamily classification, structure prediction, molecule design, involve optimization of different criteria (like energy, alignment score and overlap strength), while requiring robust, fast and close approximate solutions. The utility of genetic learning, including the aforementioned classifiers, for solving some of these problems is described in a part of this book.

GAs have been applied for some tasks in Web mining involving search and optimization over complex fitness landscapes. They have been used in searching and retrieving documents from the Web, selecting features in text mining, query optimization and document representation. Such applications of GAs to the domain of Web intelligence and mining are also discussed in this treatise.

Chapter 1 provides an introduction to pattern recognition, different research issues, challenges, and applications, and various classical and modern methods, including the relevance of GA-based approaches. Chapter 2 describes the basic principles of GAs, their theoretical analysis, applications to learning systems, and multiobjective GAs. Chapters 3–5 deal with the problems of pattern classification using GAs where the number of surfaces used for approximating the class boundaries could either be fixed a priori, or may be evolved automatically. Theoretical analyses of the genetic classifiers are provided. Based on an analogy between the working principle of the genetic classifiers and the multilayer perceptron (MLP), an algorithm for determining the architecture and connection weights of an MLP is also described. Chapter 6 describes how the incorporation of the concept of chromosome differentiation in GAs can increase its effectiveness in learning the class boundaries for designing a classifier. In Chap. 7 the classification problem is posed as one of optimizing multiple objectives, and the application of MOGA for this task is described. Several criteria for comparing the performance of multiobjective optimization algorithms are discussed. Chapter 8 describes some GA-based approaches for clustering data into both crisp and fuzzy partitions, when the number of clusters may or may not be fixed a priori. Chapter 9 provides a detailed discussion of several bioinformatic tasks, and how GAs are often applied for solving the related problems. Finally, Chap. 10 deals with several issues in Web mining, and the relevance of GAs in this domain.

The volume, which is unique in its character, will be useful to graduate students and researchers in computer science, electrical engineering, system science, information technology and related areas, both as a text and a reference book for some parts of the curriculum. Researchers and practitioners in industry and R & D laboratories working in fields such as system design, control, pattern recognition, data mining, soft computing, bioinformatics and Web intelligence will also benefit. Most parts of the text are based on the published results of the authors. However, while explaining other related work to provide a unified framework, some references might have been inadvertently omitted.

The authors gratefully acknowledge the support provided by Mrs. Maya Dey, Ms. Sriparna Saha, Mr. Praveen Tripathi, Mr. Santanu Santra, Mr. Ramkrishna Mitra and Mr. Shubhra Sankar Ray in typing and editing a part of the manuscript, and Dr. Swati Choudhury in proof reading a few chapters. The initiative and support of the project by Mr. Alfred Hofmann and Mr. Ronan Nugent of Springer is also acknowledged. Sanghamitra Bandyopadhyay specially acknowledges Prof. S. K. Pal whose inspiration gave shape to this book.

Kolkata, India   *Sanghamitra Bandyopadhyay*
June 29, 2006   *Sankar K. Pal*

Contents

1

Introduction

1.1 Introduction

Human beings are very adept in recognizing patterns, a task that they continually execute. In contrast, getting a machine to learn and recognize patterns can be extremely difficult and challenging. Machine recognition of patterns can be viewed as a twofold task – learning the invariant and common properties of a set of samples characterizing a class, and deciding that a new sample is a possible member of the class by noting that it has properties common to those of the set of samples. It is generally conjectured that animal brains process the sensory data obtained from different senses, apply rules that are embedded in the brain (which are usually thought to be learnt starting from birth) and recognize patterns. Pattern recognition deals with the theories and methodologies involved in making a machine recognize patterns which are characterized by numerical and nonnumerical attributes. The major hurdle in this task is that the functioning of the brain is much less understood. The mechanisms with which it stores huge amounts of information, processes them at lightning speeds and infers meaningful rules, and retrieves information as and when necessary have till now eluded the scientists. Attempts have therefore been made to develop algorithms and methodologies that are motivated by theoretical considerations and speculations about the working of the brain.

Objects in the real world are captured using some measurements on them. For example, a camera captures the scene in front of it as an array of pixels. Each pixel value may range from 0 to 255 if it is a gray-scale image. For color images, each pixel value will be a vector of, say, three values, indicating the intensities in the red, blue and green bands. Figure 1.1 shows an example of how a scene is converted to image data that can be stored in a computer for further processing. Similarly, a flower may be characterized by certain measurements taken on it. For example, the famous Iris flower data, described in Appendix B, measures the petal length, petal width, sepal length and sepal width of Iris flowers which belong to three classes — Setosa, Versicolor and Virginica. Thus a flower may be represented using a vector in $I\!R^4$ of the above

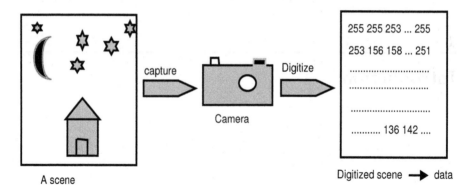

Fig. 1.1. Example of object representation

four measurement values, and it may be labelled by one of the three classes. In other words, a flower may be mapped to a point in a four-dimensional space, and this point may have an associated class label. The task of any pattern recognition system essentially involves taking some measurements of the objects to be recognized, thereby mapping them into points in some space. Thereafter, if the system has some labelled points then it has to infer some rules about the "belongingness" of points to the classes. Given an unknown point, the system will then have to decide about its classification, based on the rules that it has already learnt. On the other hand, if the system is not provided any labelled information, then its task may be to find some relationship among the unlabelled points that it has collected.

Machine learning is an area of artificial intelligence that deals with the task of making machines capable of learning from their environment. Evolution is probably the most powerful learning mechanism in existence. Over millions of years, different species have evolved in such a way that they have become more suited to their environment. In biological evolution, changes occur at the genetic level through crossover and mutation. These changes are propagated to future generations, where the process gets repeated. Usually, these changes are such that they provide some survival advantage to an individual with respect to the local environments. Changes that are disadvantageous tend to decrease in frequency, while those that are beneficial are more likely to be preserved because they aid survival.

Inspired by the powerful evolutionary learning mechanism, scientists attempted to design computer algorithms that mimic these processes. The result was a class of computer algorithms called evolutionary computation (EC). EC is a computing paradigm comprising population-based metaheuristic optimization techniques that are based on the principles of biological evolution. The essential components of EC are a strategy for representing or encoding a solution to the problem under consideration, a criterion for evaluating the fitness or goodness of an encoded solution and a set of biologically inspired operators applied on the encoded solutions. Genetic algorithms (GAs) are the most well-known and widely used techniques in this computing paradigm. Other commonly known techniques are evolutionary strategies and genetic programming. Because of the robustness and effectiveness of GAs, they find widespread applications in various engineering and scientific communities like pattern recognition, image processing, bioinformatics, data mining, Web mining, mechanical engineering, telecommunications, mobile computing, VLSI design, and embedded and real-time systems.

In particular, many tasks involved in the process of recognizing a pattern need appropriate parameter selection and efficient search in complex and large spaces in order to attain optimal solutions. This makes the process not only computationally intensive, but also leads to a possibility of losing the exact solution. Therefore, the application of GAs for solving certain problems of pattern recognition that require optimization of computation requirements, and robust, fast and close approximate solution, seems appropriate and natural. Additionally, the existence of the proof that GAs will definitely provide the global optimal solution as the number of iterations goes to infinity [60] further strengthens the theoretical basis of its use in search problems.

The basic concepts related to machine recognition of patterns is first discussed in Sect. 1.2. The different approaches to pattern recognition are mentioned in Sect. 1.3. These are followed by brief descriptions of the connectionist approach in Sect. 1.4, genetic approach in Sect. 1.5, fuzzy set-theoretic approach in Sect. 1.6 and other hybrid approaches in Sect. 1.7. Section 1.8 mentions some real-life applications of pattern recognition and learning. Finally, a summary of the chapter and the scope of the book are mentioned in Sect. 1.9.

1.2 Machine Recognition of Patterns: Preliminaries

Pattern recognition and machine learning [131, 152, 313, 384, 458] form a major area of research and development activity that encompasses the processing of pictorial and other nonnumerical information obtained from the interaction between science, technology and society. A motivation for the spurt of activity in this field is the need for people to communicate with computing machines in their natural mode of communication. Another important motivation is that the scientists are also concerned with the idea of designing and making intelligent machines that can carry out certain tasks that we human beings do. The most salient outcome of these is the concept of future-generation computing systems.

Machine recognition of patterns, as mentioned before, can be viewed as a twofold task, comprising learning the invariant properties of a set of samples characterizing a class, and of deciding that a new sample is a possible member of the class by noting that it has properties common to those of the set of samples. A typical pattern recognition system consists of three phases (Fig. 1.2). These are *data acquisition, feature extraction* and *classification*. In the data acquisition phase, depending on the environment within which the objects are to be classified, data are gathered using a set of sensors. These are then passed on to the feature extraction phase, where the dimensionality of the data is reduced by measuring and retaining only some characteristic features or properties. In a broader perspective, this stage significantly influences the entire recognition process. Finally, in the classification phase, the extracted features are passed on to the classifier that evaluates the incoming information and makes a final decision. This phase basically establishes a transformation between the features and the classes. Therefore pattern recognition can be described as a transformation from the measurement space M to the feature space F and finally to the decision space D, i.e.,

$$M \to F \to D.$$

Here the mapping $\delta : F \to D$ is the decision function, and the elements $d \in D$ are termed as decisions.

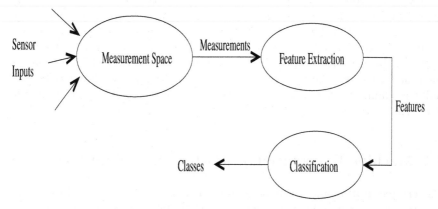

Fig. 1.2. A typical pattern recognition system

In the following sections we describe, in brief, some tasks commonly undertaken in typical pattern recognition systems. These are data acquisition, feature selection, and some classification and clustering techniques.

1.2.1 Data Acquisition

Pattern recognition techniques are applicable in a wide domain, where the data may be qualitative, quantitative or both; they may be numerical, linguistic, pictorial or any combination thereof. Generally, the data structures that are used in pattern recognition systems are of two types: *object data vectors* and *relational data*. Object data, sets of numerical vectors of Q features, are represented in what follows as $\mathcal{Y} = \{Y_1, Y_2, \ldots, Y_t\}$, a set of t feature vectors in the Q-dimensional measurement space Ω_Y. The ith object, $i = 1, 2, \ldots, t$, observed in the process has vector Y_i as its numerical representation; y_{ij} is the jth $(j = 1, 2, \ldots, Q)$ feature associated with the object i.

Relational data is a set of t^2 numerical relationships, say $\{r_{ii'}\}$, between pairs of objects. In other words, $r_{ii'}$ represents the extent to which object i and i' are related in the sense of some binary relationship ρ. If the objects that are pairwise related by ρ are called $O = \{o_1, o_2, \ldots, o_t\}$, then $\rho : O \times O \rightarrow \mathbb{R}$.

Sometimes, the data acquisition phase includes some preprocessing tasks like noise reduction, filtering, encoding and enhancement for extracting pattern vectors. For example, if the input pattern is an image, these preprocessing operations play a crucial role for extracting salient features for its recognition.

1.2.2 Feature Selection

Feature selection is one of the major tasks in any automatic pattern recognition system. The main objective of *feature selection* [47, 127] is to retain the optimum salient characteristics necessary for the recognition process and to reduce the dimensionality of the measurement space Ω_Y so that effective and easily computable algorithms can be devised for efficient classification. The problem of feature selection has two aspects: the formulation of a suitable criterion to evaluate the goodness of a feature and the selection of optimal subsets from the available features.

The major mathematical measures so far devised for the estimation of feature quality are mostly statistical in nature, and can be broadly classified into two categories:

- feature selection in the measurement space, and
- feature extraction in the transformed space.

The techniques in the first category generally reduce the dimensionality of the feature set by discarding redundant information carrying features. On the other hand, those in the second category utilize all the information contained in the pattern vectors, and map a higher-dimensional pattern vector to a lower-dimensional one.

Feature selection is the process of selecting a map of the form $X = f(Y)$, by which a sample $Y (=y_1, y_2, \ldots, y_Q)$ in a Q-dimensional measurement space Ω_Y is transformed into a point $X (=x_1, x_2, \ldots, x_N)$ in an N-dimensional feature space Ω_X, where $N < Q$.

The pioneering research on feature selection mostly deals with statistical tools. Later, the thrust of the research shifted to the development of various other approaches to feature selection, including fuzzy, neural and genetic approaches [133, 342, 351, 401, 434, 465].

1.2.3 Classification

The problem of classification basically establishes a transformation $F \rightarrow D$ between the features and the classes (Fig. 1.2). In other words, it provides a partitioning of the feature space into regions, one region for each category of input. That is, it attempts to assign every data point in the entire feature space to one of the possible (say K) classes. Different forms of transformation can be a Bayesian rule of computing a posteriori class probability, nearest neighbor rule, linear discriminant functions, perceptron rule, nearest proto-type rule, etc. [13, 127, 130, 153, 155, 161, 172, 364, 461]. Classifiers are usually, but not always, designed with labelled data, in which case these problems are sometimes referred to as *supervised classification* (where the parameters of a classifier function D are learned). Some common examples of the *supervised* pattern classification techniques are the nearest neighbor rule, Bayes' maximum likelihood classifier and the perceptron rule. Some of the commonly used classification (supervised) methods, which have been used in different chapters of this book for the purpose of comparison, are now described here.

Let $C_1, C_2, \ldots, C_i, \ldots, C_K$ be the K possible classes in an N-dimensional feature space. Let $\mathbf{x} = [x_1, x_2, \ldots, x_j, \ldots, x_N]'$ be an unknown pattern vector. In a deterministic classification approach, it is assumed that there exists only one unambiguous pattern class corresponding to each of the unknown pattern vectors. If the pattern $\mathbf{x}$ is a member of class C_i, the *discriminant (decision) function* $D_i(\mathbf{x})$ associated with the class C_i, $i = 1, 2, \ldots, K$, must then possess the largest value. In other words, a classificatory decision would be as follows:

$$\text{Decide} \quad \mathbf{x} \in C_i, \qquad \text{if } D_i(\mathbf{x}) > D_l(\mathbf{x}), \tag{1.1}$$

$(\forall \ i, l)$ in $(1, \ldots, K)$ and $i \neq l$. Ties are resolved arbitrarily. The decision boundary in the feature space between regions associated with the classes C_i and C_l would be governed by the expression

$$D_i(\mathbf{x}) - D_l(\mathbf{x}) = 0. \tag{1.2}$$

Many different forms satisfying Eq. (1.2) can be selected for $D_i(\mathbf{x})$. The functions that are often used are linear discriminant functions, quadratic discriminant functions and polynomial discriminant functions. One commonly used piecewise linear discriminant function, which is used in the nearest neighbor (NN) classifier, involves distance as a similarity measure for classification. This is described below.

NN rule

Let us consider a set of n patterns of known classification $\{\mathbf{x}_1, \mathbf{x}_2, \ldots, \mathbf{x}_n\}$, where it is assumed that each pattern belongs to one of the classes $C_1, C_2, \ldots, C_i, \ldots, C_K$. The NN classification rule [130, 155, 461] then assigns a pattern $\mathbf{x}$ of unknown classification to the class of its nearest neighbor, where $\mathbf{x}_i \in \{\mathbf{x}_1, \mathbf{x}_2, \ldots, \mathbf{x}_n\}$ is defined to be the nearest neighbor of $\mathbf{x}$ if

$$D(\mathbf{x}_i, \mathbf{x}) = \min_l\{D(\mathbf{x}_l, \mathbf{x})\}, \qquad l = 1, 2, \ldots, n. \tag{1.3}$$

Here D is any distance measure definable over the pattern space.

Since the aforesaid scheme employs the class label of only the nearest neighbor to $\mathbf{x}$, this is known as the 1-NN rule. If k neighbors are considered for classification, then the scheme is termed as the k-NN rule. The k-NN rule assigns a pattern $\mathbf{x}$ of unknown classification to class C_i if the majority of the k nearest neighbors belongs to class C_i. The details on the k-NN rule along with the probability of error are available in [130, 155, 461].

In most of the practical problems, the features are usually noisy and the classes in the feature space are overlapping. In order to model such systems, the feature values $x_1, x_2, \ldots, x_j, \ldots, x_N$ are considered as random values in the probabilistic approach. The most commonly used classifier in such probabilistic systems is the *Bayes' maximum likelihood classifier* [9, 461], which is now described.

Bayes' Maximum Likelihood Classifier

Let P_i denote the a priori probability and $p_i(\mathbf{x})$ denote the class conditional density corresponding to the class C_i $(i = 1, 2, \ldots, K)$. If the classifier decides $\mathbf{x}$ to be from the class C_i, when it actually comes from C_l, it incurs a loss equal to L_{li}. The expected loss (also called the conditional average loss or risk) incurred in assigning an observation $\mathbf{x}$ to the class C_i is given by

$$r_i(\mathbf{x}) = \sum_{l=1}^{K} L_{li} \, p(C_l/\mathbf{x}), \tag{1.4}$$

where $p(C_l/\mathbf{x})$ represents the probability that $\mathbf{x}$ is from C_l. Using Bayes' formula, Eq. (1.4) can be written as

$$r_i(\mathbf{x}) = \frac{1}{p(\mathbf{x})} \sum_{l=1}^{K} L_{li} \, p_l(\mathbf{x})P_l, \tag{1.5}$$

where

$$p(\mathbf{x}) = \sum_{l=1}^{K} p_l(\mathbf{x})P_l.$$

The pattern $\mathbf{x}$ is assigned to the class with the smallest expected loss. The classifier which minimizes the total expected loss is called the *Bayes' classifier*.

Let us assume that the loss (L_{li}) is zero for a correct decision and greater than zero but the same for all erroneous decisions. In such situations, the expected loss, Eq. (1.5), becomes

$$r_i(\mathbf{x}) = 1 - \frac{P_i p_i(\mathbf{x})}{p(\mathbf{x})}. \tag{1.6}$$

Since $p(\mathbf{x})$ is not dependent upon the class, the Bayes' decision rule is nothing but the implementation of the decision functions

$$D_i(\mathbf{x}) = P_i p_i(\mathbf{x}), \qquad i = 1, 2, \ldots, K, \tag{1.7}$$

where a pattern $\mathbf{x}$ is assigned to class C_i if $D_i(\mathbf{x}) > D_l(\mathbf{x})$, $\forall l \neq i$. This decision rule provides the minimum probability of error. It is to be noted that if the a priori probabilities and the class conditional densities are estimated from a given data set, and the Bayes' decision rule is implemented using these estimated values (which may be different from the actual values), then the resulting classifier is called the *Bayes' maximum likelihood classifier*.

Assuming normal (Gaussian) distribution of patterns, with mean vector μ_i and covariance matrix $\sum_i$, the Gaussian density $p_i(\mathbf{x})$ may be written as

$$p_i(\mathbf{x}) = \frac{1}{(2\pi)^{\frac{N}{2}} |\Sigma_i|^{\frac{1}{2}}} \exp\left[-\frac{1}{2} (\mathbf{x} - \mu_i)' \Sigma_i^{-1} (\mathbf{x} - \mu_i) \right], \tag{1.8}$$
$$i = 1, 2, \ldots, K.$$

Then, $D_i(\mathbf{x})$ becomes (taking the natural logarithm)

$$D_i(\mathbf{x}) = \ln P_i - \tfrac{1}{2} \ln |\Sigma_i| - \frac{1}{2} (\mathbf{x} - \mu_i)' \Sigma_i^{-1} (\mathbf{x} - \mu_i), \tag{1.9}$$
$$i = 1, 2, \ldots, K.$$

Note that the decision functions in Eq. (1.9) are hyperquadrics, since no terms higher than the second degree in the components of $\mathbf{x}$ appear in it. It can thus be stated that the Bayes' maximum likelihood classifier for normal distribution of patterns provides a second-order decision surface between each pair of pattern classes. An important point to be mentioned here is that if the pattern classes are truly characterized by normal densities, on an average, no other surface can yield better results. In fact, the Bayes' classifier designed over known probability distribution functions provides, on an average, the best performance for data sets which are drawn according to the distribution. In such cases, no other classifier can provide better performance on an average, because the Bayes' classifier gives the minimum probability of misclassification over all decision rules.

1.2.4 Clustering

As already mentioned, when only unlabelled data are available, then the classification methodology adopted is known as *unsupervised classification* or *clustering*. Many clustering algorithms are used as precursors to the design of a classifier in such situations. In clustering [8, 127, 187, 212, 461] a set of patterns, usually vectors in a multidimensional space, are grouped into clusters in such a way that patterns in the same cluster are similar in some sense and patterns in different clusters are dissimilar in the same sense. Clustering in N-dimensional Euclidean space $I\!R^N$ is the process of partitioning a given set of n points into a number, say K, of groups (or clusters) based on some similarity/dissimilarity metric. Let the set of n points $\{\mathbf{x}_1, \mathbf{x}_2, \dots, \mathbf{x}_n\}$ be represented by the set S, and the K clusters be represented by $C_1, C_2, \dots, C_K$. Then

$$
\begin{aligned}
&C_i \neq \emptyset, \qquad for\ i = 1, \dots, K, \\
&C_i \cap C_j = \emptyset,\ for\ i = 1, \dots, K,\quad j = 1, \dots, K,\ \text{and}\ i \neq j,\ \text{and} \\
&\bigcup_{i=1}^{K} C_i = S.
\end{aligned}
$$

For this it is necessary to first define a measure of similarity which will establish a rule for assigning patterns to the domain of a particular cluster center. One such measure of similarity may be the Euclidean distance $\mathbf{D}$ between two patterns $\mathbf{x}$ and $\mathbf{z}$ defined by $\mathbf{D} = \|\mathbf{x} - \mathbf{z}\|$. The smaller the distance between $\mathbf{x}$ and $\mathbf{z}$, the greater is the similarity between the two, and vice versa.

Clustering techniques have been broadly categorized into partitional and hierarchical methods. One commonly used clustering technique in the partitional category is the K-means algorithm [461], where the number of clusters K is assumed to be known a priori. The K-means algorithm has been widely recommended after extensive studies dealing with comparative analysis of different clustering methods [129, 155, 307]. Popular techniques in the hierarchical category are the single linkage, complete linkage and average linkage algorithms. A distinguishing feature of the hierarchical clustering techniques from the partitional ones is that while the former provide a valid clustering of the data at each iteration of the algorithm, the latter do not do so (they provide a valid clustering only on termination of the algorithm). Minimal spanning tree-based graph-theoretic clustering techniques are also quite popular in the pattern recognition community. A description of some existing techniques like the K-means and single linkage algorithms are described in Chap. 8 of this book.

1.3 Different Approaches

Pattern recognition, by its nature, admits many approaches, sometimes complementary, sometimes competing, to the approximate solution of a given problem. These include the decision-theoretic approach (both deterministic and probabilistic), syntactic approach, connectionist approach, genetic approach and fuzzy set-theoretic approach.

In the decision-theoretic approach [127, 131, 155, 488], once a pattern is transformed, through feature selection, to a vector in the feature space, its characteristics are expressed only by a set of numerical values. Classification can be done using deterministic or probabilistic techniques. Classification techniques described in Sect. 1.2.3 fall under the decision-theoretic approach.

On the other hand, when a pattern is rich in structural information (e.g., picture recognition, character recognition, scene analysis), i.e., the structural information plays an important role in describing and recognizing the patterns, syntactic approaches, which deal with representation of structures via sentences, grammars and automata, are usually adopted. In the syntactic method [153, 461], the ability to select and classify the simple pattern primitives and their relationships represented by the composition operations is a vital criterion for making a system effective. Since the techniques of composition of primitives into patterns are usually governed by the formal language theory, the approach is often referred to as the linguistic approach. An introduction to a variety of approaches based on this idea can be found in [153].

For any pattern recognition system, one desires to achieve robustness with respect to random noise and failure of components. Another important requirement is to have output in real time. It is also desirable for the system to be adaptive to changes in the environment. Moreover, a system can be made artificially intelligent if it is able to emulate some aspects of the human processing system. Connectionist (neural network-based) approaches [250, 406] to pattern recognition [64, 420] are attempts to achieve these goals. An artificial neural network (ANN) is a system composed of several simple processing elements that operate in parallel and that are connected by some interconnection structure. The behavior of a neural network is determined by its structure, connection weights and the function that is implemented at each node. Some ANN models are closely related to statistical pattern recognition [213, 223], with each approach benefiting from the theories/applications developed in the other. For example, many classical pattern recognition algorithms like principal component analysis, K-means clustering, etc., can be mapped to ANNs for faster hardware implementation. Similarly, ANNs derive benefit from well-known results in statistical pattern recognition, such as Bayes' decision theory, and nearest neighbor rules.

A pattern recognition problem can be mapped to one of search and optimization of certain criteria (e.g., search for a suitable decision boundary that provides the maximum abstraction of the data, search for a suitable subset of features that provides the most compact representation of the data). Often the search spaces in such problems are extremely huge, complex and multimodal. Hence genetic approaches to pattern recognition have been attempted by several researchers for constructing more accurate and efficient systems.

Methods and technologies developed under the above-mentioned categories may, again, be fuzzy set-theoretic in order to handle uncertainties arising from vague, incomplete, linguistic, overlapping patterns, at various stages of pattern recognition systems. This approach is developed based on the realization that a pattern may belong to more than one class, with varying degrees of class membership. Accordingly, fuzzy decision-theoretic, fuzzy syntactic and fuzzy neural approaches are developed [57, 59, 224, 225, 351, 368].

Some of the classification methods under the aforesaid approaches are briefly described, in the following sections, along with their relevance and characteristic features.

1.4 Connectionist Approach: Relevance and Features

Neural networks can be formally defined as *massively parallel interconnections of simple (usually adaptive) processing elements that interact with objects of the real world in a manner similar to biological systems.* Their origin can be traced to the work of Hebb [189], where a local learning rule is proposed. The benefit of neural nets lies in the high computation rate provided by their inherent massive parallelism. This allows real-time processing of huge data sets with proper hardware backing. All information is stored distributed among the various connection weights. The redundancy of interconnections produces a high degree of robustness, resulting in a *graceful degradation* of performance in the case of noise or damage to a few nodes or links. The characterizing features of ANNs are

- simple processing units
- high degree of interconnection
- simple scalar messages
- adaptive interaction within elements

In [353] the relevance of neural networks is summarized as follows: Neural networks are natural classifiers having resistance to noise, tolerance to distorted images or patterns (or the ability to generalize), superior ability to recognize partially occluded or degraded images or overlapping pattern classes or classes with highly nonlinear boundaries, and potential for parallel processing. Neural network models have been studied for many years with the hope of achieving human-like performance (artificially), particularly in the field of pattern recognition, by capturing the key ingredients responsible for the remarkable capabilities of the human nervous system. Note that these models are extreme simplifications of the actual human nervous system.

The perceptron [115, 188, 357, 406] is the earliest development in this direction. It may be broadly categorized into two classes: *single-layer perceptrons* and *multilayer perceptrons*. The single-layer perceptron, given two classes of patterns, attempts to find a linear decision boundary between them. If the classes are linearly separable, the perceptron algorithm is guaranteed to find a separating hyperplane in a finite number of steps. However, as shown by Minsky and Papert [312], if the pattern space is not linearly separable, the perceptron algorithm will not converge. In fact, it is shown to fail for the simple XOR problem, which is linearly inseparable.

This limitation was overcome by Rumelhart, Hinton and Williams [405], who developed the generalized delta rule for training of multilayer perceptrons

(MLP). The multilayer perceptron, known for discriminating nonlinearly separable classes as well, has since then been widely studied and applied in various
fields [279]. Below we briefly explain the working principle of an MLP.

Multilayer Perceptron (MLP)

An MLP [115, 188, 357] consists of several layers of simple neurons with full
connectivity existing between neurons of adjacent layers. Figure 1.3 shows an
example of a three-layer MLP which consists of an input layer (*layer 1*), one
hidden layer (*layer 2*) and an output layer (*layer 3*).

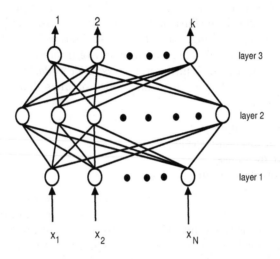

Fig. 1.3. Multilayer perceptron

The neurons in the input layer serve the purpose of fanning out the input
values to the neurons of *layer 2*. Let

$$w_{ji}^{(l)}, \qquad l = 2,\ 3 \tag{1.10}$$

represent the connection weight on the link from the ith neuron in layer $l-1$
to the jth neuron in layer l. Let $\theta_j^{(l)}$ represent the threshold of the jth neuron
in layer l. The total input, $x_j^{(l)}$, received by the jth neuron in layer l is given
by

$$x_j^{(l)} = \sum_i y_i^{(l-1)}\, w_{ji}^{(l)} + \theta_j^{(l)}, \tag{1.11}$$

where $y_i^{(l-1)}$ is the output of the ith neuron in layer $l-1$. For the input layer

$$y_i^{(1)} = x_i, \tag{1.12}$$

where x_i is the ith component of the input vector. For the other layers

$$y_i^{(l)} = f(x_i^{(l)}) \qquad l = 2, \ 3. \tag{1.13}$$

Several functional forms like threshold logic, hard limiter and sigmoid can be used for $f(.)$.

There are several algorithms for training the network in order to learn the connection weights and the thresholds from a given training data set. Back-propagation (BP) is one such learning algorithm, where the least mean square error of the network output is computed, and this is propagated in a top-down manner (i.e., from the output side) in order to update the weights. The error is computed as the difference between the actual and the desired output when a known input pattern is presented to the network. A gradient descent method along the error surface is used in BP. More information on MLP and other neural methods are available in [115, 188, 279, 357].

1.5 Genetic Approach: Relevance and Features

Genetic algorithms (GAs) [19, 164, 195, 475], first proposed by Holland, are a class of computational models that mimic the principles of natural evolution for performing search and optimization. They use two important features from natural evolution: handing down of information from one generation to another, or *inheritance*, and competition for survival, or *survival of the fittest*. The main advantages of GAs that make them suitable for solving real-life problems are

- They are adaptive.
- They possess inherent parallelism.
- They are efficient for solving complex problems where the main focus is on obtaining good, not necessarily the best, solutions quickly.
- They are easy to parallelize, without much communication overhead.

GAs are particularly suited for applications involving search and optimization, where the space is huge, complex and multimodal, and the need for finding the exact optimal solution is not all important. Several tasks in the domain of pattern recognition meet these requirements. For example, consider the problem of clustering n observations into m groups such that some clustering measure is optimized to the largest possible extent. The number of possibilities is $\frac{1}{m!} \sum_{k=0}^{k=m} (-1)^{m-k} \binom{m}{k} k^n$. This number becomes immensely large even for small problems. It is expected that application of an effective search strategy like GAs is likely to provide significantly superior performance compared to schemes that are ad hoc or mostly local in nature.

Moreover, in many pattern recognition problems, one is on the lookout for an acceptable good solution rather than the exact optimal one. Therefore, using genetic algorithms for pattern recognition problems that need optimization of computation requirements, and robust, fast, and close approximate solutions appear to be natural and appropriate. The following characteristics of GAs make them adequate for use in pattern recognition problems:

(a) Because of implicit parallelism [164], GAs can search huge spaces in less time.
(b) Being stochastic in nature they have the ability to come out of local optima.
(c) They make no assumptions (e.g., differentiability requirement) about the search space, except that it is bounded.

Note that, unlike ANNs, which were developed keeping pattern classification tasks in mind, GAs were developed for the purpose of optimizing a given criterion. However, another way to look at classification is as a problem of searching for the best, or optimum, distinguishing boundary between the different classes. Use of GAs for performing this search will lead to the development of a nonparametric classifier that is applicable to a wide variety of data sets.

In addition to pattern classification and feature selection [160, 356, 434], GAs find widespread applications in solving different problems in image processing and scene recognition [14, 193, 347, 424], rule generation and classifier systems [65, 207, 208, 209, 210, 216], neural network design [66, 185, 290, 316, 321, 346, 407, 417, 491], scheduling problems [21, 92, 249, 277, 278, 390], VLSI design [85, 108, 302, 480], path planning [91, 109], the travelling salesman problem [176, 196, 218, 306, 333], graph coloring [113], and numerical optimization [309].

1.6 Fuzzy Set-Theoretic Approach: Relevance and Features

Uncertainty can arise either implicitly or explicitly in each and every phase of a pattern recognition system. It results from the incomplete, imprecise or ambiguous input information, the ill-defined and/or overlapping boundaries among the classes or regions, and the indefiniteness in defining/extracting features and relations among them. Any decision taken at a particular level will have an impact on all other higher-level activities. It is therefore required for a pattern recognition system to have sufficient provision for representing these uncertainties involved at every stage, so that the ultimate output of the system can be obtained with the least uncertainty.

Fuzzy sets were introduced in 1965 by Zadeh [509] as a way to represent vagueness in everyday life. Since this theory is a generalization of the classical set theory, it has greater flexibility to capture various aspects of incompleteness, impreciseness or imperfection in information about a situation.

The relevance of fuzzy set theory in the realm of pattern recognition [57, 59, 224, 225, 351, 352, 367, 368] is adequately justified in

- representing input patterns as an array of membership values denoting the degree of possession of certain properties
- representing linguistically phrased input features for processing
- providing an estimate (representation) of missing information in terms of membership values
- representing multiclass membership of ambiguous patterns and in generating rules and inferences in linguistic form
- extracting ill-defined image regions, primitives, and properties and describing relations among them as fuzzy subsets

1.7 Other Approaches

In the previous sections we have described, in addition to some conventional approaches to pattern recognition, three modern approaches based on tools like neural networks, genetic algorithms and fuzzy logic. These three tools are essential components of the soft computing paradigm. Soft computing is a relatively recent term that is used to denote a consortium of methodologies that works synergistically and provides in one form or another flexible information processing capability for handling real-life ambiguous situations. Its aim is to exploit the tolerance for imprecision, uncertainty, approximate reasoning and partial truth in order to achieve tractability, robustness and low-cost solutions. The guiding principle is to devise methods of computation that lead to an acceptable solution at low cost by seeking for an approximate solution to an intractable problem.

There have been several attempts over the last decade to derive hybrid methods by combining the merits of several existing techniques. Moreover, other pattern recognition paradigms like support vector machines, case-based reasoning and rough sets are also being extensively studied and developed. Let us now discuss some such attempts in this section.

An integration of neural network and fuzzy theory, commonly known as neuro-fuzzy computing, is one such hybrid paradigm [215, 353] that enables one to build more intelligent decision-making systems. In the context of pattern recognition, it incorporates the capability of neural networks in generating required decision boundaries, which may be linearly nonseparable, along with the capability of fuzzy logic for handling uncertain, imprecise, linguistic and incomplete data.

Application of rule-based systems in pattern recognition has also gained popularity in the recent past. By modelling the rules and facts in terms of fuzzy sets, it is possible to make interfaces using the concept of approximate reasoning [45, 128]. An approach to classifying unknown patterns by combining the k-NN rule with the Dempster–Shafer theory of evidence [427] has been formulated in [125]. In this regard, GAs have also been used, under the genetic-fuzzy integration, to evolve the fuzzy membership functions as well as the rule base [192, 197, 207, 208, 370, 508]. Tettamanzi [457] describes a way in which the features of fuzzy sets can be used for improving the performance of genetic algorithms. A new method of fuzzy coding making use of linguistic terms modelled as a collection of fuzzy sets is proposed in [372]. This fuzzy coding establishes a relevant level of information granularity and provides some specific search orientation (guidance). However, calculation of the fitness function now needs to be performed prudently. Under genetic-neural integration, there have been several attempts at using GAs to train a given network, determine the network topology, or perform both determining and training an appropriate network for a given problem. Branke [71] provides an overview of some such attempts.

Integration of neural networks, GAs and fuzzy logic have also been studied in some systems [305, 346, 389, 503]. Some other approaches for pattern recognition are based on case-based reasoning [349], support vector machines [105], rough sets [365] and their hybridizations.

1.8 Applications of Pattern Recognition and Learning

Pattern recognition research is mostly driven by the need to process data and information obtained from the interaction between human, society, science and technology. As already mentioned, in order to make machines more intelligent and human-like, they must possess automatic pattern recognition and learning capabilities. Some of the typical application areas of such intelligent machines are:

- *medicine:* medical diagnosis, image analysis, disease classification
- *computation biology:* gene identification, protein modelling, rational drug design
- *natural resource study and estimation:* agriculture, forestry, geology, the environment
- *human-machine communication:* automatic speech recognition and understanding, natural language processing, image processing, script recognition, signature analysis
- *biometry:* face recognition, gesture recognition
- *vehicular:* automobile, airplane, train and boat controllers
- *defense:* automatic target recognition, guidance and control
- *police and detective:* crime and criminal detection from analysis of speech, handwriting, fingerprints, photographs
- *remote sensing:* detection of manmade objects and estimation of natural resources
- *industry:* CAD, CAM, product testing and assembly, inspection, quality control
- *domestic systems:* appliances
- *space science:* Mars rover control, camera tracking
- *investment analysis:* predicting the movement of stocks, currencies, etc., from historical data

1.9 Summary and Scope of the Book

Pattern recognition and learning are two important issues that need to be addressed for making machines that can emulate some human behavior. In this chapter, an introduction to pattern recognition has been provided. The different tasks are described briefly. Various approaches to pattern recognition are mentioned, along with their relevance.

In the past, several computational algorithms have been developed that are based on the principles of biological systems. Effectiveness of integrating these computing techniques with classical pattern recognition and learning tasks has time and again been established by various researchers. GAs, an efficient search and optimization tool based on the mechanism of natural genetics and evolution, are such an important and relatively recent computing paradigm. Remaining chapters of this book describe different methodologies, theories, and algorithms, developed with various real-life applications using different variants of GAs for efficient pattern recognition and learning in both supervised and unsupervised modes.

Chapter 2 provides a detailed discussion on GAs. The different steps of GAs are described followed by the derivation of the schema theorem. The proof of convergence of GAs to the optimal string under limiting conditions is stated. Some implementation issues and various genetic operators are mentioned. A discussion on multiobjective GAs, a relatively recent development in this field, is provided in a part of this chapter. Finally, different applications of GAs are mentioned.

Chapters 3–7 deal with several single and multiobjective GA-based classifiers. Detailed comparative results on artificial and real-life data sets are provided. The basic classification principle used in the genetic classifiers is the approximation of the class boundaries using a fixed or variable number of hyperplanes (as well as higher-order surfaces) such that one or more objective criteria are optimized. While Chap. 3 describes a *GA-classifier* using a fixed number of hyperplanes (and hence, fixed chromosome length), Chap. 5 considers a variable number of hyperplanes (and hence, variable chromosome length) for designing the *VGA-classifier*. The latter facilitates better approximation of the decision boundary with less a priori information. Theoretical analysis of the two genetic classifiers are provided in Chaps. 4 and 5, respectively. A new model of GA, viz., GA with chromosome differentiation, keeping analogy with the sexual differentiation in nature, is described in Chap. 6 for reducing the computation time and improving performance. Its incorporation in the *VGA-classifier* to provide the *VGACD-classifier* is described, along with a real-life application for pixel classificaton in remote sensing imagery. In Chap. 7, the classification problem is modelled as one of multiobjective optimization, and a multiobjective genetic classifier, *CEMOGA-classifier*, is discussed. Different quantitative indices are defined to measure the quality of Pareto front solutions.

The application of genetic algorithms for clustering, both crisp and fuzzy, is demonstrated in Chap. 8. The number of clusters can be fixed or variable, thereby necessitating the use of both fixed- and variable-length genetic representation schemes.

Bioinformatics and Web intelligence are two upcoming areas where the use of genetic learning holds promise. Some studies have already been carried out that demonstrate that the integration of the search capability of GAs in these domains leads to the production of more effective and accurate solutions, more efficiently. Chapters 9 and 10 are devoted to these two important application areas of GAs.

In Chap. 9, the basic concepts of bioinformatics are first explained, followed by a listing of some of the tasks in bioinformatics. The relevance of GAs in bioinformatics is discussed. Thereafter, a detailed description of the bioinformatics tasks is provided along with the applications of GAs for solving them. Some experimental results demonstrating the effectiveness of GAs for designing small ligands are also provided.

Finally, Chap. 10 starts with an introduction to Web intelligence and mining, and the various components and tasks in Web mining. The existing limitations and challenges in this regard are mentioned. Subsequently, the different applications of genetic search for solving Web mining-related problems are discussed.

2

Genetic Algorithms

2.1 Introduction

Nature has a wonderful and powerful mechanism for optimization and problem solving through the process of evolution. Emulating the power of natural selection and evolution in computer programs therefore appears to be natural for solving hard problems. Based on the principles of natural genetic systems, evolutionary algorithms (EA) were developed for performing search and optimization in complex landscapes. The important components of EAs are genetic algorithms (GAs) [164, 166, 195], genetic programming (GP) [255] and evolutionary strategies (ES) [56].

This chapter provides a detailed description of genetic algorithms, their working principles, a theoretical analysis and some empirical studies. A brief description of some search techniques is first provided in Sect. 2.2. Section 2.3 provides an overview of GAs. This includes a description of the basic principles and features of GAs, the representation strategy, fitness evaluation and genetic operators, algorithmic parameters and schema theorem. A theoretical analysis of GAs, including Markov chain modelling and the proof of convergence of GAs to the optimal string under limiting conditions, is described in Sect. 2.4. (The derivation of the ϵ-optimal stopping time is provided in Appendix A for the convenience of readers.) The next section highlights some empirical aspects and implementation issues of GAs. Sections 2.6 and 2.7 provide a description of multiobjective genetic algorithms and a discussion on the different applications of genetic algorithms, respectively. Finally, Sect. 2.8 summarizes the chapter.

2.2 Traditional Versus Nontraditional Search

Search techniques can be broadly classified into three classes (Fig. 2.1):

- numerical methods

- enumerative techniques
- guided random search techniques

Numerical methods, also called calculus-based methods, use a set of necessary and sufficient conditions that must be satisfied by the solution of the optimization problem. They can be further subdivided into two categories, viz., direct and indirect. Direct search methods perform a hill climbing on the function space by moving in a direction related to the local gradient. In indirect methods, the solution is sought by solving a set of equations resulting from setting the gradient of the objective function to zero. The calculus-based methods are local in scope and also assume the existence of derivatives. These constraints severely restrict their application in many real-life problems, although they can be very efficient in a small class of unimodal problems.

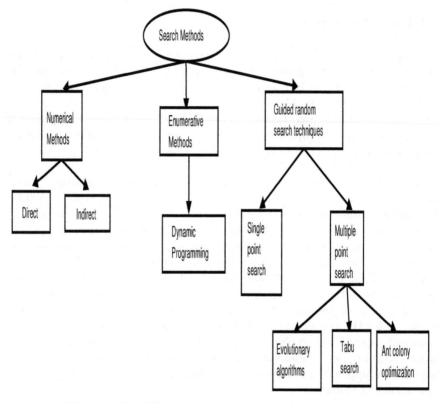

Fig. 2.1. The different search and optimization techniques

Enumerative techniques involve evaluating each and every point of the finite, or discretized infinite, search space in order to arrive at the optimal solution. Dynamic programming is a well-known example of enumerative search. It is obvious that enumerative techniques will break down even on problems of moderate size and complexity because it may become simply impossible to search all the points in the space.

Guided random search techniques are based on enumerative methods, but they use additional information about the search space to guide the search to potential regions of the search space. These can be further divided into two categories, namely, single-point search and multiple-point search, depending on whether it is searching just with one point or with several points at a time. Simulated annealing is a popular example of the single-point search technique that uses thermodynamic evolution to search for the minimum energy states. Evolutionary algorithms like genetic algorithms are a popular examples of multiple-point search, where random choice is used as a tool to guide a highly explorative search through a coding of the parameter space. The guided random search methods are useful in problems where the search space is huge, multimodal and discontinuous, and where a near-optimal solution is acceptable. These are robust schemes, and they usually provide near-optimal solutions across a wide spectrum of problems.

2.3 Overview of Genetic Algorithms

Genetic algorithms (GAs)[113, 164, 308] are efficient, adaptive and robust search and optimization processes that are usually applied in very large, complex and multimodal search spaces. GAs are loosely modelled on the principles of natural genetic systems, where the genetic information of each individual or potential solution is encoded in structures called *chromosomes*. They use some domain- or problem-dependent knowledge to compute the *fitness function* for directing the search in more promising areas. Each individual or chromosome has an associated fitness value, which indicates its degree of goodness with respect to the solution it represents. GAs search from a set of points, called a population. Various biologically inspired operators like *selection, crossover* and *mutation* are applied on the chromosomes in the population to yield potentially better solutions.

2.3.1 Basic Principles and Features

GAs emulate biological principles to solve complex optimization problems. They essentially comprise a set of individual solutions or chromosomes (called the population) and some biologically inspired operators that creates a new (and potentially better) population from an old one. According to the theory of evolution, only those individuals in a population who are better suited to the environment are likely to survive and generate offspring, thereby transmitting their superior genetic information to new generations. GAs are different from most of the normal optimization and search procedures in four ways [164]:

- GAs work with the coding of the parameter set, not with the parameters themselves.

- GAs work simultaneously with multiple points, not with a single point.
- GAs search via sampling (a blind search) using only the payoff information.
- GAs search using stochastic operators, not deterministic rules.

Since a GA works simultaneously on a set of coded solutions, it has very little chance of getting stuck at a local optimum when used as an optimization technique. Again, the search space need not be continuous, and no auxiliary information, like the derivative of the optimizing function, is required. Moreover, the resolution of the possible search space is increased by operating on coded (possible) solutions and not on the solutions themselves. In general, GAs perform best when potential solutions can be represented in a way which exposes important components of potential solutions, and operators to mutate and hybridize these components are available. Conversely, GAs are hampered if the chosen representation does not expose key features of potential solutions, or operators do not generate "interesting" new candidates. The essential components of GAs are the following:

- A representation strategy that determines the way in which potential solutions will be coded to form string-like structures called *chromosomes*.
- A population of *chromosomes*.
- Mechanism for evaluating each chromosome.
- Selection/reproduction procedure.
- Genetic operators.
- Probabilities of performing genetic operations.

It operates through a simple cycle of

(a) evaluation of each chromosome in the population to get the fitness value,
(b) selection of chromosomes, and
(c) genetic manipulation to create a new population of chromosomes,

over a number of iterations (or, generations) till one or more of the following termination criteria is satisfied:

- The average fitness value of a population becomes more or less constant over a specified number of generations.
- A desired objective function value is attained by at least one string in the population.
- The number of iterations is greater than some predefined threshold.

A schematic diagram of the basic structure of a genetic algorithm is shown in Fig. 2.2. These components are now described in detail.

2.3.2 Encoding Strategy and Population

To solve an optimization problem, GAs start with the chromosomal representation of a parameter set. The parameter set is to be coded as a finite-length string over an alphabet of finite length. Usually,

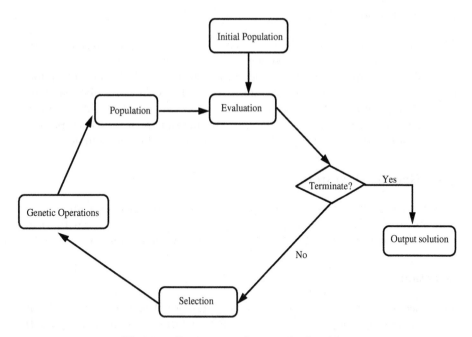

Fig. 2.2. Basic steps of a genetic algorithm

the chromosomes are strings of 0's and 1's. For example, the string

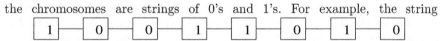

is a binary chromosome of length 8. It is evident that the number of different chromosomes (or strings) is 2^l, where l is the string length. Each chromosome actually refers to a coded possible solution. A set of such chromosomes in a generation is called a *population*, the size of which may be constant or may vary from one generation to another. A general practice is to choose the initial population randomly.

A commonly used principle for coding is known as the *principle of minimum alphabet* [164]. It states that for efficient coding, the smallest alphabet set that permits a natural expression of the problem should be chosen. In general, it has been found that the binary alphabet offers the maximum number of schemata per bit of information of any coding [164]. Hence, binary encoding is one of the commonly used strategies, although other techniques like floating point coding [118, 308] are also quite popular.

2.3.3 Evaluation

The fitness/objective function is chosen, depending on the problem to be solved, in such a way that the strings (possible solutions) representing good points in the search space have high fitness values. This is the only information (also known as the payoff information) that GAs use while searching for possible solutions. In the commonly used *elitist* model of GAs, thereby providing an *elitist GA* (EGA), the best chromosome seen up to the present generation is retained either in the population or in a location outside it.

2.3.4 Genetic Operators

The frequently used genetic operators are the selection, crossover and mutation operators. These are applied on a population of chromosomes to yield potentially new offspring. These operators are described below.

Selection

The selection/reproduction process copies individual strings (called parent chromosomes) into a tentative new population (known as the mating pool) for genetic operations. The number of copies that an individual receives for the next generation is usually taken to be directly proportional to its fitness value, thereby mimicking the natural selection procedure to some extent. This scheme is commonly called the *proportional selection scheme. Roulette wheel parent selection, stochastic universal selection* and *binary tournament selection* are some of the most frequently used selection procedures [164]. Figure 2.3 demonstrates the roulette wheel selection. The wheel has as many slots as the popualtion size P, where the size of a slot is proportional to the relative fitness of the corresponding chromosome in the population. An individual is selected by spinning the roulette and noting the position of the marker when the roulette stops. Therefore, the number of times that an individual will be selected is proportional to its fitness (or, the size of the slot) in the population.

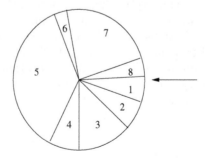

Fig. 2.3. Roulette wheel selection

In the stochastic universal selection, P equidistant markers are placed on the wheel. All the P individuals are selected by spinning the wheel, the number of copies that an individual gets being equal to the number of markers that lie within the corresponding slot. Figure 2.4 demostrates this selection procedure.

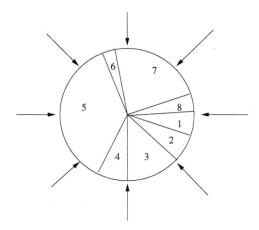

Fig. 2.4. Stochastic universal selection

In binary tournament selection, two individuals are taken at random, and the better of the two is selected. This process is repeated till the mating pool is full. The binary tournament selection procedure may be implemented with and without replacement of the competing individuals.

Crossover

The main purpose of crossover is to exchange information between randomly selected parent chromosomes by recombining parts of their genetic materials. This operation, performed probabilistically, combines parts of two parent chromosomes to produce offspring for the next generation. *Single-point crossover* is one of the most commonly used schemes. Here, first of all, the members of the selected strings in the mating pool are paired at random. Then for performing crossover on a pair, an integer position k (known as the crossover point) is selected uniformly at random between 1 and $l - 1$, where l is the string length. Two new strings are created by swapping all characters from position $(k + 1)$ to l. For example, let the two parents and the crossover point be as shown in Fig. 2.5(a). Then after crossover the offsping formed are shown in Fig. 2.5(b). Some other common crossover techniques are the two-point crossover, multiple-point crossover, shuffle-exchange crossover and uniform crossover [113].

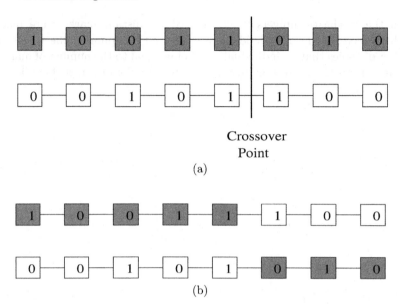

Crossover
Point

(a)

(b)

Fig. 2.5. Example of single-point crossover operation: (a) before crossover; (b) after crossover

The success of GAs depends significantly on the coding technique used to represent the problem variables [230, 388]. The *building block hypothesis* indicates that GAs work by identifying good building blocks, and by eventually combining them to get larger building blocks [164, 176, 195]. Unless good building blocks are coded tightly, the crossover operation cannot combine them together [168, 169]. Thus coding-crossover interaction is important for the successful operation of GAs. The problem of tight or loose coding of problem variables is largely known as the *linkage problem* [167]. Some work on linkage learning GAs that exploits the concept of gene expression can be found in [30, 228, 229].

Mutation

Mutation is the process by which a random alteration in the genetic structure of a chromosome takes place. Its main aim is to introduce genetic diversity into the population. It may so happen that the optimal solution resides in a portion of the search space which is not represented in the populations' genetic structure. The process will therefore be unable to attain the global optima. In such a situation, only mutation can possibly direct the population to the optimal section of the search space by randomly altering the information in a chromosome. Mutating a binary gene involves simple negation of the bit, while mutations for real coded genes are defined in a variety of ways [136, 308]. Here, we discuss the binary bit-by-bit mutation, where every bit in a chromosome is subjected to mutation with a (usually low) probability μ_m. An example of bit-by-bit binary mutation is shown in Fig. 2.6. Here, the positions 3 and 7 of the chromosome shown above have been subjected to mutation.

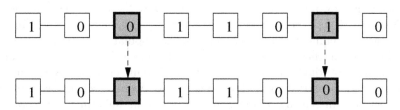

Fig. 2.6. Example of bit-by-bit mutation

2.3.5 Parameters of Genetic Algorithms

There are several parameters in GAs that have to be tuned and/or fixed by the programmer. Some among these are the population size, the length of a chromosome, the probabilities of performing crossover and mutation, the termination criteria, and the population replacement strategy. For example, one needs to decide whether to use the generational replacement strategy, where the entire population is replaced by the new population, or the steady-state replacement policy, where only the less fit individuals are replaced. Such parameters in GAs are mostly problem dependent, and no guidelines for their choice exist in the literature. Several researchers have therefore also kept some of the GA parameters variable and/or adaptive [22, 25, 450].

2.3.6 Schema Theorem

This section describes the *schema theorem*, which is one of the most fundamental theoretical results of GAs. In the following discussion, available in detail in [164], the schema theorem for the conventional GAs is derived. A *schema h* is a similarity template describing a subset of strings with similarities at certain string positions. For binary representation, it is composed of 0, 1 and the # (don't care) symbols. For example, # 1 # 1 # # 0 # # # is a schema of length 10. This schema will be subsequently referred to as h'. A schema indicates the set of all strings that match it in the positions where the schema has either a 0 or a 1.

The *defining position* of a schema is a position in it which has either a 1 or a 0. The *defining length* of a schema h, denoted by $\delta(h)$, is defined as the distance between the last defining position and the first defining position of the schema, and is obtained by subtracting the first defining position from the last defining position. For the schema h' given above, the first defining position (counting from the left) is 2, and the last defining position is 7. Hence

$\delta(h') = 7 - 2 = 5$. The *order* of a schema h, denoted by $O(h)$, is the number of defining positions in the schema. For the schema h', $O(h') = 3$.

A schema h_1 is said to be contained in another schema h_2 if for each defining position in h_2, the position is defined in h_1, the defining bit being the same. For example, let $h_1 = \# 1 1 1 0 1 0 \# \#$, then h_1 is contained in h'. Note that if h_1 is contained in h_2, then $m(h_2, t) \geq m(h_1, t)$, where $m(h, t)$ represents the number of instances of h in the population at time t.

The schema theorem [164] estimates the lower bound of the number of instances of different schemata at any point of time. According to this theorem, a short-length, low-order, above-average schema will receive exponentially increasing number of instances in subsequent generations at the expense of below-average ones. This is now discussed in detail. The notations that will be used for this purpose are first explained.

h : a short-length, low-order, above-average schema
$m(h, t)$: number of instances of schema h in a population at generation t
$\delta(h)$: the defining length of schema h
$O(h)$: order of schema h
l : length of a chromosome
$\overline{f_h}$: average fitness value of schema h
$\overline{f}$: average fitness value of the population

As already mentioned, the number of copies of each string that go into the mating pool is proportional to its fitness in the population. Accordingly, the expected number of instances of the schema h that go into the mating pool after selection, $m(h, t + 1)$, is given by

$$m(h, t + 1) = m(h, t) * \frac{\overline{f_h}}{\overline{f}}. \tag{2.1}$$

Hence the number of above-average schemata will grow exponentially, and below-average ones will receive decreasing numbers of copies.

If a crossover site is selected uniformly at random among $(l - 1)$ possible sites, a schema h may be destroyed when the crossover site falls within its defining length $\delta(h)$. Therefore, the probability of schema disruption due to crossover is

$$p_{d1} = \frac{\delta(h)}{(l - 1)}.$$

Hence the survival probability (p_s) (when the crossover site falls outside the defining length) is

$$p_s = 1 - p_{d1},$$
$$= 1 - \frac{\delta(h)}{(l - 1)}.$$

If μ_c is the crossover probability, then

$$p_s \geq 1 - \mu_c * \frac{\delta(h)}{(l-1)}.$$

Again, for survival of a schema h, all the fixed positions of h, $(O(h))$, should remain unaltered. If μ_m is the mutation probability, the probability that one fixed position of the schema will remain unaltered is $(1-\mu_m)$. Hence, for $O(h)$ number of fixed positions of a schema h to survive, the survival probability is

$$(1-\mu_m)^{O(h)}.$$

If $\mu_m \ll 1$, the above value becomes $(1 - O(h) * \mu_m)$. Therefore,

$$m(h, t+1) \geq m(h, t) * \frac{\overline{f_h}}{\overline{f}} * \left(1 - \mu_c * \frac{\delta(h)}{(l-1)}\right) * (1 - \mu_m * O(h)). \quad (2.2)$$

Neglecting the smaller term, we have

$$m(h, t+1) \geq m(h, t) * \frac{\overline{f_h}}{\overline{f}} * \left(1 - \mu_c * \frac{\delta(h)}{(l-1)} - \mu_m * O(h)\right). \quad (2.3)$$

Equation (2.3) indicates that short (small $\delta(h)$), low-order (small $O(h)$) and above-average ($\overline{f_h} > \overline{f}$) schemata will receive exponentially increasing numbers of instances in subsequent generations [164, 195]. Note that the schema theorem does not guarantee the convergence of the process to the global optimal solution. Vose extended this work and interpreted GAs as constrained random walk and generalized the concept of schemata [476]. Some results on the analysis of schema distribution and deceptive problems, i.e., the class of problems which mislead the GA, are available in [106, 204, 276, 387, 492].

2.4 Proof of Convergence of GAs

This section deals with some of the theoretical studies that have been carried out to study the convergence properties of GAs. In this regard an investigation of the limiting behavior of GAs is provided, where it is shown that under limiting conditions, the elitist model of GAs will surely provide the optimal string.

2.4.1 Markov Chain Modelling of GAs

Several researchers have analyzed GAs by modelling them as states of a Markov chain [60, 114, 171, 199, 452, 476]. The states of the chain are denoted by the populations of GAs. A population Q is a collection of strings of length l generated over a finite alphabet $\mathcal{A}$ and is defined as follows:

$$Q = \quad \{S_1, S_1, \cdots, (\sigma_1 \; times), S_2, S_2, \cdots, (\sigma_2 \; times), \cdots,$$
$$S_\xi, S_\xi, \cdots, (\sigma_\xi \; times) : S_i \in \Sigma;$$
$$\sigma_i \geq 1 \text{ for } i = 1, 2, \cdots \xi; S_i \neq S_j \; \forall i \neq j \text{ and } \sum_{i=1}^{\xi} \sigma_i = P\}.$$

Here Σ denotes the search space of the GA consisting of 2^l strings. Let $\mathcal{Q}$ denote the set of all populations of size P. The number of populations or states in a Markov chain is finite and is given by

$$\mathcal{N} = \binom{2^l + P - 1}{P}.$$

Goldberg and Segrest [171] assume proportional selection without mutation. Since single-bit chromosomes are considered in [171], for a population size P, there are $P + 1$ possible states i, where i is the population with exactly i ones and $(P - i)$ zeros. A $(P + 1) \times (P + 1)$ transition matrix $\tau[i, j]$ is defined that maps the current state i to the next state j. The major focus has been the study of the pure genetic drift of proportional selection using the transition probability expression, where they investigated the expected times to absorption for the drift case. Horn [199] extended this work by modelling a niched GA using finite, discrete time Markov chains, limiting the niching operator to *fitness sharing*, introduced in [170].

Vose and Liepins [476] studied the transient and asymptotic behavior of the genetic algorithm by modelling it as a Markov chain. In [114], Davis and Principe attempted to provide a Markov chain model and an accompanying theory for the simple genetic algorithm by extrapolating the existing theoretical foundation of the simulated annealing algorithm [111, 243, 471]. Rudolph has also found the asymptotic behavior of elitist GAs to the optimal string [403].

Suzuki [452] presented a Markov chain analysis of GAs using a modified elitist strategy, by considering selection, crossover and mutation for generational change. By evaluating the eigenvalues of the transition matrix of the Markov chain, the convergence rate of GAs has been computed in terms of the mutation probability μ_m. It is shown that the probability of the population including the individual with the highest fitness value is lower-bounded by $1 - O(|\lambda_*|^n)$, $|\lambda_*| < 1$, where n is the number of generation changes and λ_* is a specified eigenvalue of the transition matrix.

In [477], Vose and Wright surveyed a number of recent developments concerning the simple genetic algorithms, its formalism, and the application of the Walsh transformation to the theory of genetic algorithms. It is mainly concerned with the mathematical formalization of the simple genetic algorithm, and how the Walsh transform appertains to its theoretical development.

The issue of asymptotic behavior of GAs has also been pursued in [60], where GAs are again modelled as Markov chains having a finite number of states. A state is represented by a population together with a potential string.

Irrespective of the choice of initial population, GAs have been proved to approach the optimal string after infinitely many iterations, provided the conventional mutation operation is incorporated. The details of the proof are provided in the following section. Murthy et al. [324] provided a stopping criterion, called ϵ-optimal stopping time, for the elitist model of the GAs. Subsequently, they derived the ϵ-optimal stopping time for GAs with elitism under a "practically valid assumption". Two approaches, namely, pessimistic and optimistic, have been considered to find out the ϵ-optimal stopping time. It has been found that the total number of strings to be searched in the optimistic approach to obtain an ϵ-optimal string is less than the number of all possible strings for sufficiently large string length. This observation validates the use of genetic algorithms in solving complex optimization problems. The derivation of ϵ-optimal stopping time is provided in Appendix A.

2.4.2 Limiting Behavior of Elitist Model of GAs

This section derives a proof for convergence of an elitist model of GAs. It is available in detail in [60]. A string over a finite set of alphabet $\mathcal{A} = \{\alpha_1, \alpha_2, \cdots, \alpha_a\}$ is used as a chromosomal representation of a solution x. The length (l) of the strings depends on the required precision of the solution. Each string S corresponds to a value x in D and is of the form

$$S = (\beta_1 \beta_2 \cdots \beta_{l-1} \beta_l); \quad \beta_i \in \mathcal{A}, \ \forall i.$$

For example, if we consider $\mathcal{A} = \{0, 1\}$, then a binary string of length l is a chromosomal or string representation of a solution. Note that the total number of strings (i.e., the number of different values for the variable x) is 2^l. To generate an initial population of size P, generally, a random sample of size P is drawn from 2^l strings.

In each iteration t, a genetic algorithm starts with a population of potential solutions in the form of chromosomes or strings, $Q^{(t)} = \{S_1^{(t)}, S_2^{(t)}, \cdots, S_P^{(t)}\}$. Each string $S_i^{(t)}$ is evaluated to give some measure of its fitness in terms of fitness function fit. A mating pool (a tentative new population) is formed by selecting the potential (more fit) individuals from $P^{(t)}$. Members of this population undergo reproduction by means of crossover (with probability μ_c) and mutation (with probability μ_m) operations to form a new population of solutions for the next iteration.

Crossover operation on the mating pool of size P (P is even) is generally performed in the following way:

- Select $\frac{P}{2}$ pairs of strings randomly from the mating pool so that every string in the mating pool belongs to exactly one pair of strings.
- For each pair of strings, generate a random number rnd from $[0, 1]$. If $rnd \leq \mu_c$ perform crossover; otherwise no crossover is performed and the two parent strings will become a part of the new population generated.

If P is odd then one string in the mating pool will not take part in the crossover operation, i.e., $\frac{P-1}{2}$ pairs of strings are selected for crossover. For the purpose of this proof, single-point crossover operation was considered.

During bit-by-bit mutation, every character β_i, $i = 1, 2, \cdots, l$, in each chromosome (generated after crossover) may undergo mutation depending on the mutation probability value. Note that any string can be generated from any given string by mutation operation. This result is stated in the form of the following lemma.

Lemma 2.1. *Probability of generating any string S_1 from a given string S_2 is greater than zero and its value is $\mu_m^\nu (1 - \mu_m)^{l-\nu}$, where ν ($0 \leq \nu \leq l$) is the number of places where those two strings have distinct characters.*

Proof. Let $S_1 = \beta_1 \beta_2 \cdots, \beta_{l-1}\beta_l$ and $S_2 = \gamma_1 \gamma_2 \cdots \gamma_{l-1}\gamma_l$ be two strings such that $\beta_i \neq \gamma_i$ for $i = i_1, i_2, \cdots, i_\nu$ ($\nu \leq l$); $\beta_i = \gamma_i$ otherwise. Then from S_2 one can obtain S_1 if γ_i is mutated as β_i for $i = i_1, i_2, \cdots, i_\nu$, and no mutation occurs at all the other places. Given that μ_m is the probability of mutation and it has to be performed in each bit position $i_1, i_2, \cdots, i_\nu$, the probability of obtaining S_1 from S_2 is evidently $\mu_m^\nu \cdot (1 - \mu_m)^{l-\nu} > 0$.

Note that the probability of mutating i bit positions is more than that of mutating $i + 1$ bit positions, i.e.,

$$\mu_m^i (1 - \mu_m)^{l-i} \geq \mu_m^{i+1}(1 - \mu_m)^{l-i-1} \quad \forall i = 0, 1, 2, \cdots, l - 1.$$

This results in $\mu_m \leq 0.5$. Hence the probability of mutation may be kept in the range of $(0, 0.5]$.

Genetic algorithms search over a space Σ of 2^l strings and eventually provide the best with respect to the fitness function fit. The strings can be classified into a set of s classes depending on their fitness function values. The classes are defined as

$$\Sigma_i = \{S : fit(S) = F_i\},$$

where F_i denotes the ith highest fitness function value. Thus $F_1 > F_2 > \cdots > F_s$. Let us also assume without loss of generality that $F_s > 0$. Note that $1 \leq s \leq 2^l$. The fitness function value $fit(Q)$ of a population is defined as $fit(Q) = \max_{S \in Q} fit(S)$. Then the populations can be partitioned into s sets of populations where $E_i = \{Q : Q \in \mathcal{Q} \text{ and } fit(Q) = F_i\}$, $i = 1, 2, \cdots, s$, is a set of populations having the same fitness function value F_i. Let e_k be the number of populations in E_k, $k = 1, 2, \ldots, s$. Let the jth population in E_i be denoted by Q_{ij}, $j = 1, 2, \ldots, e_i$, $i = 1, 2, \ldots, s$. In an iteration or generation, the genetic operators (selection, crossover and mutation) create a population $Q_{kl} \in E_k$; $l = 1, 2, \cdots, e_k$, and $k = 1, 2, \cdots, s$, from some $Q_{ij} \in E_i$. Here $k \leq i$ since the elitist model is used. The generation of a population Q_{kl} from Q_{ij} is considered as a transition from Q_{ij} to Q_{kl}, and let $p_{ij.kl}$ denote the transition probability. Then the probability of transition from Q_{ij} to any population in E_k can be calculated as

$$p_{ij.k} = \sum_{l=1}^{e_k} p_{ij.kl}; j = 1, 2, \cdots, e_i; \ i, \ k = 1, 2, \cdots, s.$$

For all $j = 1, 2, \cdots, e_i$, and $i = 1, 2, \cdots, s$, the following holds:

$$\begin{aligned} p_{ij.k} &> 0, \quad \text{if} \quad k \leq i, \\ &= 0, \quad \text{otherwise.} \end{aligned} \tag{2.4}$$

Let $p_{ij.kl}^{(n)}$ denote the probability of reaching Q_{kl} from Q_{ij} in n iterations. Note that $p_{ij.kl}^{(1)} = p_{ij.kl}$. Let $p_{ij.k}^{(n)}$ denote the probability of reaching a population in E_k from Q_{ij} in n iterations. So

$$p_{ij.k}^{(n)} = \sum_{l=1}^{e_k} p_{ij.kl}^{(n)}.$$

It can be easily shown that

$$\begin{aligned} p_{ij.k}^{(n)} &> 0 \text{ if } k \leq i \ \forall n, \\ &= 0 \text{ if } k > i \ \forall n. \end{aligned}$$

To show the limiting nature of a GA with elitist model (EGA), the following theorem is proved.

Theorem 2.2. *For an EGA with probability of mutation $\mu_m \in (0, \frac{1}{2}]$,*

$$\lim_{n \longrightarrow \infty} p_{ij.k}^{(n)} = 0 \ \text{for } 2 \leq k \leq s; \ \forall j = 1, 2, \cdots, e_i, \ \text{and } i = 1, 2, \cdots, s.$$
Hence $\lim_{n \longrightarrow \infty} p_{ij.1}^{(n)} = 1 \ \forall j = 1, 2, \cdots, e_i, \ \text{and } i = 1, 2, \cdots, s.$

Proof. It is evident from Eq. (2.4) that $p_{ij.1} > 0$ for all $j = 1, 2, \cdots, e_i$ and $i = 1, 2, \cdots, s$. Let

$$\max_{i,j} (1 - p_{ij.1}) = \delta. \tag{2.5}$$

Note that $\delta < 1$ since $\min_{i,j} p_{ij.1} > 0$. Now,

$$\sum_{k \neq 1} p_{ij.k}^{(1)} = \sum_{k=2}^{s} p_{ij.k} = 1 - p_{ij.1} ; \tag{2.6}$$

$$\sum_{k\neq 1} p_{ij.k}^{(2)} = \sum_{k=2}^{s} \sum_{i_1 \neq 1} \sum_{j_1=1}^{e_{i_1}} p_{ij.i_1 j_1} p_{i_1 j_1 .k} \quad \text{(since, } p_{1j_1 .k} = 0 \text{ for } k > 1\text{)},$$

$$= \sum_{i_1 \neq 1} \sum_{j_1=1}^{e_{i_1}} p_{ij.i_1 j_1} \sum_{k=2}^{s} p_{i_1 j_1 .k},$$

$$= \sum_{i_1 \neq 1} \sum_{j_1=1}^{e_{i_1}} p_{ij.i_1 j_1} (1 - p_{i_1 j_1 .1}),$$

$$\leq \delta \sum_{i_1 \neq 1} \sum_{j_1=1}^{e_{i_1}} p_{ij.i_1 j_1},$$

$$= \delta \sum_{i_1 \neq 1} p_{ij.i_1} \quad \text{(from Eq. (2.5))},$$

$$= \delta (1 - p_{ij.1}),$$

$$\leq \delta^2.$$

(2.7)

Similarly,

$$\sum_{k\neq 1} p_{ij.k}^{(3)} = \sum_{k=2}^{s} \sum_{i_1 \neq 1} \sum_{j_1=1}^{e_{i_1}} p_{ij.i_1 j_1} p_{i_1 j_1 .k}^{(2)},$$

$$= \sum_{i_1 \neq 1} \sum_{j_1=1}^{e_{i_1}} p_{ij.i_1 j_1} \sum_{k=2}^{s} p_{i_1 j_1 .k}^{(2)},$$

$$\leq \sum_{i_1 \neq 1} \sum_{j_1=1}^{e_{i_1}} p_{ij.i_1 j_1} \delta (1 - p_{i_1 j_1 .1}) \quad \text{(from Eq. (2.7))},$$

$$\leq \delta^2 \sum_{i_1 \neq 1} \sum_{j_1=1}^{e_{i_1}} p_{ij.i_1 j_1},$$

$$= \delta^2 \sum_{i_1 \neq 1} p_{ij.i_1},$$

$$= \delta^2 (1 - p_{ij.1}),$$

$$\leq \delta^3.$$

In general,

$$\sum_{k\neq 1} p_{ij.k}^{(n+1)} = \sum_{k=2}^{s} \sum_{i_1 \neq 1} \sum_{j_1=1}^{e_{i_1}} p_{ij.i_1 j_1}^{(n)} p_{i_1 j_1 .k},$$

$$\leq \delta^n (1 - p_{ij.1}),$$

$$\leq \delta^{n+1}.$$

(2.8)

Note that $\delta^{n+1} \longrightarrow 0$ as $n \longrightarrow \infty$ since $0 \leq \delta < 1$. Hence $\sum_{k\neq 1} p_{ij.k}^{(n+1)} \longrightarrow 0$ as $n \longrightarrow \infty$. This immediately implies that $\lim_{n \to \infty} p_{ij.k}^{(n)} = 0$ for $2 \leq k \leq s$ for all i and j.

It is clear that,

$$\lim_{n \longrightarrow \infty} p_{ij.1}^{(n)} = \lim_{n \longrightarrow \infty} (1 - \sum_{k \neq 1} p_{ij.k}^{(n)})$$
$$= 1.$$

The following conclusions can be made from the above theorem:

- For any state (or population) Q_{ij}; $i \geq 2$, $p_{ij.k}^{(n)} \longrightarrow 0$ as $n \longrightarrow \infty$, $\forall k \geq 2$. In other words, for sufficiently large number of iterations, Q_{ij} will result in Q_{1l} for some $l = 1, 2, \cdots, e_1$, i.e., the convergence to optimal string is assured with any initial population.
- The proof is independent of the crossover operation, but mutation should be performed with probability $\mu_m > 0$. Moreover, in each iteration, the knowledge of the best string obtained in the previous iteration is preserved within the population.
- Q_{ij} will eventually result in a population containing the best string (or in a population containing a best string if the number of best strings is more than one) for all $i \geq 2$.
- Note that the nature of the proof does not change even if mutation probability μ_m is changed from iteration to iteration as long as it lies in the interval $(0, \frac{1}{2}]$. In the later chapters, μ_m is varied from iteration to iteration.

The stopping time for GAs have been derived theoretically using the above proof. This is provided in Appendix A.

2.5 Some Implementation Issues in GAs

Here we describe some important issues which need attention while implementing GAs in practical problems. Some of the developments of different components of GAs for enhancing its performance are also mentioned.

One important issue with respect to the effectiveness of genetic algorithms is the population size. The result of Holland [195] in this regard states that the number of schemata processed effectively is proportional to the cube of the population size. However, a study by Goldberg [165] suggests that this result should not lead one to assume very large population sizes. He has, in fact, indicated that relatively small populations are appropriate for serial implementations of GAs, while large populations are appropriate for perfectly parallel implementations. There have been a few attempts regarding the initialization of population. These include *random initialization* [164], where the population is initialized randomly, *extended random initialization* [70], where each member of the initial population is chosen as the best of a number of randomly chosen individuals, as well as knowledge-based initialization [175, 400]. In [391], a case-based method of initializing genetic algorithms that are used to guide search in changing environments was proposed. This was incorporated in anytime learning systems, which are a general approach to continuous learning in a changing environment. Domain knowledge and/or knowledge gained by previous runs of the algorithm can be used at any stage of the GA, not only during initialization. In [283, 284] a case-injected genetic algorithm for combinational logic design is studied, and the effect of injection percentage on the performance is investigated. A case-injected genetic algorithm is a genetic algorithm augmented with a case-based memory of past problem solving attempts which learns to improve performance on sets of similar design problems. In this approach, rather than starting anew on each design, appropriate intermediate design solutions to similar, previously solved problems are periodically injected into a genetic algorithm's population. Experimental results on a configuration design problem of a parity checker demonstrate the performance gains from this approach and show that this system learns to take less time to provide quality solutions to a new design problem as it gains experience from solving other similar design problems.

Designing a good representation strategy in GAs is more of an art, and is a crucial aspect in the success of their application. In [157], the effects that a choice of the representation primitives and the encapsulation of the primitives into modules can have on the size of the search space and bias towards a solution are analyzed. The concept of modularity-preserving representation is introduced in GAs, where the existence of modularity in the problem space is translated into a corresponding modularity in the search space. The choice of the representation strategy is usually guided by and is intuitively natural to the problem under consideration. For example, a bit-array representation is used in structural topology optimization since it is natural to the problem domain, and the decoding step is quite simple and intuitive [455, 483, 484].

In practical implementations of GAs, it is often necessary to scale the fitness values in order to keep appropriate levels of competition throughout the generations. Different mechanisms for scaling the fitness values are linear scaling, sigma truncation, power law scaling [164], etc. In [194] the effectiveness of an inversion operator in a basic GA and in a GA using fitness scaling has been compared. It is found that at higher levels of epistasis, inversion is more useful in a basic GA than a GA with fitness scaling. In [107], a new type of genetic algorithm called the Structured Genetic Algorithm is designed for function optimization in nonstationary environments. The novelty of this genetic model lies primarily in its redundant genetic material and a gene activation mechanism which utilizes a multilayered structure for the chromosome (e.g., a directed graph or a tree). The genes at any level can be active or passive. Moreover, genes at a higher level control the activity of the genes at a lower level, by either activating or deactivating them. Experimental results are provided for the nonstationary 0–1 knapsack problem where the weight constraint was varied in time as a periodic step function. In [72], a number of techniques are surveyed for assessing the suitability of evolutionary algorithms for dynamic optimization problems where the goal is to track the progression of the extrema through the space as closely as possible.

An attempt to incorporate the ancestors influence into the fitness of individual chromosomes has been made in [116]. This is based on the observations in nature where an individual is not an independent entity, but is highly influenced by the environment.

Several alternate selection schemes have been suggested in the literature for reducing the stochastic errors associated with roulette wheel selection [22, 164]. Some of these are deterministic sampling, remainder stochastic sampling without replacement, stochastic sampling without replacement, remainder stochastic sampling with replacement, stochastic sampling with replacement, stochastic tournament selection, ranking, etc. A study where the selection scheme itself is made adaptive can be found in [22]. A selection method called *disruptive selection* has been proposed in [263]. This method adopts a nonmonotonic fitness function and favors both superior and inferior individuals. It has been shown here that this scheme can be used to solve nonstationary search problems, where conventional GAs fare poorly. In [240] a new selection strategy is proposed based on the natural characteristics and behaviors in genetics. Recently a self-adaptive selection has been proposed in [3] as a second level of selection in GAs. The experimental results, for very large problem dimensions, are found to be encouraging. In [80] a comparison of several selection strategies, namely, linear ranking, exponential ranking, Boltzmann selection, truncation selection and tournament selection, is provided using the cumulants of the fitness distribution of the selected individuals. Experimental results reported in [80] suggest that while Bolzmann selection preserves the original fitness variance more than the other methods, tournament selection results in the largest reduction of the fitness variance. Although selection of the operator that results in large fitness variance (and hence more diverse solutions) is expected to be beneficial, severely reducing the diversity may be advantageous if the variation operators produce very diverse individuals.

There have been several investigations on developing various crossover operators and studying their interaction with some other operators. In order to improve the performance, Syswerda [453] introduced a new crossover operator called *uniform crossover*, where two offspring are generated from two parents by selecting each component from either parent with a given probability. The comparative utility of single-point, two-point and uniform crossover operators is also investigated, where uniform crossover is found to provide a generally superior performance. This has been further confirmed by an analysis in [444], where a theoretical framework for understanding the virtues of uniform crossover is discussed. A new dynamic, knowledge-based, nonuniform crossover technique has been proposed in [288], which generalizes the uniform crossover, and constantly updates the knowledge extracted from the environment's feedback on previously generated chromosomes. Some additional information and new operators in this line are available in [119, 442, 443, 478].

To sustain diversity (which may be lost due to crossover and very low mutation rates) in the population, Whitley and Starkweather [490] proposed a technique called *adaptive mutation*, where instead of a fixed mutation rate the probability to perform mutation operation increases with increase of genetic homogeneity in the population. Bhandari et al. [61] have proposed a new mutation operator known as *directed mutation* which follows from the concept of induced mutation in biological systems [320]. This operation uses the information acquired in the previous generations rather than probabilistic decision rules. In certain environments, directed mutation will deterministically introduce a new point in the population. The new point is directed (guided) by the solutions obtained earlier, and therefore the technique is called *directed mutation*. Ghannadian et al. suggested the use of random restart in place of the mutation operation in [162]. The expected time for the method to reach an optimal solution is also derived using the Markov chain method.

Schaffer and Eshelman [418] compared crossover and mutation on binary function optimization problems. They observed that crossover generally dominates the population, which may not be necessarily beneficial to the problem. The results indicated that the utility of specific crossover and mutation operations is problem dependent.

In [174], a metalevel GA is used to set the control parameters in order to search for an optimal GA for a given set of numerical optimization problems. A study regarding the optimal mutation rates has been presented in [18]. In another investigation to improve the performance of GAs, concepts of adaptive probabilities of crossover and mutation based on various fitness values of the population are recommended [450]. Here, high solutions are protected and the subaverage solutions are completely disrupted. Besides these, some other techniques for appropriate control parameter selection and investigations regarding the interaction between different operators can be found in [17, 308, 444]. In [240] a new GA called maturing assortative GA is proposed, and an adaptive method for determining the parameters is provided.

A nontraditional genetic algorithm, called cross-generational elitist selection, heterogeneous recombination, and cataclysmic mutation (CHC), that combines a conservative selection strategy, which always preserves the best individuals found so far, with a highly disruptive recombination operator, which produces offspring that are maximally different from both the parents, has been developed by Eshelman [134]. It incorporates an incest prevention technique where individuals are randomly paired for mating, but are only mated if their Hamming distance is above a certain threshold. Another version of genetic algorithm, known as delta coding, where the diversity of the population is constantly monitored, and the algorithm restarted when it show signs of losing diversity, is described in [493].

Mathias et al. [296] compared several types of genetic algorithms against a mutation-driven stochastic hill climbing algorithm, on a standard set of benchmark functions which had Gaussian noise added to them. The genetic algorithms used in these comparisons include *elitist simple genetic algorithm*, *CHC adaptive search algorithm* [134] and *delta coding genetic algorithm* [493], where *CHC* is found to perform better than the others. One reason for this, as pointed out in [296], is that as the noise component of the objective functions becomes a significant factor, the algorithms that sample the immediate neighborhood of solutions in the current populations may have an advantage in finding the global optimum.

Gender differentiation, which is largely encountered in natural systems and results in the preservation of genetic diversity among species, has been one of the approaches for enhancing the performance of GAs [164]. In [43] a way of incorporating differentiation of the population into two subpopulations, M and F, and a form of restrictive mating are described. In [479], four gender types are considered, namely, male, female, self-fertilizing and hermaphrodite. Each of the four genders have different types of mating restrictions. A spontaneous sex change from male to female and vice versa is also allowed. The offspring inherits gender type from one randomly chosen parent. Results are demonstrated on both standard and deceptive problems. Another technique for incorporating chromosome differentiation in GAs is discussed in detail in Chap. 6 of this book.

Messy genetic algorithms (mGAs) [168, 169], another development in the area of GAs, use variable length strings that may be under- or overspecified with respect to the problem being solved. mGAs divide the genetic processing into two distinct phases, a primordial phase and a juxtapositional phase. In the primordial phase, the population is first initialized to contain all possible building blocks of a specific length, where the characteristic length is chosen to encompass possibly misleading building blocks. Thereafter, the portion of good building blocks is enriched through a number of generations of reproduction only. At the same time, the population size is usually reduced by halving the number of individuals in the population at specified intervals. The juxtapositional phase proceeds with a fixed population size and the invocation of reproduction, cut and splice, and other genetic operators. It has been found that mGAs solve problems of bounded deception to global optimality in a time that grows no faster than a polynomial function of the number of decision variables on a serial machine. The use of partial specification (or variable-length genomes) in mGA allows the individuals themselves, rather than the ordering of genes within an individual, to represent which genes go together during recombination. In [487] this critical feature of the mGA is examined, and the impact that partial specification has on recombination is illustrated. An Incremental Commitment GA (ICGA) is formulated that uses partially specified representations and recombination inspired by the mGA but separates these features from the moving-locus aspects and many of the other features of the existing algorithm. The superiority of the ICGA over mGA is demonstrated for the "Shuffled-Hierarchical-if-and-only-if" problem [487] that is characterized by random linkage (hence not solvable by conventional GA since genetic and epistatic linkages do not correspond), and is not delineable into separate subfunctions (hence not solvable by mGA).

Inspired by mGAs, Kargupta developed the gene expression messy genetic algorithm (GEMGA), which explores the computational role of gene expression (or, DNA $\rightarrow$ RNA $\rightarrow$ Protein transformation) in evolutionary linkage learning [226]. This version with quadratic sample complexity is further improved by Bandyopadhyay et al. in [30] to provide linear sample complexity. Further developments in this regard may be found in [228, 229].

Recently Hu et al. [203] have proposed a hierarchical fair competition framework in evolutionary algorithms, particularly genetic programming, which defined hierarchical levels with fitness value boundaries. Individuals undergo selection and recombination within these levels and move between them based on their fitness. Higher fitness levels undergo higher selection pressure to simulate added exploitation. The resulting system is scalable, sustainable and robust. This work was motivated by the observation that current evolutionary computation systems fail to achieve the desired scalability for solving many complex problems like an electric circuit with several thousands of components, even given huge numbers of computers.

The limitation of uniprocessor computing systems and the increase in availability of multiprocessors have led to investigations of parallel genetic algorithms (PGAs). In PGAs, a single large population is divided into smaller subpopulations, where each subpopulation evolves independently on different processors. While obtaining higher speed-up using multiple processors is one motivation for this division of populations, the other and more interesting motivation is that in sequential GAs the diversity may soon be lost because of selection pressure. After some generations, the majority of the chromosomes in a single population become very similar, and crossover thereafter may not be productive. However, if a number of independent populations evolves simultaneously, they will most likely search different areas of the search space. Occasional interaction between subpopulations will allow exchange of information among widely different information carrying groups. Depending on the amount of parallelism, and the level of interaction, PGAs may be broadly classified into coarse-grain and fine-grain. More details can be found in [78, 79, 274, 294, 318, 319]. The island model of PGA has been widely studied where isolated populations evolved in different geographical location and individuals migrated according to a mesh and hyper-cube topology after a number of generations. Martin et al. describe the results of using the island model on a VLSI design problem [292]. Several other models have been proposed for various types of problems such as multiobjective optimization [518], multimodal optimization [55] and agent-based evolutionary algorithms [437].

2.6 Multiobjective Genetic Algorithms

In many real-world situations there may be several objectives that must be optimized simultaneously in order to solve a certain problem. This is in contrast to the problems tackled by conventional GAs, which involve optimization of just a single criterion. The main difficulty in considering multiobjective optimization is that there is no accepted definition of optimum in this case, and therefore it is difficult to compare one solution with another one. One approach for solving such multiobjective problems is to optimize each criterion separately and combine the solutions thus obtained. However, this method is seldom likely to provide a solution where each criterion is optimally balanced. In fact, it may so happen that optimizing one objective may lead to unacceptably low performance of another objective. For example, consider the case of planning a trip from place X to place Y. The objectives here are to minimize the time as well as the cost of the trip. It is evident that the two objectives are conflicting in nature, i.e., if time is minimized, the cost goes up and vice versa. Therefore, there cannot be a single optimum in this case.

Thus, for solving multiobjective problems all the objectives need to be treated together. In general, these problems admit multiple solutions, each of which is considered acceptable and equivalent when the relative importance of the objectives is unknown. The best solution is subjective and depends on the need of the designer or decision maker.

The multiobjective optimization is formally stated as follows [93]: Find the vector $\overline{x}^* = [x_1^*, x_2^*, \ldots, x_n^*]^T$ of decision variables which will satisfy the m inequality constraints:

$$g_i(\overline{x}) \geq 0, \quad i = 1, 2, \ldots, m, \tag{2.9}$$

the p equality constraints

$$h_i(\overline{x}) = 0, \quad i = 1, 2, \ldots, p, \tag{2.10}$$

and optimizes the vector function

$$\overline{f}(\overline{x}) = [f_1(\overline{x}), f_2(\overline{x}), \ldots, f_k(\overline{x})]^T. \tag{2.11}$$

The constraints given in Eqs. (2.9) and (2.10) define the feasible region $\mathcal{F}$ which contains all the admissible solutions. Any solution outside this region is inadmissible since it violates one or more constraints. The vector $\overline{x}^*$ denotes an optimal solution in $\mathcal{F}$. In the context of multiobjective optimization, the difficulty lies in the definition of optimality, since it is only rarely that we will find a situation where a single vector $\overline{x}^*$ represents the optimum solution to all the objective functions. The concept of *Pareto optimality* comes in handy in the domain of multiobjective optimization. A formal definition of Pareto optimality from the point of view of minimization problem may be given as follows [151]:

A decision vector $\overline{x}^*$ is called Pareto optimal if and only if there is no $\overline{x}$ that dominates $\overline{x}^*$, i.e., there is no $\overline{x}$ such that

$$\forall i \in 1, 2, \ldots, k, f_i(\overline{x}) \leq f_i(\overline{x}^*) \text{ and } \exists i \in 1, 2, \ldots, k, \text{ where } f_i(\overline{x}) < f_i(\overline{x}^*).$$

In other words, $\overline{x}^*$ is Pareto optimal if there exists no feasible vector $\overline{x}$ which causes a reduction on some criterion without a simultaneous increase in at least one other. A solution $\overline{x}^*$ strongly dominates a solution $\overline{x}$ if $\overline{x}^*$ is strictly better than $\overline{x}$ in all the objectives. In general, Pareto optimality usually admits a set of solutions called *nondominated* solutions.

A characteristic of a genetic algorithm is that it tends to converge to a single solution when the population size is finite [123]. This may be acceptable when the aim is to find the global optimum of a single criterion. However, in the case of multiobjective optimization, one is interested in obtaining a set of nondominated solutions. Therefore, it is imperative that genetic diversity is maintained in the population. Several attempts in this direction may be found in [117, 123, 134, 170, 378, 438].

Traditional search and optimization methods such as gradient descent search, and other nonconventional ones such as simulated annealing are difficult to extend to multiobjective case, since their basic design precludes such situations. For these methods, the multiobjective problems have to be reformulated as single-objective ones using appropriate combination techniques like weighting, etc. On the contrary, population-based methods like evolutionary algorithms are well suited for handling several criteria at the same time. The different approaches for solving multiobjective optimization problems may be categorized as follows [93, 121]:

(a) Aggregating techniques:
- Weighted sum approach, where the different objectives are combined using weighting coefficients w_i, $i = 1, 2, \ldots, k$. The objective to minimize becomes $\sum_{i=1}^{k} w_i f_i(\overline{x})$.
- Goal programming-based approach, where the user is required to assign targets $(T_i, i = 1, 2, \ldots, k)$ or goals for each objective. The aim then becomes the minimization of the deviation from the targets to the objectives, or $\sum_{i=1}^{k} |f_i(\overline{x}) - T_i|$.
- Goal attainment-based approach, where the user is required to provide, in addition to the goal vector, a vector of weights w_i, $i = 1, 2, \ldots, k$, relating the relative under- or over-attainment of the desired goals.
- ϵ-constraint approach, where the primary objective function is minimized and the other objectives are treated as constraints bound by some allowable levels ϵ_i.

(b) Population-based non-Pareto techniques:
- Vector evaluated genetic algorithm (VEGA) [415, 419] that incorporated a special selection operator in which a number of subpopulations were generated by applying proportional selection according to each objective function in turn.
- Lexicographic ordering [264], where the objectives are ranked in order of importance by the user, and optimization is carried out on these objectives according to this order.
- Use of game theory [373], where it is assumed that a player is associated with each objective.

- Use of gender for identifying the objectives [280], where panmitic reproduction, in which several parents combine to produce a single child, is allowed.
- Use of contact theorem for detecting Pareto-optimal solutions [336], where the fitness of an individual is set according to its relative distance with respect to the Pareto set.
- Use of nongenerational GA [470], where a multiobjective problem is transformed to a single-objective one through a set of appropriate transformations and the fitness of an individual is calculated incrementally. Genetic operations are utilized to produce a single individual that replaces the worst individual in the population.

(c) Pareto-based nonelitist strategies:
- Multiple objective GA (MOGA) [150] in which an individual is assigned a rank corresponding to the number of individuals in the current population by which it is dominated plus 1. All nondominated individuals are ranked 1. Fitness of individuals with the same rank are averaged so that all of them are sampled at the same rate. A niche formation method is used to distribute the population over the Pareto-optimal region.
- Niched Pareto GA [200], where a Pareto dominance-based tournament selection with a sample of the population is used to determine the winner between two candidate solutions. Around ten individuals are used to determine dominance, and the nondominated individual is selected. If both the individuals are either dominated or nondominated, then the result of the tournament is decided through fitness sharing.
- Nondominated sorting GA (NSGA) [449], where all nondominated individuals are classified into one category, with a dummy fitness value that is proportional to the population size. Then this group is removed and the process repeated on the remaining individuals iteratively till all the individuals are classified. Stochastic-remainder-proportionate selection is used in this technique. With this scheme, any number of objective functions, and both maximization and minimization problems can be solved.

(d) Pareto-based elitist strategies:
- Strength Pareto Evolutionary Algorithm (SPEA), proposed by Zitzler and Thiele [522], introduces elitism explicitly by maintaining an external population called an archive. At any generation t, two populations co-exist:
  - Population P_t of size N
  - Archive (External Population) P_t' of maximum size N'

  All the nondominated solutions of P_t are stored in P_t'. Fitness is assigned to all the individuals in population P_t and archive P_t' as follows: Individuals in the archive P_t' are assigned strength S_i using the equation

$$S_i = \frac{n}{N+1}, \tag{2.12}$$

where n is the number of population members dominated by individual i of P_t' in the population, and N is the total number of the individuals in the population. The fitness of the members in the archive is taken to be equal to their strength values. The fitness of the individual i in P_t is calculated as:

$$f_i = 1 + \sum_{j \in P_t', j \geq i} S_j, \tag{2.13}$$

where $j \geq i$ indicates that member j of the archive dominates member i of the population P_t. Fitness is determined relative to the individuals stored in the archive, irrespective of the relative dominance between the members of P_t. Binary tournament selection with assigned fitness followed by crossover and a mutation operator creates the new population P_{t+1} of size N. The nondominated members of P_{t+1} are copied into the archive, which is updated by deleting any dominated solutions if present. In case the size of the archive exceeds N', clustering-based niching is used to reduce it.

- SPEA2: Two potential weaknesses of SPEA were
  - Fitness assignment is determined entirely on the basis of the strength of archive members. This results in individuals having the same fitness value in P_t, if the corresponding set of dominating members in the archive is same. In the worst case if the archive contains one member then all the members of the population will have the same rank.
  - The clustering technique used for ensuring diversity may lose the outer solutions, which should be kept in the archive to obtain good spread of the nondominated solutions.

SPEA2 [521] was developed to avoid the situation where individuals dominated by the same archive members have the same fitness values. SPEA2 considers both the population and the archive to determine the fitness. Strength S_i of every individual i (belonging to either P_t or P_t') is set equal to the number of individuals it dominates. A raw fitness $R(i)$ is then assigned to an individual i as

$$R(i) = \sum_{j \in \{P_t \bigcup P_t'\},\ j \geq i} S_j. \tag{2.14}$$

Here $R(i) = 0$ corresponds to the nondominated members. The final fitness of i is computed as

$$F_i = R(i) + D(i), \tag{2.15}$$

where $D(i)$ is the density of individual i, computed based on its distance to the kth nearest neighbor in the objective space. A different scheme for updating the archive that prevents the loss of the boundary solutions is adopted.

- Elitist Nondominated Sorting Genetic Algorithm (NSGA-II) NSGA-II was proposed by Deb and Srinivas to resolve the weaknesses of NSGA, specially its nonelitist nature. Here the chromosomes in a population are first sorted based on their *domination* status using the procedure *nondominated sort* [121, 122], which results in all the chromosomes being assigned a rank. Thereafter selection, using the *crowded tournament selection*, crossover and mutation are performed to generate a new child population. The parent and the child population are combined, and elitism is applied to generate the next population. Details of these techniques are available in [121, 122]. They are also described in some detail in Chap. 7 of this book.

Some other related and promising techniques in this regard are based on other methods such as evolutionary strategies [246], tabu search [49], particle swarm optimization [48], integration of tabu search and evolutionary algorithm [456], and ant colony optimization [156, 304]. Multiobjective GAs have evolved as a viable alternative for solving problems where the purpose is to optimize several objectives simultaneously. An important aspect of such techniques is that the decision maker is provided with a set of possible alternative solutions, as well as an intermediate solution, which the decision maker may subsequently refine.

A study establishing the effectiveness of some multiobjective optimization techniques over their single-objective counterparts for the multiobjective set covering problem with varying degrees of difficulty has been carried out in [217]. Theoretical analysis characterizing multiobjective optimization algorithms [267], and investigating the possibility of generating a unified framework for them, such that individual multiobjective evolutionary algorithms could be generated by varying some parameter within this framework [67], have been taken up in recent studies.

Research on multiobjective GAs is far from complete. Future investigation should include the development of enhanced techniques and new operators such that the Pareto front is explored more efficiently. Improvements in the techniques for incorporation of elitism should be studied. Moreover, handling of constraints in an efficient manner needs to be studied. Some attempts in this direction may be found in [121, 122, 191] as well as in Chap. 7. Extensive study related to ways of measuring the quality of solutions as well as devising standard benchmark problems needs to be undertaken. Finally, theoretical analysis of multiobjective GAs should be conducted so that the working principles of such techniques may be formalized and the effects of different parameter settings as well as operators may be explained. In this regard, appropriate stopping times also need to be devised.

2.7 Applications of Genetic Algorithms

As mentioned before, GAs find widespread applications in various business, scientific and engineering circles. A review of some of their applications is presented here. These include design of classifier systems and knowledge-based systems, automatic determination of neural network architecture, and development of image processing and pattern classification methodologies. Application of GA in bioinformatics and Web intelligence, two emerging areas, are discussed in Chaps. 9 and 10.

A classifier system with linguistic if–then rules is designed in [207], where genetic algorithms are used to optimize the number of rules and to learn the grade of certainty associated with each rule. In [98], Congdon describes an application of genetic algorithms to classify epidemiological data where the user may tune a parameter in order to evolve a smaller set of more generalized rules. A comparison of the GA-based scheme with the decision tree-based approaches is also provided. Other applications of GAs for developing classifier systems can be found in [65, 208, 209, 210, 275, 396].

Janikow [216] utilized GAs for optimizing the fuzzy knowledge components of a fuzzy decision tree both prior to and during its construction. The suggested method also finds application in optimizing fuzzy rule-based systems. An approach of integrated genetic learning in construction of fuzzy relational structures is suggested in [369]. Another application where a GA is used for auto-fuzzy tuning in fuzzy neural networks is described in [211]. In [431], methods of self-tuning fuzzy modelling based on GAs are proposed. Several issues of integration of fuzzy sets with evolutionary computation are addressed in [371, 410].

In [154], self-adaptive GAs are used for training a recurrent neural network, which is then used for learning the best among the several stability configurations for a biped locomotion robot. Integration of GAs with neural networks has been investigated for the problem of vehicle routing [379]. A method for the automatic generation of fuzzy controllers using GAs is described in [101]. A new encoding scheme for the rules is also suggested, which results in a more compact rule base. Ishibuchi et al. have used GAs for selecting a small number of significant fuzzy if–then rules in [210]. In [260], an evolutionary concept is proposed which aims at generating rules that are interpretable and represent a relevant aspect of the system behavior or of the control strategy, with acceptable calculation times. Xue et al. [502] have also studied the application of genetic algorithms for optimization of fuzzy systems in prediction and control. Genetic algorithms are used in [285] to optimize the membership functions of a fuzzy logic controller for smart structure systems. A new encoding method and a fitness function with variable factors are proposed. The effectiveness of the genetic algorithm is demonstrated for a cantilever beam that is attached with piezoelectric materials.

An approach to develop general heuristics, for solving problems in knowledge-lean environments using GAs, is developed in [480]. The generalization is attempted over the problem space which is not seen during learning. The method described in [480] uses a new statistical measure called probability of win, which assesses the performance of heuristics in a distribution independent manner. To validate the proposed method, experimental results on generalizing heuristics learned for sequential circuit testing, VLSI cell placement and routing, and branch and bound search have been demonstrated.

Dev and Ram Murthy [126] presented a GA solution to the problem of optimal assignment of production rules in a knowledge base to a variable number of partitions. Partitioning of a knowledge base is an important problem since an efficient solution can increase its performance both in the compilation and the execution phases. The authors demonstrated that the solutions obtained using GAs compare favorably with those obtained with a clustering algorithm. There have been many attempts [356] for learning rules of classification. Some of them use decision trees [76, 383, 393]. A procedure has been suggested by integrating genetic algorithms with similarity based learning for acquiring rules for classification [435].

Methods for designing architectures of neural networks using GAs have also been reported. These are primarily divided into two parts: in one part GAs replace the learning to find appropriate connection weights of some predefined architecture. In another part, GAs themselves are used to find the architecture (connectivity) and it is evaluated using some learning algorithm.

Romanuik [397] has described an approach for the automatic construction of neural networks for pattern classification. It utilizes GAs to locally train network features using the perceptron rule. Transdimensional learning is introduced which can automatically adjust the learning bias inherent to all learning systems. Whitley et al. [491] suggested a way to apply GAs to neural network problems, and they showed how genetic search can be improved to achieve more accurate and consistent performance. The method is able to optimize the weighted connections for small neural network problems.

In [346], Pal and Bhandari incorporated GAs to find out the optimal set of weights (biases) in a layered network. Weighted mean square error over the training examples has been used as the fitness measure. They introduced a new concept of selection, called *nonlinear selection,* which enhances genetic homogeneity of the population and speeds up searching. Implementation results on both linearly separable and nonlinearly separable pattern sets are reported.

Harp and Samad suggested an approach [185] to evolve the architecture of neural networks and their learning parameters using GAs. The network is trained separately by backpropagation algorithm. Each chromosome contains two parts: (1) *fixed-length area parameter specification,* which corresponds to the layer number, the number of units in it, how the units are organized, learning parameters and threshold values; (2) *projection specification field,* which shows the way of connection between layers. Since the total number of layers may vary, the chromosomal length is kept variable. They also introduced a modified crossover operator which exchanges homologous segments of two chromosomes (representing two networks). Architecture, connection strengths and thresholds of individual neurons of feed-forward neural networks have been simultaneously found out in another investigation [407], without using any error propagation algorithm. In [198], the parameters of type-2 fuzzy neural networks (FNN) are learned using real-coded genetic algorithms. The type-1 FNN has the properties of parallel computation scheme, easy to implement, fuzzy logic inference system, and parameters convergence. And, the membership functions (MFs) and the rules can be designed and trained from linguistic information and numeric data. However, there is uncertainty associated with information or data. These can be treated by type-2 fuzzy sets that can model and minimize the effects of uncertainties in rule-based fuzzy logic systems. Here the fitness function is defined as

$$fitness = \frac{1}{\sum_t \sum_i (d_i(t) - y_i(t))^2},   \tag{2.16}$$

where $d_i(t)$ and $y_i(t)$ are the desired and the type-2 FNN system outputs, respectively. Some other applications of GAs for determination of the architectures of neural networks are presented in [232, 348, 411, 412].

Application of GAs to image processing problems is described in [202, 347]. A method for determining the optimal enhancement operator for both bimodal and multimodal images is described by Pal et al. in [347]. The algorithm does not need iterative visual interaction and prior knowledge of image statistics for this purpose. The fuzziness measures are used as fitness function. A modified scheme of chromosome encoding is suggested in [202], which utilizes trees and two-dimensional bit maps as its structures. This is useful in the framework of image processing, where ordinary genetic algorithms using one-dimensional strings cannot be applied in the natural way because of the loss of two-dimensional correlation in the process. In [337], the authors have applied GAs for partial shape matching, a challenging task when the image contains overlapping, noisy, occluded and partial shapes. For this purpose, they have used the attributed strings [464] for representing outline features of shapes. Another application of GAs to automatic scene labelling of remote sensing images, where the domain constraints are represented as semantic nets, is presented in [77]. Variable string length GAs have been used in [299] for pixel classification using genetic fuzzy clustering scheme. This work is further extended in [27] to take into account the spatial information in the pixels. These are described in detail in Chap. 8.

An early attempt towards using GAs for pattern recognition and subroutine selection problems is reported by Cavicchio [83], who stored the detectors from known images, along with the associated class in the training phase. In the testing phase, the detectors of the unknown image are compared to the stored ones, and an appropriate classification is performed. The detectors are encoded as genes of a chromosome, and conventional genetic operators are applied with several variants of the mutation operation.

Kelly and Davis [237] made an attempt to hybridize the k-NN classification algorithm with GAs. The k-NN algorithm uses the training data set to compute the classification of the test data set based on a metric of nearness or similarity. Ideally, it assigns equal importance to all the attributes. But in practice, this should not be the case. Consequently, a weighting factor is assigned to each attribute which determines its contribution to the overall similarity measure. In addition, the neighbors are assigned weights, that control their effect in determining the class of an unknown data point. Obviously, the nearest neighbor is given more importance and so on. GA is used to determine the weights, thereby providing the GA-based weighted k-NN (GA-WKNN) algorithm. In order to make the performance of the GA-WKNN invariant to rotation of the data set, the rotated GA-WKNN (GA-RWKNN) algorithm was proposed in [238], where GAs are again employed to search for a suitable rotation and weight vectors. The algorithm GA-RWKNN is found to significantly outperform both k-NN and Iterative Dichotomiser 3 (ID3) [383] algorithms for two data sets.

Application of GAs for editing the set of prototypes for the k-NN rule, thereby providing a reference set, is addressed by Kuncheva [261]. Knight and Sen [245] have also used GAs for developing a prototype learning system (PLEASE), where each chromosome encodes a set of class prototypes. Results are demonstrated on several two-dimensional simulated data sets, where it is found that PLEASE produces accurate solutions to most of the problems considered. In a similar fashion, GAs have been applied for selecting the initial seed points for a Radial Basis Function classifier [262]. Another attempt in this direction has been made by Zhao and Higuchi [517], where an evolutionary algorithm, called individual evolutionary algorithm (IEA) (which uses four basic operations viz., competition, gain, loss and retraining), is used for reducing the set of prototypes for the k-NN rule. IEA has also been used by them for obtaining reduced NN-MLP (a nearest neighbor-based MLP) in terms of the number of neurons [516].

Srikanth et al. [448] described a genetic algorithmic approach to pattern classification, both crisp and fuzzy, where clusters in pattern space are approximated by ellipses or sets of ellipses in two dimensions and ellipsoids in general. A variable number of ellipsoids is searched for, which collectively classify a set of objects. Since the number of ellipsoids is not fixed a priori, a variable-length chromosome is considered. The performance of the classifier is illustrated using two two-dimensional data sets. Performance on a four-dimensional data set is compared with those of neural networks for several architectures. The comparison is marginally in favor of the genetic technique presented. In [343] a similar work is described where the boundaries between classes are approximated using a number of hyperplanes. This work has been subsequently extended to variable length representation [36, 38] and incorporation of the concept of chromosome differentiation in GAs [39, 41]. These are described in Chaps. 3–6 of this book.

Murthy and Chowdhury [325] have used GAs for finding optimal clusters, without the need for searching all possible clusters. The experimental results show that the GA-based scheme may improve the final output of the K-means algorithm [461], where an improvement is possible. Another application of GA for clustering, feature selection and classification is reported in [465]. Sarkar et al. [413] have proposed a clustering algorithm using evolutionary programming-based approach, when the clusters are crisp and spherical. The algorithm automatically determines the number and the center of the clusters, and is not critically dependent on the choice of the initial clusters. A series of work using GAs for clustering in both crisp and fuzzy domains is reported in [27, 31, 32, 34, 298, 299]. Some of these will be discussed in detail in Chap. 8.

Piper [376] has described the use of GAs in chromosome classification. The fitness function takes into account constraints imposed by the context of a metaphase cell, as well as similarity of homologues. Comparison with previous classification methods demonstrates similar classification accuracy. Incorporation of the homologue similarity constraint does not substantially improve the error rate. Selection of a subset of principal components for classification using GAs is made in [380]. Since the search space depends on the product of the number of classes and the number of original features, the process of selection by conventional means may be computationally very expensive. Results on two data sets with small and large cardinalities are presented. Another application of GAs for searching for error correcting graph isomorphism, an important issue in pattern recognition problems, is described in [485].

In [498], application of GAs to the problem of learning how to control advanced life support systems, a crucial component of space exploration, is studied. Performance is compared with a stochastic hill-climber (SH) for two different types of problem representations, binary and proportional representations. Results indicate that although proportional representation can effectively boost GA performance, it does not necessarily have the same effect on other algorithms such as SH. Results also indicate that multivector control strategies are an effective method for control of coupled dynamical systems. Besides these, several applications of GAs for solving a variety of problems arising in industry are discussed in [473, 496].

Hybridization of genetic algorithms with other components of the soft computing paradigm like fuzzy sets and neural networks have proved effective. Different hybrid genetic algorithms have been developed for various application domains. Geophysical static corrections of noisy seismic data using a hybrid genetic algorithm has been described in [296]. An iterative rule learning using a fuzzy rule-based genetic classifier to provide an intelligent intrusion detection system is described in [338]. Several advancements in genetic fuzzy systems are also available in [102].

A hybrid technique involving artificial neural network and GA is described in [358] for optimization of DNA curvature characterized in terms of the reliability value. In this approach, first an artificial neural network (ANN) approximates (models) the nonlinear relationship(s) existing between its input and output example data sets. Next, the GA searches the input space of the ANN with a view to optimize the ANN output. The method of prediction of eukaryotic PolII promoters from DNA sequence [248], takes advantage of the hybridizations of neural networks and genetic algorithms to recognize a set of discrete subpatterns with variable separation as a promoter. The neural networks use, as input, a small window of DNA sequence, as well as the output of other neural networks. Through the use of genetic algorithms, the weights in the neural networks are optimized to discriminate maximally between promoters and nonpromoters. Several other applications of GAs for solving problems related to bioinformatics as well as Web intelligence are discussed in Chaps. 9 and 10, respectively, of this book.

2.8 Summary

Genetic algorithms and their variants have attracted the attention of researchers for well over the last two decades. The defining characteristics of GAs are their population-based nature and encoding of the search space. In contrast to classical optimization techniques, GAs make no assumptions about the nature of the solution space, except that it must be bounded. They have been fairly successful in solving complex, multimodal problems in extremely large search spaces.

An important characteristic of GAs is their flexibility. Since there are no hard and fast guidelines regarding the operators, their definitions, probabilities of their use, etc., new operators can be defined that may be more suited to the problem under consideration. Domain-specific knowledge may also be incorporated in the process. GA parameters can be tuned appropriately. The downside of this amount of flexibility is that new users are often confused and intimidated. Parameter values are often set using trial and error, leading to wasted time. Moreover, GAs also tend to be computationally time consuming, though they have been found to provide superior results as compared to other heuristic methods that are faster. A possible compromise is to incorporate results of heuristic and/or local search in GA. This has given rise to a new branch of study, namely memetic algorithms [186].

The following chapters of this book are devoted to the application of GAs in classification and clustering problems, bioinformatics and Web intelligence. Both experimental and theoretical analyses are provided.

3

Supervised Classification Using Genetic Algorithms

3.1 Introduction

The problem of classification, as mentioned earlier, involves taking a pattern, characterized by a set of features, as input, and making a decision about how well it belongs to one (or more) of the classes as output. In case the classifier is designed using a set of labelled patterns, it is called a supervised classifier. The classification problem can be modelled in a variety of ways, e.g., by generating a set of discriminatory if–then rules, learning optimal decision trees, generating class boundaries capable of distinguishing among the different classes. Most of these approaches may, in turn, be modelled as problems of optimization of some criterion. Hence genetic algorithms appear to be a natural choice for developing supervised classifiers.

In this chapter, a brief discussion on some such applications of GAs for designing supervised classification systems by generating fuzzy if–then rules and optimizing decision trees are provided. Subsequently, the application of GAs in searching for a number of linear segments which can approximate the nonlinear class boundaries of a data set while providing minimum misclassification of training sample points is described in detail. A distinguishing feature of this approach is that the boundaries (approximated by piecewise linear segments) need to be generated explicitly for making decisions. This is unlike the traditional methods or the multilayered perceptron (MLP) based approaches, where the generation of boundaries is a consequence of the respective decision making processes. The issue of fitting higher-order surfaces, instead of linear ones, is addressed. The criteria of selecting the genetic parameters are discussed. The merits of this GA-based scheme over other conventional classifiers, e.g., those based on k-NN rule, Bayes' maximum likelihood ratio and MLP, are explained.

This chapter is organized as follows: Sects. 3.2 and 3.3 provide a description of the use of GAs for designing a classifier system using linguistic if–then rules and decision trees, respectively. A detailed description of the genetic classifier where the class boundaries are approximated using a number of hyperplanes is provided in Sect. 3.4. The important subtasks are explained though examples. Section 3.5 provides the experimental results. Finally Sect. 3.6 summarizes the chapter.

3.2 Genetic Algorithms for Generating Fuzzy If–Then Rules

This section describes a classifier system with linguistic if–then rules, where genetic algorithms are used to optimize the number of rules and to learn the grade of certainty (CF) associated with each rule. It is described in detail in [207]. Here, a small number of linguistic rules are selected by a GA to construct a compact fuzzy classification system. To improve the performance of the system, a hybrid approach that incorporates learning procedures into the GA is used.

Here initially the set of all possible rules is considered [207]. For example, for a two-class, two-dimensional problem with six linguistic states, the rule R_{ij} is written as:

$$R_{ij}: \quad \text{If } x_{p1} \text{ is } A_i \text{ and } x_{p2} \text{ is } A_j,$$
$$\text{then } (x_{p1}, \ x_{p2}) \text{ belongs to class } C_{ij} \text{ with } CF = CF_{ij},$$
$$i = 1, 2, \ldots, 6; \ j = 1, 2, \ldots, 6.$$

Here, A_i and A_j are the linguistic labels. The consequent class C_{ij} and the grade of certainty of the rule CF_{ij} are computed as follows. First of all, obtain

$$\beta_{C1} = \sum_{x_p \in C1} \mu_i(x_{p1})\mu_j(x_{p2}), \qquad (3.1)$$

$$\beta_{C2} = \sum_{x_p \in C2} \mu_i(x_{p1})\mu_j(x_{p2}), \qquad (3.2)$$

where $\mu_i(x_{p1})$ and $\mu_j(x_{p2})$ are the membership functions of the linguistic labels A_i and A_j, respectively. β_{C1} and β_{C2} can be viewed as the indices that measure the number of compatible patterns with the antecedent of the linguistic rule R_{ij}.

The consequent class C_{ij} is then computed as

$$C_{ij} = \begin{cases} C1, & \text{if } \beta_{C1} > \beta_{C2}, \\ C2, & \text{if } \beta_{C1} < \beta_{C2}, \\ \emptyset, & \text{if } \beta_{C1} = \beta_{C2}. \end{cases} \qquad (3.3)$$

In case $C_{ij} = \emptyset$, the rule is called a dummy rule. Such rules may also result if none of the patterns matches the consequent class.

The grade of certainty CF_{ij} is computed as

$$CF_{ij} = \begin{cases} (\beta_{C1} - \beta_{C2})/(\beta_{C1} + \beta_{C2}), & \text{if } \beta_{C1} > \beta_{C2}, \\ (\beta_{C2} - \beta_{C1})/(\beta_{C1} + \beta_{C2}), & \text{if } \beta_{C1} < \beta_{C2}, \\ 0, & \text{if } \beta_{C1} = \beta_{C2}. \end{cases} \qquad (3.4)$$

The entire set of rules is denoted by S_{ALL}, and a subset of S_{ALL} by S. The problem of rule selection is formulated as finding a rule set S with high classification power and a small number of significant linguistic rules from S_{ALL}. Hence, the objectives are to (i) maximize the number of correctly classified training patterns by S, and (ii) minimize the number of linguistic rules in S. These two objectives are combined into a single function as follows:

$$\text{Maximize } W_{NCP}.NCP(S) - W_S.|S|, \text{ subject to } S \subseteq S_{ALL}. \qquad (3.5)$$

Here, $NCP(S)$ denotes the number of correctly classified training patterns by S, $|S|$ denotes the cardinality of S, and W_{NCP} and W_S are the weights associated with the two terms. In general, $0 < W_S << W_{NCP}$.

In order to classify a pattern $x_p = (x_{p1}, x_{p2})$, α_{C1} and α_{C2} are calculated as:

$$\alpha_{C1} = \max\{\mu_i(x_{p1}).\mu_j(x_{p2}).CF_{ij}|C_{ij} = C1 \text{ and } R_{ij} \in S\}, \qquad (3.6)$$

$$\alpha_{C2} = \max\{\mu_i(x_{p1}).\mu_j(x_{p2}).CF_{ij}|C_{ij} = C2 \text{ and } R_{ij} \in S\}. \qquad (3.7)$$

If $\alpha_{C1} > \alpha_{C2}$, then x_p is classified to class 1. If $\alpha_{C1} < \alpha_{C2}$, then x_p is classified to class 2. Otherwise, when $\alpha_{C1} = \alpha_{C2}$, the pattern is left unclassified.

For designing the genetic fuzzy classification system, the fitness function may be taken as in Eq. (3.5). However, in [207] an additional term called the fineness of a rule is incorporated in the fitness function, which increases the fitness value of a rule that is more general. An additional linguistic label, called "don't care", is considered, and the fineness values are defined as follows:

$$fineness(A_i) = 5, \text{ where } A_i \text{ is a valid linguistic label,}$$
$$fineness(A_i) = 1, \text{ if } A_i = don't\ care.$$

The fineness of a linguistic rule R_{ij}, mentioned earlier, is therefore defined as

$$fineness(R_{ij}) = fineness(A_i) + fineness(A_j). \qquad (3.8)$$

This term is used as a penalty function by replacing $|S|$ in Eq. (3.5) with $\sum_{R_{ij} \in S} fineness(R_{ij})$. The modified function therefore becomes

$$W_{NCP}.NCP(S) - W_S. \sum_{R_{ij} \in S} fineness(R_{ij}). \qquad (3.9)$$

Each individual in the GA encodes a rule set S. A rule set S is denoted by a string $s_1 s_2 \ldots s_N$, where

$\qquad N = K^2$, the number of $K \times K$ rules in S_{ALL}, K denoting the number of linguistic labels,

$\qquad s_r = 1$, if rth rule belongs to S,

$\qquad s_r = -1$, if rth rule does not belong to S, and

$\qquad s_r = 0$, if rth rule is a dummy rule.

A rule is called dummy if the consequent part of the rule is $\emptyset$, and if there are no patterns that match its antecedent part. Such rules play no part in the inferencing process. The index r of a rule R_{ij} is computed as

$$r = K(i-1) + j. \tag{3.10}$$

A chromosome $s_1 s_2 \ldots s_N$ is decoded as

$$S = \{R_{ij} | s_r = 1; r = 1, 2, \ldots, N\}, \tag{3.11}$$

and its fitness is computed with Eq. (3.9).

For initializing the population of GAs, all the dummy rules are first assigned zero values in the corresponding positions of the chromosomes. Each nondummy rule is included in the chromosome (by setting the corresponding position to 1) with a probability of 0.5. Otherwise, the position is set to -1, indicating that the corresponding rule is absent from the rule set encoded in the chromosome.

The probability $P(S)$ of selecting a chromosome with rule set S in a population ψ is computed as

$$P(S) = \frac{f(S) - f_{min}(\psi)}{\sum_{S' \in \psi} f(S') - f_{min}(\psi)}, \tag{3.12}$$

where

$$f_{min}(\psi) = \min\{f(S) | S \in \psi\}. \tag{3.13}$$

Uniform crossover [453] is applied on pairs of strings. For mutation, 1 is changed to -1 and vice versa, probabilistically. In [207], the probability of changing from 1 to -1 (i.e., deselecting a rule) is kept higher than the converse. Elitism is incorporated by randomly replacing one chromosome of the current population by the best chromosome of the previous generation (i.e., one with maximum fitness value). The process is continued until some prespecified stopping criterion is attained.

Results on an artificially generated data set, and the well-known Iris data are provided in [207] to demonstrate the effectiveness of the genetic pattern classification system. As an extension to this work, a learning procedure, which appropriately modifies the certainty factor of the rules based on the patterns that they correctly and incorrectly classify, is incorporated to yield a hybrid genetic classifier. After the application of the genetic operators in an iteration, a learning phase is applied on all the rule sets generated. The learning algorithm is as follows:

Repeat for a specified number of iterations

    For each pattern x_p, $p = 1, 2, \ldots, m$ do the following;

        Classify x_p

        Identify the rule that is responsible for classification of x_p

        (from Eq. (3.6) or (3.7)).

        If x_p is correctly classified, increase the corresponding CF_{ij} as

$$CF_{ij} = CF_{ij} + \eta_1(1 - CF_{ij}),$$

where $0 < \eta_1 < 1$ is a positive learning constant.

Otherwise, decrease the corresponding CF_{ij} as

$$CF_{ij} = CF_{ij} - \eta_2(1 - CF_{ij}),$$

where $0 < \eta_2 < 1$ is also a positive learning constant.

Since there are many more correctly classified patterns than incorrectly classified ones, a larger value is assigned to η_2 than to η_1. Ishibuchi et al. [207] report a classification rate of 100% on Iris data with the hybrid genetic classification system, which indicates a gain over the version without the learning phase.

3.3 Genetic Algorithms and Decision Trees

Induction of decision trees constitutes an important approach to classification by learning concepts from labelled examples. Quinlan [383] initiated the development of ID3, which is an iterative method for constructing decision trees from a limited number of training instances. Initially, ID3 was able to cope with categorical data only. Thereafter, several researchers tried to overcome this limitation of ID3. In this regard, Quinlan developed the C4.5 algorithm [384, 385], which is widely used for supervised pattern classification. Hybridization of genetic algorithms and genetic programming with decision trees for classification and rule generation has also been attempted [82, 254, 340, 341, 466]. Here we discuss briefly one such attempt reported in [466].

In [466], Turney describes an algorithm called Inexpensive Classification with Expensive Tests (ICET) that uses GAs for evolving the biases of a decision tree induction algorithm for cost-sensitive classification. Cost-sensitive classification takes into account the cost of classification errors as well as the cost of the tests in order to decide whether to go in for a test or not. If tests are inexpensive relative to the cost of classification errors, then it is better to perform all the tests. However, if this is not true, it may be unwise to perform all the tests. Moreover, in many situations, the cost of a test is conditional to the prior test on the path in the decision tree. It may so happen that two or more tests on the path share some common costs. The shared cost should logically be considered only once. For example, a set of blood tests share the cost of collecting blood. This cost is charged only once, when the first blood test is carried out. In such situations, the optimal decision tree is the one that optimizes the cost of the tests as well as the classification errors, given both these costs and the conditional probabilities of test results and the classes. Searching for an optimal decision tree is, in general, infeasible. ICET attempts to find a good tree in this regard.

In ICET, the cost of classification of a case is computed as the sum of (i) the classification error (which is provided in a $c \times c$ matrix, c being the number of distinct classes) and (ii) the total cost of all the tests in the corresponding path from the root to the leaf (if in this path the same test appears more than once, its cost is considered only once, assuming that the result of the test when performed first is available later as well).

Another practical issue in this regard that has been taken care of in [466], albeit in a crude way, is the fact that test results may not be immediately available; i.e., the results may be delayed. The solution suggested is as follows. Let T be such a test whose result is delayed. Then all the tests in the tree rooted at T are performed, and the cost of all these tests are taken into account while computing the cost of classification.

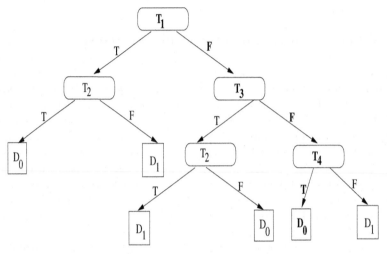

Fig. 3.1. Decision tree [466]

A simple example in the medical domain can be considered to illustrate the process of cost computation used in [466]. Let there be four tests T_1, T_2, T_3 and T_4, such that the results of T_1 and T_2 are immediately available, while those of T_3 and T_4 are delayed. Let the last two tests share a common cost of $2.00. Table 3.1 shows the above mentioned situation. Assume that each test has two possible outcomes, True (T) or False (F), and there are two disease classes D_0 and D_1. Figure 3.1 shows a decision tree corresponding to this problem. The classification cost matrix is shown in Fig. 3.2. Consider the case shown in Table 3.2 (the corresponding path in the decision tree is indicated by boldface letters in Fig. 3.1). The first step is to do the test at the root of the tree (T_1). In the second step, a delayed test (T_3) is encountered, so the cost of the entire subtree rooted at this node is computed. Note that T_4 only costs $8.00, since T_3 has already been selected, and T_3 and T_4 have a common cost. In the third step, T_4 is performed, but no payment is required, since it was already paid for in the second step. In the last step, the class of the case is guessed, incorrectly, as D_0. So a penalty of $50.00 is paid. Therefore, the overall cost computed for this case is $80.00.

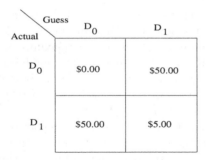

Fig. 3.2. Classification cost matrix [466]

Table 3.1. Test costs for an example [466]

Test	T_1	T_2	T_3	T_4
Whether delayed	No	No	Yes	Yes
Cost	$5.00	$10.00	$7.00 if T_3 or T_4 not performed earlier $5.00 otherwise	$10.00 if T_3 or T_4 not performed earlier $8.00 otherwise

Table 3.2. Computing the cost for a particular case [466]

Test	Delayed?	Result	Next tests	Reason	Cost
T_1	No	F	T_3	Right branch from root	$5.0
T_3	Yes	F	T_2, T_4	T_3 is delayed test T_3 T_4 share costs	$7.00 +$10.00 + $8.00 = $25.00
T_4	Yes	T	–	Reached a leaf node	Already paid earlier
–	–	guess = D_0	actual = D_1	Wrong guess	$50.00
Total cost	–	–	–	–	$80.00

ICET uses a hybrid of GAs and a decision tree induction algorithm, EG2 [332]. This is a top down induction of decision trees, which uses the Information Cost Function (ICF) for selection of attributes to test at a particular level of the decision tree. The ICF is a function of both the information gain (ΔI), and cost (C) of measuring the attribute. ICF for the ith attribute is computed as

$$ICF_i = \frac{2^{\Delta I_i} - 1}{(C_i + 1)^\omega}, \quad \text{where } 0 \leq \omega \leq 1. \tag{3.14}$$

Note that C4.5 [384] selects the attribute that maximizes the information gain ratio which is a function of the information gain ΔI_i. On the contrary, EG2 selects the attribute for which ICF_i is maximized. In the experiments reported by Turney, ICET is implemented by modifying the source code of C4.5 so that it selects the attribute for testing on the basis of ICF_i, rather than the information gain ratio. Also, ω was set to 1, thereby making it strongly biased by the cost of measuring an attribute.

ICET uses a two-tiered search strategy. On the bottom tier, EG2 performs greedy search through the search space of decision trees. On the top, a GA performs the genetic search through a space of biases, where the costs C_is are assumed to be the bias parameters, in contrast to cost parameters as originally modelled in [332]. The C_is are encoded in a chromosome. With each such chromosome, EG2 is executed in order to generate a decision tree, based on some training data (which is taken as a portion of the entire data, while the remaining data is taken as the test data). The fitness of the chromosome is taken as the average classification cost of the resultant decision tree computed over the test data. The genetic operators, selection, crossover and mutation are applied as usual, and the process repeats for a number of generations before termination. At the end, the fittest chromosome provides the final bias values (C_is). These are then used, along with the entire data set, for a new run of EG2 in order to generate the final decision tree that is output to the user. The author compares the performance of the ICET algorithm with four other techniques, three of which belong to the category of cost sensitive classification like ICET, while the fourth, C4.5, ignores the costs. It is found that the performance of ICET is significantly better when the cost of the tests is of the same order of magnitude as the cost of misclassification errors, than the performance of other competing methods, although ICET requires considerably more execution time.

3.4 *GA-classifier*: Genetic Algorithm for Generation of Class Boundaries

Pattern classification can be viewed as a problem of generating decision boundaries that can successfully distinguish the various classes in the feature space. In real-life problems, the boundaries between the different classes are usually nonlinear. In this section we describe a classifier where the characteristics of GAs are exploited in searching for a number of linear segments which can approximate the nonlinear boundaries while providing minimum misclassification of training sample points. Such a classifier, designed using GA, is subsequently referred to as the *GA-classifier* [35, 343].

3.4.1 Principle of Hyperplane Fitting

The *GA-classifier* [343] attempts to place H hyperplanes in the feature space appropriately such that the number of misclassified training points is minimized. From elementary geometry, the equation of a hyperplane in N-dimensional space $(X_1 - X_2 - \cdots - X_N)$ is given by

$$x_N \cos \alpha_{N-1} + \beta_{N-1} \sin \alpha_{N-1} = d, \tag{3.15}$$

where $\beta_{N-1} = x_{N-1} \cos \alpha_{N-2} + \beta_{N-2} \sin \alpha_{N-2}$,
$\qquad \beta_{N-2} = x_{N-2} \cos \alpha_{N-3} + \beta_{N-3} \sin \alpha_{N-3}$,

$$\vdots$$

$\qquad \beta_1 = x_1 \cos \alpha_0 + \beta_0 \sin \alpha_0$.

The various parameters are as follows:

X_i: the ith feature (or dimension).

$(x_1, x_2, \ldots, x_N)$: a point on the hyperplane.

α_{N-1}: the angle that the unit normal to the hyperplane makes with the X_N axis.

α_{N-2}: the angle that the projection of the normal in the $(X_1 - X_2 - \cdots - X_{N-1})$ space makes with the X_{N-1} axis.

$$\vdots$$

α_1: the angle that the projection of the normal in the $(X_1 - X_2)$ plane makes with the X_2 axis.

α_0: the angle that the projection of the normal in the (X_1) plane makes with the X_1 axis $= 0$. Hence, $\beta_0 \sin \alpha_0 = 0$.

d: the perpendicular distance of the hyperplane from the origin.

Thus the N-tuple $< \alpha_1, \alpha_2, \ldots, \alpha_{N-1}, d >$ specifies a hyperplane in an N-dimensional space.

Each angle α_j, $j = 1, 2, \ldots, N - 1$ is allowed to vary in the range of 0 to 2π. If b_1 bits are used to represent an angle, then the possible values of α_j are

$$0, \delta * 2\pi, 2\delta * 2\pi, 3\delta * 2\pi, \ldots, (2^{b_1} - 1)\delta * 2\pi,$$

where $\delta = \frac{1}{2^{b_1}}$. Consequently, if the b_1 bits contain a binary string having the decimal value v_1, then the angle is given by $v_1 * \delta * 2\pi$.

Once the angles are fixed, the orientation of the hyperplane becomes fixed. Now only d must be specified in order to specify the hyperplane. For this purpose the hyper-rectangle enclosing the training points is considered. Let (x_i^{min}, x_i^{max}) be the minimum and maximum values of feature X_i as obtained from the training points. Then the vertices of the enclosing hyper-rectangle are given by

$$(x_1^{ch_1}, x_2^{ch_2}, \ldots, x_N^{ch_N}),$$

where each ch_i, $i = 1, 2, \ldots, N$ can be either *max* or *min*. (Note that there will be 2^N vertices.) Let *diag* be the length of the diagonal of this hyper-rectangle given by

$$diag = \sqrt{(x_1^{max} - x_1^{min})^2 + (x_2^{max} - x_2^{min})^2 + \ldots + (x_N^{max} - x_N^{min})^2}.$$

A hyperplane is designated as the *base hyperplane* with respect to a given orientation (i.e., for some $\alpha_1, \alpha_2, \ldots, \alpha_{N-1}$) if

(i) It has the same orientation.

(ii) It passes through one of the vertices of the enclosing rectangle.

(iii) Its perpendicular distance from the origin is minimum (among the hyperplanes passing through the other vertices). Let this distance be d_{min}.

If b_2 bits are used to represent d, then a value of v_2 in these bits represents a hyperplane with the given orientation and for which d is given by $d_{min} + \frac{diag}{2^{b_2}} * v_2$.

Thus, each chromosome is of a fixed length of $l = H((N-1) * b_1 + b_2)$, where H denotes the number of hyperplanes. These are initially generated randomly for a population of size Pop. The following example illustrates the principles of the chromosome encoding strategy.

Example 3.1. Let $n = 4$, $b_1 = 3$, $b_2 = 3$, $N = 3$ and $H = 1$. Then a string

$$\overset{\alpha_1}{\overbrace{0\ 0\ 0}} \quad \overset{\alpha_2}{\overbrace{0\ 1\ 0}} \quad \overset{d}{\overbrace{1\ 0\ 0}}$$

indicates a plane whose

(i) normal makes an angle $(2 * \frac{2\pi}{8} = \frac{\pi}{2})$ with the X_3 axis, since $\alpha_2 = 2$, and $2^{b_1} = 2^3 = 8$,

(ii) projection of the normal in the $X_1 - X_2$ plane makes an angle 0 with the X_2 axis, since $\alpha_1 = 0$, and

(iii) the perpendicular distance from the origin is $d_{min} + 4 * \frac{diag}{2^3}$, since $d = 4$.

In other words, the plane lies parallel to the $X_1 - X_3$ plane. The equation of the plane is

$$x_2 = d_{min} + 4 * \frac{diag}{2^3} = d_{min} + \frac{diag}{2}.$$

3.4.2 Region Identification and Fitness Computation

The computation of the fitness is done for each string in the population. The fitness of a string is characterized by the number of points it misclassifies. A string with the lowest misclassification is therefore considered to be the fittest among the population of strings. Note that every string str_i, $i = 1, 2, \ldots, Pop$, represents H hyperplanes denoted by pln_j^i, $j = 1, 2, \ldots, H$, in the N-dimensional space.

For each pln_j^i, the parameters $\alpha_1^{ij}, \alpha_2^{ij}, \ldots, \alpha_{N-1}^{ij}$ and d^{ij} are retrieved (see Sect. 3.4.1). For each training pattern point $(x_1^m, x_2^m, \ldots x_N^m)$, $m = 1, 2, \ldots, n$, the sign with respect to the hyperplane pln_j^i, $(j = 1, 2, \ldots, H)$, i.e., the sign of the expression

$$x_N^m \cos \alpha_{N-1}^{ij} + \beta_{N-1}^{ijm} \sin \alpha_{N-1}^{ij} - d^{ij} \qquad (3.16)$$

is found where β_{N-1}^{ijm} corresponds to the jth hyperplane of the ith chromosome for the mth data point, and its form is as defined in Eq. (3.15). The sign is digitized as 1(0) if the point lies on the positive (negative) side of the hyperplane. The process is repeated for each of the hyperplanes, at the end of which a string $sign_i^m$, subsequently referred to as the *sign_string*, is obtained. This string, of length H, corresponds to the classification yielded by the string str_i of the population, for the mth training pattern. Note that, each unique *sign_string* corresponds to a unique region provided by the classifier.

In order to compute the misclassification associated with a particular arrangement of the lines (or a particular string), use of an auxiliary matrix (*class_matrix*) is made. This is a $2^H * K$ matrix, where each row corresponds to a unique decimal value of the *sign_string*, and each column, 1 to K, corresponds to a class. This matrix is initialized with zeros. For each training point, the *sign_string* is formulated, and its decimal value, *dec(sign_string)*, is computed. If the class of the point is i, then the entry in the location $[dec(sign\_string), i]$ of matrix *class_matrix* is incremented by 1. At the same time, the entry in the 0th column of this row is set to 1, indicating that some training data points lie in the region represented by the *sign_string*. This process is repeated for all the n training points.

In order to associate each region of the search space corresponding to sign string Reg_i with a particular class (denoted by $Class(Reg_i)$), each row of *class_matrix* (indicating a specific region) having a 1 in its 0th column is considered. Its maximum value is computed. Then the corresponding column number (class) is the label that is associated with the said region. In other words,

$$Class(Reg_i) = \text{argmax}_{p=1,2,...,K} class\_matrix[dec(Reg_i), p]$$
$$\text{such that } class\_matrix[dec(Reg_i), 0] \neq 0.$$

All other points in this region are misclassified. The steps for computing the number of misclassified points are given in Fig. 3.3.

It is to be noted that although *sign_string* can take on at most 2^H possible values (since H lines will yield a maximum of 2^H possible classifications), all of them may not occur in practice. Also, it may so happen that the maximum entries for two (or more) different *sign_strings* may correspond to the same class. In that case, all these strings (correspondingly, union of all the different regions) are considered to provide the region for the said class. A tie is resolved arbitrarily. The example stated below will clarify this method.

Example 3.2. Let there be 8 training patterns belonging to two classes, 1 and 2, in a two-dimensional feature space $X_1 - X_2$. Let us assume H to be 3, i.e., 3 lines will be used to classify the points. Let the training set and a set of three lines be as shown in Fig. 3.4. Each point i_j, $i = 1, 2, \ldots, 8$, and $j = 1, 2$ indicates that it is the ith training point and that it belongs to class j. Let

---

Begin
Initialize *class_matrix* with 0's
for $i = 1$ to n /** each training data point **/
Begin

       $m \leftarrow$ class of point i.

       Compute the *sign_string*

       Compute $dec =$ decimal value of *sign_string*

       Increment *class_matrix*[dec, m]

       class_matrix[$dec, 0$] $= 1$ /** Some points lie in this region **/

End
$miss = 0$ /** To hold the total misclassification **/
for $i = 0$ to 2^H
Begin

       If *class_matrix*[$i, 0$] $= 1$

       Begin

              Compute $max =$ maximum of *class_matrix*[i]

              Let m be the corresponding class

              Output $\rightarrow$ Region i associated with class m

              Compute $miss = miss + \sum_{j=1}^{k} class\_matrix[i, j] - max$

       End

End
Output $\rightarrow$ Total misclassification $= miss$.
End

Fig. 3.3. Steps for computing the number of misclassified training points

the positive and the negative sides of the lines be as shown in Fig. 3.4. Then, point 1_1 yields a *sign_string* 111 since it lies on the positive side of all the three lines $Line_1$, $Line_2$ and $Line_3$. The corresponding *class_matrix* formed for all the eight points is shown in Fig. 3.5. It is to be noted that one region (denoted by *sign_string* 110 or row 6) is empty, while two regions (100 and 101, or rows 4 and 5, respectively) do not exist. The number of misclassifications here is found to be $1 + 1 = 2$, one each for *sign_strings* 001 (row 1) and 111 (row 7). Note that in this example both the strings 000 (row 0) and 001 (row 1) are providing the regions for class 2 (assuming that the tie for region 111 is resolved in favor of class 1).

If the number of misclassified points for a string is denoted by *miss*, then the fitness of the string is computed as $(n - miss)$, where n is the number of training data points. The best string of each generation or iteration is the one which has the fewest misclassifications. This string is stored after each iteration. If the best string of the previous generation is found to be better than the best string of the current generation, then the previous best string replaces the worst string of the current generation. This implements the *elitist strategy*, where the best string seen up to the current generation is propagated to the next generation. The flowchart for the *GA-classifier* is given in Fig. 3.6.

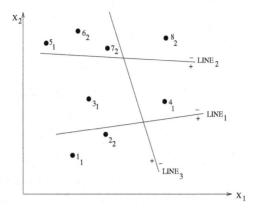

Fig. 3.4. Region identification for $H = 3$ and $n = 8$

	0	1	2
0	1	0	1
1	1	1	2
2	1	1	0
3	1	1	0
4	0	0	0
5	0	0	0
6	0	0	0
7	1	1	1

Fig. 3.5. *class_array* for the example in Fig. 3.4

3.4.3 Genetic Operations

The *roulette wheel* selection procedure has been adopted here to implement a *proportional selection* strategy. Single-point crossover is applied according to a fixed crossover probability μ_c. Mutation is done on a bit-by-bit basis (for binary strings) [147, 164] according to a mutation probability μ_m. The process of fitness computation, selection, crossover and mutation continues for a fixed number of iterations *max_gen* or till the termination condition (a string with misclassification number reduced to zero, i.e., *miss* = 0 or $fit_m = n$ in Fig. 3.6) is achieved.

3.5 Experimental Results

The effectiveness of the *GA-classifier* in classifying both overlapping, and nonoverlapping, nonconvex regions is extensively demonstrated on several artificial and real-life data sets having widely varying characteristics [25, 343]. Here, some results are provided on an artificial data, viz., ADS 1 (Fig. 3.7), and two real-life data sets, viz., Iris data and a speech data, *Vowel* (Fig. 3.8). The

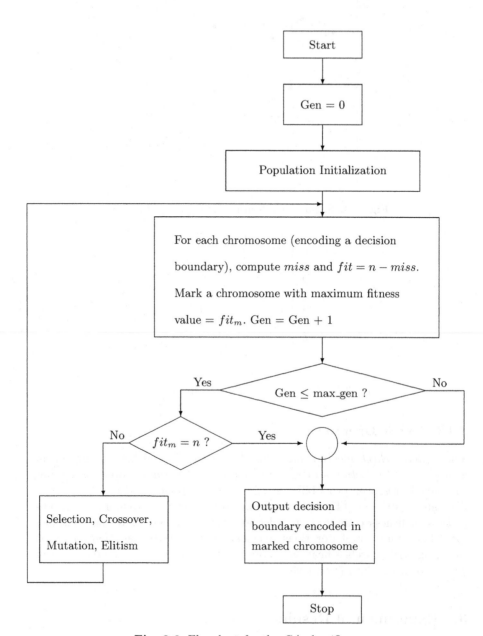

Fig. 3.6. Flowchart for the *GA-classifier*

generalization ability of the decision boundary provided by the *GA-classifier* is demonstrated pictorially. The results are compared with those of Bayes' maximum likelihood classifier (which is known to provide optimal performance from a statistical viewpoint, when the a priori probabilities and class conditional densities are known exactly), k-NN classifier and MLP (where both the k-NN classifier and the MLP with hard limiters are known to provide piecewise linear boundaries) [155, 461]. These existing classifiers are described in Chap. 1.

Data Sets

ADS 1

This artificial data set is shown in Fig. 3.7. It has 557 points in 2 classes, of which 460 points are in class 1 and 97 points are in class 2. These are also mentioned in Appendix B. The class boundary of the data sets is seen to be highly nonlinear, although the classes are separable.

Vowel Data

As mentioned in Appendix B, this speech data (Indian Telugu vowel sounds) consists of 871 patterns having 3 features F_1, F_2 and F_3, and 6 vowel classes $\{\delta, a, i, u, e, o\}$. Figure 3.8 shows the distribution of the six classes in the $F_1 - F_2$ plane. (It is known [350] that these two features are more important in characterizing the classes than F_3.) Note that, the boundaries of the classes are seen to be ill-defined and overlapping.

Iris Data

This data, described in Appendix B, has 4 feature values, and 3 classes Setosa, Versicolor and Virginica (labelled 1, 2 and 3), with 50 samples per class. It is known that two classes Versicolor and Virginica have a large amount of overlap, while the class Setosa is linearly separable from the other two.

Selection of Control Parameters

The population size and crossover probability are kept fixed at 20 and 0.8, respectively. The other parameter chosen is the mutation probability value, the importance of which has been stressed in [324, 416]. In [60], it is shown that the mutation probability should be chosen in the range of $(0, \frac{a-1}{a}]$, where a is the cardinality of the set of alphabets for the strings (cardinality= 2 for binary strings). In [324], it is proved that in many situations, the mutation probability value could be at least equal to $\frac{c}{l}$, where c is the number of bit positions of a string in a generation that needs to be changed for arriving at the optimal solution for binary strings. However, such a priori knowledge of c is almost impossible to acquire.

Here, the mutation probability value is considered to vary approximately in the range of [0.015,0.333] as shown in Fig. 3.9. The range is divided into

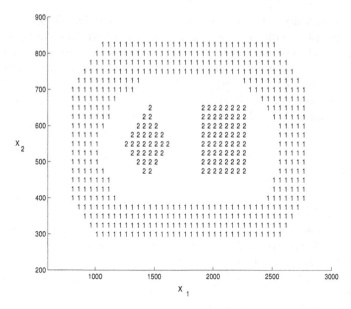

Fig. 3.7. ADS 1

eight equispaced values. Initially μ_m has a high value ($=0.333$), which is slowly decreased in steps to 0.015. Then μ_m is again increased in steps to 0.333. One hundred iterations are executed with each value of μ_m, thereby giving a maximum of 1500 iterations. Given that the number of bits in a string is l, the number of different possible strings is 2^l. The objective is to search this space and arrive at the optimal string as fast as possible. The initial high value of the mutation probability ensures sufficient diversity in the population, which is desired, as at this stage the algorithm knows very little about the nature of the search space. As generations pass, the algorithm slowly moves towards the optimal string. It is therefore necessary that the space be searched in detail without abrupt changes in population. Consequently, the mutation probability value is decreased gradually until it is sufficiently small. It may so happen that in spite of this, the optimal string obtained so far has a large Hamming distance from the actual optimal string. This may very well happen for deceptive problems [106, 164]. Thus, if one continues with the small mutation probability value, it may be too long before the optimal string is found. So to avoid being stuck at a local optima, the value is again gradually increased. Even if the optimal string had been found earlier, one loses nothing since the best string is always preserved in subsequent generation of strings. Ideally the process of decreasing and then increasing the mutation probability value should continue, but here the cycle was restricted to just one due to practical limitations. The process is terminated if the maximum number of iterations has been executed or a string with zero misclassification is attained.

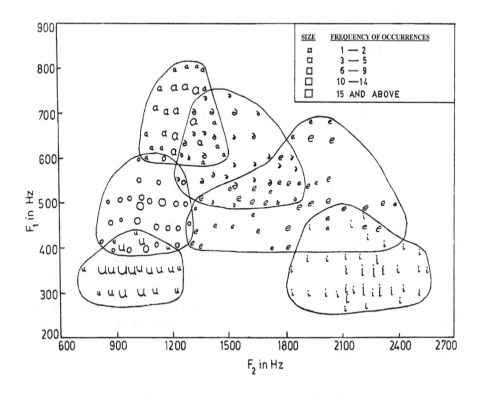

Fig. 3.8. *Vowel* data in the F_1–F_2 plane

3.5.1 Results

Using *GA-Classifier*

The effectiveness of the *GA-classifier* is demonstrated for the aforementioned three data sets. Other results are available in [25, 35, 343]. Table 3.3 presents the test recognition scores corresponding to the artificial data set (*ADS 1*) for different percentages of the training samples (e.g., *perc* = 5, 10, 50) and *H* = 6. The results shown are the average values computed over five runs of the algorithm from different initial populations.

It is seen from Table 3.3 that for *perc* = 50 and 10, the recognition ability of the classifier is considerably high for class 1 of the data set *ADS 1*. Class 2,

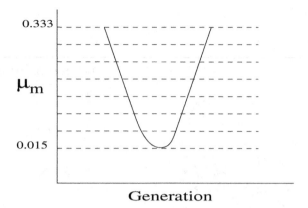

Fig. 3.9. Variation of mutation probability with the number of generations

on the other hand, is recognized relatively poorly. This disparity is due to the fact that the region for class 2 is of a relatively smaller size compared to the region of class 1, and it is totally surrounded by region 1 (see Fig. 3.7). The reversed situation for $perc = 5$ appears to be due to some error in sampling. As expected, the recognition scores are seen to improve with the value of $perc$, since, in general, increasing the size of the training data implies improving the representation of the entire data set. However, an important point to be noted here is that the performance of the *GA-classifier* remains respectable even for $perc = 5$, thereby indicating its effectiveness.

Table 3.3. Recognition scores (%) for *ADS 1* for $H = 6$

Class	$perc=50$	$perc = 10$	$perc = 5$
1	98.26	94.44	91.07
2	85.71	87.50	93.54
Overall	96.05	93.22	91.51

The variation of recognition score of the *GA-classifier* with H is demonstrated in Tables 3.4–3.7 for the different data sets. For *ADS 1* (Table 3.4), the recognition score improves with H only up to a specific value. Then the score degrades, since further increase in the number of lines makes the resulting decision boundary greatly dependent on the training data set. In other words, when a large number of lines is given for constituting the decision boundary, the algorithm can easily place them to approximate the distribution of the training set and hence the boundary closely. (This is also demonstrated in Table 3.5, which shows, for the purpose of illustration, the variation of the recognition scores with H during training of the *GA-classifier* for *ADS 1*. As seen from the table, the scores consistently improve with increase in H, thereby providing more accurate abstraction of the data.) This may not necessarily be beneficial (in the sense of generalization) for the test data set as the results in Table 3.4 demonstrate. For the *Vowel data*, the score is found to improve with the value of H up to $H = 7$, which provides the maximum score. On investigation, two out of these seven hyperplanes were found to be redundant. However, $H = 5$ provided a comparatively lower score. This may be due to the fact that with $H = 7$, the classifier has more flexibility in placing the five hyperplanes appropriately than for $H = 5$. It is also to be noted that the *GA-classifier* will be able to place the hyperplanes appropriately even with $H = 5$ if it is executed for sufficiently more iterations. (A similar observation was made in case of *ADS 1* for $H = 6$. It was found that three lines were actually required to provide the class boundaries. However, it is evident from Tables 3.5 and 3.4 that consideration of $H = 3$ provides neither good training nor test performance.) For the *Iris* data, $H = 2$ and 3 are found to provide the best score.

Table 3.4. Variation of recognition scores (%) during testing with H for *ADS 1*, with $perc = 10$

Class	Lines						
	8	7	6	5	4	3	2
1	93.71	93.71	94.44	99.03	93.96	94.68	93.23
2	77.27	77.27	87.50	55.68	57.95	64.77	22.72
Overall	90.83	90.83	93.22	91.43	87.65	89.42	80.87

Table 3.5. Variation of recognition scores (%) during training with H for *ADS 1*, with $perc = 10$

Class	Lines						
	8	7	6	5	4	3	2
1	100.00	100.00	100.00	100.00	100.00	100.00	97.82
2	100.00	100.00	100.00	66.67	60.56	66.67	44.44
Overall	100.00	100.00	100.00	94.55	93.72	94.55	89.09

Since *ADS 1* has a very small class totally surrounded by a larger class, the classwise recognition score in this case is of greater importance than the overall score. Thus, it is seen from Tables 3.4 and 3.5, $H = 2$ is not a good choice in this case since the recognition of class 2 is very poor, although the overall score is reasonably good. The reason for the good overall recognition score is that it is more affected by the score of class 1 than that of class 2 (since, as already mentioned, there are many more points belonging to the former than to the latter).

Table 3.6. Variation of recognition scores (%) during testing with H for *Vowel* data with $perc = 10$

Class	Surfaces						
	8	7	6	5	4	3	2
δ	67.69	21.53	46.15	49.23	52.30	55.38	0.00
a	75.30	86.42	32.09	30.86	0.00	0.00	0.00
i	79.35	77.42	84.51	78.06	76.12	81.29	80.00
u	88.97	90.44	75.73	77.94	84.55	56.61	0.00
e	69.31	80.75	78.07	83.95	86.63	72.19	73.26
o	66.04	67.28	41.35	74.07	70.98	48.14	97.50
Overall	74.56	74.68	71.99	71.37	69.21	57.50	53.30

Table 3.7. Variation of recognition scores (%) during testing with H for *Iris* with $perc = 10$

Class	Surfaces						
	8	7	6	5	4	3	2
1	100.00	97.78	100.00	100.00	100.00	100.00	100.00
2	75.55	82.22	84.44	82.22	82.22	93.33	93.33
3	97.78	93.33	95.55	97.78	97.78	100.00	100.00
Overall	91.11	91.11	93.33	93.33	93.33	97.78	97.78

For the *Vowel* data, since the number of classes k is equal to 6, the minimum number of planes required for its proper classification is obviously 3 ($3 \geq \lceil \log_2 6 \rceil$). This is evident from Table 3.6, where it is found that the first two classes are not recognized at all for $H = 2$. In fact, even for $H = 3$ and 4, one class is not recognized by the classifier, indicating that even four hyperplanes may not be sufficient for modelling the class boundaries of this data. In spite of this, the algorithm manages to attain a fairly respectable overall recognition score, since the other classes are well identified for these cases. Also, in general, class δ is found to be classified very poorly. This is expected since this is the class with maximum overlap. A similar observation was also found in [350], where a fuzzy set-theoretic classifier and Bayes' maximum likelihood classifier were used for vowel classification problem.

The decision boundary obtained for *ADS 1*, with $H = 6$ is shown in Fig 3.10 as an illustration. This boundary was obtained on termination of training after 538 iterations when a string with no misclassified points had been found. The dotted lines are obviously redundant.

Comparison with Existing Methods

A comparison of the performance of the *GA-classifier* with those of MLP (for three different architectures), Bayes' maximum likelihood classifier (where a priori probabilities and the class conditional densities are estimated from the

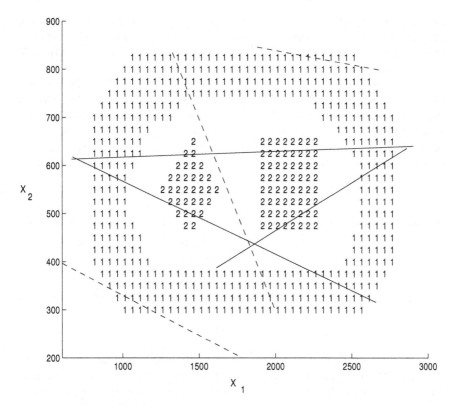

Fig. 3.10. Classification of 90% test data set using the final boundary generated with 10% training data set for $H = 6$. *Dotted lines* indicate redundant lines

training data set) and k-NN classifier is provided here for the above three data sets. For MLP, a conventional back-propagation (BP) algorithm, with the neurons executing the sigmoid function, is applied for learning. The learning rate η is fixed at 0.8, the momentum factor at 0.2 and the maximum number of generations at 3000. Online weight updating is performed, where the connection weights are updated every time an input pattern is presented to the network. The three different architectures considered consist of two hidden layers, with 5, 10 and 20 nodes in each layer, respectively. The number of nodes in the input layer is two, and that in the output layer is equal to the number of classes in the data set.

k-NN algorithm is executed taking k equal to $\sqrt{n}$, where n is the number of training samples. It is known that as number of training patterns n goes to infinity, if the values of k and k/n can be made to approach infinity and 0, respectively, then the k-NN classifier approaches the optimal Bayes' classifier [103, 155]. One such value of k for which the limiting conditions are satisfied is $\sqrt{n}$. For the Bayes' maximum likelihood classifier, a multivariate normal distribution of the samples with unequal dispersion matrices and unequal a priori probabilities ($= \frac{n_i}{n}$ for n_i patterns from class i) in each case is assumed. Although for some data sets, application of Bayes' maximum likelihood classifier, with the assumption of normal distribution may not be meaningful, the results are included to keep parity with others.

The comparative results are presented in Tables 3.8–3.10 for $H = 6$ and $perc = 10$. For the artificial data set, ADS 1, the GA-based algorithm is found to provide the best result, followed by that of the k-NN classifier (see Table 3.8). Note that k-NN classifier is reputed to partition well this type of nonoverlapping, nonlinear regions. MLP performs the worst for this data. It totally fails to recognize class 2 for ADS 1 for all the three cases; consequently, its performance on ADS 1 is poor.

Table 3.8. Comparative recognition scores (%) for ADS 1 for $perc = 10, H = 6$

Class	Bayes'	k-NN	MLP			GA
			2:5:5:2	2:10:10:2	2:20:20:2	
1	100.00	96.85	100.00	100.00	100.00	94.44
2	18.18	59.09	0.00	0.00	0.00	87.50
Overall	85.65	90.23	82.47	82.47	82.47	93.22

For Vowel (see Table 3.9), the Bayes' maximum likelihood classifier is found to provide the best performance followed by the score of the GA-classifier. Note that although the GA-classifier score is presented for $H=6$, this is much less compared to the one for $H = 7$ (74.68% from Table 3.6). MLP is again found to perform poorly for this data set.

For Iris data, as seen from Table 3.10, GA-classifier is found to provide a significantly better score than all other classifiers, even for $H=6$. (Note from Table 3.7 that the score is even better for $H = 2$ and 3). Both the Bayes' maximum likelihood classifier and MLP perform poorly for this data set.

Table 3.9. Comparative recognition scores (%) for Vowel data for $perc = 10, H = 6$

Class	Bayes'	k-NN	MLP			GA
			3:5:5:6	3:10:10:6	3:20:20:6	
δ	49.23	35.38	12.30	47.69	47.23	46.15
a	83.95	74.07	43.20	16.04	27.62	32.09
i	81.29	83.87	69.67	79.35	80.58	84.51
u	75.00	80.88	80.14	77.94	83.29	75.73
e	79.14	74.33	68.44	90.37	87.30	78.07
o	83.33	56.17	77.16	54.93	51.70	41.35
Overall	77.73	70.35	65.26	67.55	68.48	71.99

Table 3.10. Comparative recognition scores (%) for *Iris* for *perc* = 10, *H* = 6

Class	Bayes'	k-NN	MLP			GA
			4:5:5:3	4:10:10:3	4:20:20:3	
1	100.00	100.00	100.00	98.81	66.67	100.00
2	57.55	97.77	66.67	64.77	71.11	84.44
3	92.11	73.33	77.78	80.51	97.78	95.55
Overall	83.22	90.37	81.48	81.36	78.51	93.33

A comparison of the performance of the *GA-classifier* is also made with one based on simulated annealing [111, 243, 471], which is a well-known search and optimization strategy. It is found that the two perform comparably when the criterion to be minimized is the number of misclassified training data points [44]. However, if the error rate (obtained by dividing the number of misclassified points by the size of the training data) is to be minimized, then the performance of the simulated annealing based classifier degrades considerably, while that of the *GA-classifier* remains unchanged.

3.5.2 Consideration of Higher-Order Surfaces

This section describes a method of utilizing a fixed number of hyperspherical segments (i.e., using higher-order surfaces instead of hyperplanes as considered in the previous subsection) to constitute the decision surface in N-dimensional space, $N \geq 2$.

Description

The equation of a hypersphere in N dimensions, with center $(h_1, h_2, \ldots, h_N)$, and radius r is given by

$$(x_1 - h_1)^2 + (x_2 - h_2)^2 + \ldots + (x_N - h_N)^2 = r^2. \tag{3.17}$$

Thus the $N+1$ parameters $h_1, h_2, \ldots, h_N$ and r correspond to a unique hypersphere. As for the case of hyperplanes, a hyper-rectangle around the training data points is constructed. The lengths of the N sides of this hyper-rectangle, $l_1, l_2, \ldots, l_N$, are given by

$$l_1 = x_1^{max} - x_1^{min},$$
$$l_2 = x_2^{max} - x_2^{min},$$
$$\vdots$$
$$l_N = x_N^{max} - x_N^{min}.$$

This rectangle will subsequently be referred to as the *inner rectangle*. A possible encoding scheme for a number (say, C) of hyperspherical segments is to represent the parameters $h_1, h_2, \ldots, h_N$ and r as a bit string, such that $h_1, h_2, \ldots, h_N$ are constrained to lie within the *inner rectangle*, and r in the range $[0, diag]$, where *diag* is the length of the diagonal of the *inner rectangle*. Thus if one assumes the following:

$b_{h_1}, \ldots, b_{h_N}$ = number of bits for representing $h_1, \ldots, h_N$, respectively
b_r = number of bits for representing r
$v_{h_1}, \ldots, v_{h_N}$ = value in the $h_1, \ldots, h_N$ fields of a chromosome, respectively
v_r = value in the r field of a chromosome

then these parameters encode the hypersphere with the following specifications:

$$h_1 = x_1^{min} + \frac{v_{h_1}}{2^{b_{h_1}}} * l_1,$$
$$h_2 = x_2^{min} + \frac{v_{h_2}}{2^{b_{h_2}}} * l_2,$$
$$\vdots$$
$$h_N = x_N^{min} + \frac{v_{h_N}}{2^{b_{h_N}}} * l_N,$$
$$r = \frac{v_r}{2^{b_r}} * diag.$$

There is a subtle problem with this representation. Since the center of the hypersphere is constrained to lie within the *inner rectangle* and the radius in the range $[0, diag]$, the decision boundary generated by a finite number of such hyperspheres becomes restricted. Thus all kinds of complicated boundaries cannot be generated easily with this technique.

In order to overcome this limitation, the following approach can be adopted. Surrounding the *inner rectangle*, a larger rectangle is considered which is referred to as the *outer rectangle*, and whose sides are of length $l'_1, l'_2, \ldots, l'_N$, where

$$l'_1 = (2p + 1) * l_1,$$
$$l'_2 = (2p + 1) * l_2,$$
$$\vdots$$
$$l'_N = (2p + 1) * l_N,$$

and p is an integer, $p \geq 0$. Then the center of the hypersphere (which will be considered for constituting the decision boundary) is allowed to lie anywhere within the *outer rectangle*. p can be chosen sufficiently large in order to approximate any form of the decision boundary.

The center $(h_1, h_2, \ldots, h_N)$ of the hypersphere is a randomly chosen point within the outer rectangle. For computing the radius, the nearest distance of the center $(h_1, h_2, \ldots, h_N)$ from the vertices of *inner rectangle* is determined. Let this be d_1. Similarly, let the farthest distance be d_2. Now if $(h_1, h_2, \ldots, h_N)$ lies within the *inner rectangle*, then the radius can take on values in the range $[0, d_2]$. Otherwise the range is considered as $[d_1, d_2]$. Given any chromosome, the detailed steps of decoding the parameter values are given in Fig. 3.11.

The number of bits required for encoding one hypersphere is, therefore, given by

$$b_{h_1} + b_{h_2} + \ldots + b_{h_N} + b_r.$$

Step 1: Computation of the center of the hypersphere.
Let $v_{h_1}, v_{h_2}, \ldots, v_{h_N}$ and v_r be the values in the $h_1, h_2, \ldots, h_N$ and r bits, respectively.
Let p be input by the user. Compute

$h_1 = (x_1^{min} - p * l_1) + \frac{v_{h_1}}{2^{b_{h_1}}} * (2 * p + 1) \, l_1,$

$h_2 = (x_2^{min} - p * l_2) + \frac{v_{h_2}}{2^{b_{h_2}}} * (2 * p + 1) \, l_2,$

$\vdots$

$h_N = (x_N^{min} - p * l_N) + \frac{v_{h_N}}{2^{b_{h_N}}} * (2 * p + 1) \, l_N.$

Step 2: Computation of maximum and minimum distances.
Compute the distances of all the vertices of the *inner rectangle* from $(h_1, h_2, \ldots, h_N)$.
Let d_1 and d_2 be the minimum and maximum of these values.
Step 3: Whether center lies within the *inner rectangle* or not.
If $(h_1, h_2, \ldots, h_N)$ lies within the *inner rectangle*, go to step 5.
(i.e., if $x_1^{min} \leq h_1 \leq x_1^{max}$, $x_2^{min} \leq h_2 \leq x_2^{max}$, $\ldots$ and $x_N^{min} \leq h_N \leq x_N^{max}$, go to step 5).
Otherwise go to step 4.

Step 4: Center outside *inner rectangle* — compute radius.
Compute r in the range $[d_1, d_2]$. Or,
$r = d_1 + \frac{v_r}{2^{b_r}} * (d_2 - d_1).$
Go to step 6.

Step 5: Center inside *inner rectangle* — compute radius.
Compute r in the range $[0, d_2]$. Or, $r = \frac{v_r}{2^{b_r}} * d_2.$

Step 6: Proceed.

Fig. 3.11. Steps for decoding the parameters of hyperspheres

For a fixed number C of hyperspheres, the string length becomes

$$l = (b_{h_1} + b_{h_2} + \ldots + b_{h_N} + b_r) * C.$$

In general $b_{h_1} = b_{h_2} = \ldots = b_{h_N} = b_1$. Hence the string length will be

$$l = (N * b_1 + b_r) * C.$$

The regions corresponding to the different classes are determined from the training data set using Eq. (3.17). Fitness of a string is determined by the number of points correctly classified by the string. The conventional selection, crossover and mutation operators are applied till a termination criterion (string with zero misclassification) is achieved, or a maximum number of iterations has been executed. As before, an elitist strategy is used.

Implementation

The variation of test recognition scores with C for *ADS 1* and *Vowel* data provided by the *GA-classifier* is presented in Tables 3.11 and 3.12. Keeping the other genetic parameters the same, p is chosen to be 8.

Comparing these results with the results in Tables 3.4 and 3.6, it is found on an average that for both these cases, the use of higher-order surfaces helps to improve the recognition over those with hyperplanes. This is expected, since the class boundaries of both these data sets are nonlinear in nature. Overall, for *ADS 1*, three circular surfaces are required to model the decision boundary. The remaining surfaces for the cases with $C > 3$ are redundant, except for $C = 6$, where four surfaces are required. Note that, because of the associated overfitting, the score for $C = 6$ is also poorer.

For *Vowel* data (Table 3.12), unlike the case with hyperplanes, no redundant spherical surfaces are found, even with $C = 8$. The recognition scores are also consistently better than those with hyperplanes (Table 3.6). Thus it appears that even $C = 8$ is not sufficient for proper approximation of the class boundaries of *Vowel* data. This also seems to be appropriate, in view of the overlapping classes for *Vowel* data (Fig. 3.8).

Table 3.11. Variation of recognition scores (%) during testing with C for *ADS 1*, with $perc = 10$, $p = 8$

Class	Surfaces						
	8	7	6	5	4	3	2
1	96.37	96.37	87.43	95.18	94.20	97.58	98.79
2	96.59	96.59	97.72	87.21	87.50	86.36	55.68
Overall	96.41	96.41	89.24	93.20	93.02	95.62	91.23

Table 3.12. Variation of recognition scores (%) during testing with C for *Vowel* data with $perc = 10$

Class	Surfaces						
	8	7	6	5	4	3	2
δ	67.69	32.31	50.77	50.77	24.61	0.00	0.00
a	85.24	86.42	81.48	65.43	91.35	95.06	0.00
i	85.22	87.09	86.45	80.64	84.51	87.74	93.54
u	94.11	89.71	91.91	92.65	88.23	91.91	89.71
e	75.40	70.58	71.65	76.47	72.72	82.35	73.79
o	69.75	72.84	69.75	69.75	67.28	64.19	79.01
Overall	76.23	76.08	75.97	75.44	74.55	75.82	67.81

3.6 Summary

The problem of supervised classification using GAs has been considered in this chapter. Some of the attempts in designing classifiers using fuzzy if–then rules, and by optimizing decision trees are described in brief. A method of generating class boundaries through fitting of hyperplanes in $I\!R^N$, $N \geq 2$, using GAs has then been described in detail, along with its demonstration on both artificial and real-life data having overlapping, and nonoverlapping nonconvex regions. An extensive comparison of the methodology with other classifiers, namely the Bayes' maximum likelihood classifier (which is well-known for discriminating overlapping classes), k-NN classifier and MLP (where both the k-NN classifier and the MLP with hard limiters are known to provide piecewise linear boundaries) has been provided. The results of the *GA-classifier* are seen to be comparable to, and often better than, them in discriminating both overlapping and nonoverlapping nonconvex regions.

It is observed that the use of higher-order surfaces (i.e., hyperspherical segments) provides superior performance as compared to hyperplanes when the class boundaries are complicated. Since in most real-life situations, the class boundaries are unknown, and are generally nonlinear, the use of any higher-order surfaces is better and natural. This provides better approximation, albeit at the cost of increased computational complexity, since an additional parameter has to be encoded in this case. For example, for N-dimensional space, $N + 1$ parameters need to be encoded when using hyperspherical segments for modelling the class boundaries.

The method of classification using class boundaries described in this chapter is sensitive to rotation of the data sets due to the way of choosing the enclosing hyper-rectangle around the data points. It is also evident from the method of specifying a hyperplane that translation, dilation or contraction of the data sets would produce a similar change of the decision boundary. Discretization of the feature space, which is unbounded and continuous, poses a problem in digital pattern recognition with respect to the performance of the systems. In the *GA-classifier*, a hyper-rectangle has been constructed around the set of data points which makes the search space bounded. The possible orientations (angle values) of the hyperplanes are considered to be fixed, and parallel hyperplanes in any direction are kept separated by a small distance depending on the value of b_2. As the discretization is made finer, the performance of the classifier usually improves, but the size of the search space increases, thereby increasing the number of iterations required for GA.

Proper selection of H is obviously crucial for good performance of the *GA-classifier*, since an underestimation or overestimation of H may lead to improper modelling of the boundary or overfitting of the data, respectively. In this regard, a metalevel GA may also be used for determining H. Alternatively, the concept of variable string length in GAs [169] may be adopted where the value of H could be kept variable and could be evolved as an outcome of the GA process. Such a variable string-length GA-based classifier is described in Chap. 5 of this book.

When developing a new classifier, it is desirable to analyze it theoretically, and to show that the performance of the new classifier approaches that of the Bayes' classifier (known to be optimal from a statistical point of view) under limiting conditions. Such an analysis, with respect to the *GA-classifier*, is provided in the following chapter.

4

Theoretical Analysis of the *GA-classifier*

4.1 Introduction

Chapter 3 dealt with the development of the *GA-classifier*, where a number of linear (or higher-order) surfaces were used for approximating all kinds of decision boundaries. In this chapter, a theoretical investigation regarding the performance of the *GA-classifier* is provided. This mainly includes establishing a relationship between the Bayes' classifier (where the class a priori probabilities and conditional densities are known) and the *GA-classifier* under limiting conditions. It is known from the literature that Bayes' classifier is the best possible classifier if the class conditional densities and the a priori probabilities are known [461]. No classifier can provide better performance than Bayes' classifier under such conditions. (Let a denote the error probability associated with Bayes' classifier or the Bayes' decision rule. If any other decision rule is used and the error probability associated with that rule is b, then $a \leq b$. Bayes' classifier is the best classifier because of this property.) In practice, it is difficult to use Bayes' classifier because the class conditional densities and the a priori probabilities may not be known. Hence new classifiers are devised and their performances are compared to that of the Bayes' classifier. The desirable property of any classifier is that it should approximate or approach the Bayes' classifier under limiting conditions. Such an investigation for MLP was performed in [402] to show that the MLP, when trained as a classifier using back-propagation, approximates the Bayes' optimal discriminant function. As already mentioned in Sect. 3.5.1, k-NN classifier also approaches the Bayes' classifier under certain conditions as the size of the training data set goes to infinity.

There are many ways in which the performance of a classifier is compared to that of the Bayes' classifier. One such way is to investigate the behavior of the error rate (defined as the ratio of the number of misclassified points to the size of the data set) as the size of the training data goes to infinity, and check whether the limiting error rate is equal to a, the error probability associated with the Bayes' classifier. Such an investigation [37] is described in this chapter. It is proved here that for $n \to \infty$ (i.e., for sufficiently large training data set) and for a sufficiently large number of iterations, the error rate of the *GA-classifier* during training will approach the Bayes' error probability. It has also been shown theoretically that for n tending to infinity, the number of hyperplanes found by the *GA-classifier* to constitute the decision boundary will be equal to the optimum number of hyperplanes (i.e., which provide the Bayes' decision boundary) if exactly one partition provides the Bayes' error probability. Otherwise the number of hyperplanes found by the *GA-classifier* will be greater than or equal to the optimum value.

Experimental verification of the theoretical findings are provided for a number of training data sets following triangular and normal distributions having both linear and nonlinear boundaries using different values of H. Performance on test data sets has been studied. Instead of using hyperplanes, circular surfaces in two dimensions have also been considered as constituting the decision boundary. The generalization capability of the classifier has been studied as a function of the class a priori probabilities (for two-class problems). The empirical findings indicate that as the training data size (n) increases, the performance of the *GA-classifier* approaches that of Bayes' classifier for all the data sets.

The relationship of the *GA-classifier* with the Bayes' classifier is established in Sect. 4.2. Sect. 4.3 shows the relationship between the number of hyperplanes found by the former and the optimum value. The experimental results validating the theoretical findings are reported in Sect. 4.4. Finally, Sect. 4.5 summarizes the chapter.

4.2 Relationship with Bayes' Error Probability

In this section the relationship between the *GA-classifier* and Bayes' classifier is established [37]. The mathematical notations and preliminary definitions are described first. This is followed by the claim that for $n \to \infty$, the error rate of the *GA-classifier* during training will approach the Bayes' error probability a. Finally, some critical comments about the proof are mentioned.

Let there be K classes $C_1, C_2, \ldots, C_K$ with a priori probabilities $P_1, P_2, \ldots, P_K$ and continuous class conditional densities $p_1(\mathbf{x}), p_2(\mathbf{x}), \ldots, p_K(\mathbf{x})$. Let the mixture density be

$$p(\mathbf{x}) = \sum_{i=1}^{K} P_i p_i(\mathbf{x}). \tag{4.1}$$

(According to the Bayes' rule, a point is classified to class i iff

$$P_i p_i(\mathbf{x}) \geq P_j p_j(\mathbf{x}), \quad \forall j = 1, \ldots, K \text{ and } j \neq i.)$$

Let $X_1, X_2, \ldots, X_n, \ldots$ be independent and identically distributed (i.i.d) N-dimensional random vectors with density $p(\mathbf{x})$. This indicates that there is a probability space $(\Omega, \mathcal{F}, Q)$, where $\mathcal{F}$ is a σ field of subsets of Ω, Q is a probability measure on $\mathcal{F}$, and

$$X_i : (\Omega, \mathcal{F}, Q) \longrightarrow (I\!\!R^N, B(I\!\!R^N), P) \quad \forall i = 1, 2, \ldots, \quad (4.2)$$

such that

$$P(A) = Q(X_i^{-1}(A)), \quad (4.3)$$

$$= \int_A p(\mathbf{x}) d\mathbf{x} \quad (4.4)$$

$\forall A \in B(I\!\!R^N)$ and $\forall i = 1, 2, \ldots$. Here $B(I\!\!R^N)$ is the Borel σ field of $I\!\!R^N$. Let

$$\mathcal{E} = \{ E : E = (S_1, S_2, \ldots, S_K), S_i \subseteq I\!\!R^N, S_i \neq \emptyset,$$

$$\forall i = 1, \ldots, K, \bigcup_{i=1}^{K} S_i = I\!\!R^N, S_i \bigcap S_j = \emptyset, \forall i \neq j \}. \quad (4.5)$$

$\mathcal{E}$ provides the set of all partitions of $I\!\!R^N$ into K sets as well as their permutations, i.e.,

$$E_1 = (S_1, S_2, S_3 \ldots, S_K) \in \mathcal{E},$$

$$E_2 = (S_2, S_1, S_3, \ldots, S_K) \in \mathcal{E},$$

then $E_1 \neq E_2$. Note that the decision rule $E = (S_{i1}, S_{i2}, \ldots, S_{iK})$ denotes that each S_{ij}, $1 \leq j \leq K$, is the decision region corresponding to class C_j.

Let $E_0 = (S_{01}, S_{02}, \ldots, S_{0K}) \in \mathcal{E}$ be such that each S_{0i} is the decision region corresponding to the class C_i in $I\!\!R^N$ and these are obtained by using Bayes' decision rule. Then

$$a = \sum_{i=1}^{K} P_i \int_{S_{0i}^c} p_i(\mathbf{x}) d\mathbf{x} \leq \sum_{i=1}^{K} P_i \int_{S_{1i}^c} p_i(\mathbf{x}) d\mathbf{x}. \quad (4.6)$$

$\forall E_1 = (S_{11}, S_{12}, \ldots, S_{1K}) \in \mathcal{E}$. Here a is the error probability obtained using the Bayes' decision rule.

It is known from the literature that such an E_0 exists and it belongs to $\mathcal{E}$ because Bayes' decision rule provides an optimal partition of $I\!\!R^N$. Furthermore, for every such $E_1 = (S_{11}, S_{12}, \ldots, S_{1K}) \in \mathcal{E}$, $\sum_{i=1}^{K} P_i \int_{S_{1i}^c} p_i(\mathbf{x}) d\mathbf{x}$ provides the error probability for $E_1 \in \mathcal{E}$. One such E_0 is given below:

$$S_{01} = \{ \mathbf{x} : P_1 p_1(\mathbf{x}) \geq P_j p_j(\mathbf{x}) \ \forall j \neq 1 \},$$

$$S_{0i} = \{ \mathbf{x} : P_i p_i(\mathbf{x}) \geq P_j p_j(\mathbf{x}) \ \forall j \neq i \} \bigcap [\bigcup_{l=1}^{i-1} S_{0l}]^c, \quad i = 2, 3, \ldots, K.$$

Note that E_0 need not be unique.

Assumptions: Let H_{opt} be a positive integer and let there exist H_{opt} hyperplanes in $\mathbb{R}^N$ which provide an optimal decision rule (i.e., an optimal E_0). Let H_{opt} be known a priori.

Let

$$\mathcal{C}_n(\omega) = \{E : E = (S_1, S_2, \ldots, S_K), E \in \mathcal{E}, \qquad (4.7)$$
$$\text{E is generated by } H_{opt} \text{ hyperplanes for the training sample set}$$
$$\{X_1(\omega), X_2(\omega), \ldots, X_n(\omega)\}, \omega \in \Omega\}.$$

Note that $\mathcal{C}_n(\omega)$ is uncountable. Note also that each E in $\mathcal{C}_n(\omega)$ provides a value of the number of misclassifications for the training sample set $(X_1(\omega), X_2(\omega), \ldots, X_n(\omega))$. Let us now define a relation for the elements in $\mathcal{C}_n(\omega)$ as follows:

An element $E_1 = (S_{11}, S_{12}, \ldots, S_{1K})$ of $\mathcal{C}_n(\omega)$ is said to be related to element $E_2 = (S_{21}, S_{22}, \ldots, S_{2K})$ of $\mathcal{C}_n(\omega)$ if

$$X_i(\omega) \in S_{1j} \Rightarrow X_i(\omega) \in S_{2j} \ \forall \ i = 1, 2, \ldots, n, \ \forall \ j = 1, 2, \ldots, K. \qquad (4.8)$$

Let for an element $E_1 = (S_{11}, S_{12}, \ldots, S_{1K})$,

$$S_{E_1} = \{E : E \in \mathcal{C}_n(\omega), \ E \text{ is related to } E_1\}. \qquad (4.9)$$

Note that for any two points E_1 and E_2 in $\mathcal{C}_n(\omega)$, either $S_{E_1} = S_{E_2}$ or $S_{E_1} \bigcap S_{E_2} = \emptyset$. Note that the number of such different S_{E_i}s are finite since the training sample set is finite.

Let the *GA-classifier* be applied on the training sample set $\{X_1(\omega), X_2(\omega), \ldots, X_n(\omega)\}$ in such a way that the total number of possible rules under consideration, denoted by $\mathcal{T}_n$, gives rise to at least one element of the different S_{E_i}s. Note that the *GA-classifier* described in Chap. 3 satisfies this criterion with a suitable choice of the discretization parameters. Hence, if the number of iterations is sufficiently large, the *GA-classifier* will provide the minimum number of misclassified points [60], and correspondingly an element $E \in \mathcal{C}_n(\omega)$ (in one of the S_{E_i}s).

Let $\mathcal{A} = \{A : A$ is a set consisting of H_{opt} hyperplanes in $\mathbb{R}^N\}$. Let $\mathcal{A}_0 \subseteq \mathcal{A}$ be such that it provides all the optimal decision rules in $\mathbb{R}^N$, i.e., each element of $\mathcal{A}_0$ provides the regions which are also obtained using the Bayes' decision rule. Note that each $A \in \mathcal{A}$ generates several elements of $\mathcal{E}$ which result from different permutations of the list of K regions. Let $\mathcal{E}_A \subseteq \mathcal{E}$ denote all possible $E = (S_1, S_2, \ldots, S_K) \in \mathcal{E}$ that can be generated from A. Let

$$G = \bigcup_{A \in \mathcal{A}} \mathcal{E}_A. \qquad (4.10)$$

Achieving optimality in $\mathcal{E}$ is equivalent to achieving optimality in G since H_{opt} hyperplanes also provide the Bayes' decision regions. Thus, henceforth, we shall concentrate on G.

Note that $T_n(\omega)$ denotes all possible decision rules that are considered in the *GA-classifier* when the training sample set is $\{X_1(\omega), X_2(\omega), \ldots, X_n(\omega)\}$. Also, $T_n(\omega) \subseteq T_{n+1}(\omega)$ for every n and $\lim_{n \to \infty} T_n(\omega) \subseteq G$. It can also be seen that as $n \to \infty$, there will be some decision rules in $T_n(\omega)$ which will converge to an optimal decision rule (i.e., an optimal E_0), everywhere in ω.

Let $Z_{iE}(\omega) = 1$,   if $X_i(\omega)$ is misclassified when E is used as a decision
rule where $E \in G$, $\forall \omega \in \Omega$.

$= 0$,   otherwise.

Let $f_{nE}(\omega) = \frac{1}{n}\sum_{i=1}^{n} Z_{iE}(\omega)$, when $E \in G$ is used as a decision rule.

That is, $f_{nE}(\omega)$ is the error rate associated with the decision rule E for the training sample set $\{X_1(\omega), X_2(\omega), \ldots, X_n(\omega)\}$. In particular,

$$f_{nE}(\omega) = \frac{1}{n}\sum_{i=1}^{n} Z_{iE}(\omega) \text{ when } E \in T_n(\omega).$$

Let

$$f_n(\omega) = Inf\{f_{nE}(\omega) : E \in T_n(\omega)\}. \tag{4.11}$$

It is to be noted that the *GA-classifier* described in Chap. 3 uses $n*f_{nE}(\omega)$, the total number of misclassified points, as the objective function which it attempts to minimize. This is equivalent to searching for a suitable $E \in T_n(\omega)$ such that the term $f_{nE}(\omega)$ is minimized, i.e., for which $f_{nE}(\omega) = f_n(\omega)$. As already mentioned, it is known that for infinitely many iterations the elitist model of GAs will certainly be able to obtain such an E [60].

Theorem 4.1. *For sufficiently large n, $f_n(\omega) \not> a$, (i.e., for sufficiently large n, $f_n(\omega)$ cannot be greater than a) almost everywhere.*

Proof. Let $Y_i(\omega) = 1$, if $X_i(\omega)$ is misclassified
according to Bayes' rule $\forall \omega \in \Omega$.

$= 0$, otherwise.

Note that $Y_1, Y_2, \ldots, Y_n, \ldots$ are i.i.d. random variables. Now

$$Prob(Y_i = 1) = \sum_{j=1}^{K} Prob(Y_i = 1/X_i \text{ is in } C_j) P(X_i \text{ is in } C_j),$$

$$= \sum_{j=1}^{K} P_j Prob(\omega : X_i(\omega) \in S_{0j}^c \text{ given that } \omega \in C_j),$$

$$= \sum_{j=1}^{K} P_j \int_{S_{0j}^c} p_j(\mathbf{x}) d\mathbf{x} = a.$$

Hence the expectation of Y_i, $E(Y_i)$ is given by

$$E(Y_i) = a, \forall i.$$

Then by using Strong Law of Large Numbers [130], $\frac{1}{n}\sum\limits_{i=1}^{n}Y_i \longrightarrow a$ almost everywhere,

That is, $P(\omega : \frac{1}{n}\sum\limits_{i=1}^{n}Y_i(\omega) \not\longrightarrow a) = 0.$

Let $B = \{\omega : \frac{1}{n}\sum\limits_{i=1}^{n}Y_i(\omega) \longrightarrow a\} \subseteq \Omega.$ Then $Q(B) = 1.$

Note that $f_n(\omega) \leq \frac{1}{n}\sum_{i=1}^{n}Y_i(\omega)$, $\forall n$ and $\forall\omega$, since the set of regions $(S_{01}, S_{02}, \ldots, S_{0K})$ obtained by the Bayes' decision rule is also provided by some $A \in \mathcal{A}$, and consequently it will be included in G and $T_n(\omega)$. Note that $0 \leq f_n(\omega) \leq 1$, $\forall n$ and $\forall\omega$. Let $\omega \in B$. For every $\omega \in B$, $U(\omega) = \{f_n(\omega); n = 1, 2, \ldots\}$ is a bounded infinite set. Then by the Bolzano–Weierstrass theorem [16], there exists an accumulation point of $U(\omega)$. Let $y = \text{Sup}\{y_0 : y_0$ is an accumulation point of $U(\omega)\}$. From elementary mathematical analysis we can conclude that $y \leq a$, since $\frac{1}{n}\sum\limits_{i=1}^{n}Y_i(\omega) \longrightarrow a$ almost everywhere and $f_n(\omega) \leq \frac{1}{n}\sum_{i=1}^{n}Y_i(\omega)$. Thus it is proved that for sufficiently large n, $f_n(\omega)$ cannot be greater than the error probability a for $\omega \in B$.

Corollary 4.2. $\lim_{n\to\infty} f_n(\omega) = a$ *for* $\omega \in B$.

Proof. Consider a sequence $f_{i_k}(\omega)$ in $U(\omega)$ such that $\lim_{n\to\infty} f_{i_k}(\omega) = b$ and $b < a$. We shall show that $b < a$ is not possible. Suppose $b < a$ is possible. Let $\varepsilon = \frac{a-b}{4}$. Then for $k > k_0$, $f_{i_k}(\omega) \in (b - \varepsilon, b + \varepsilon)$. That is,

$$f_{i_k}(\omega) < a - \varepsilon, \quad \forall\, k > k_0. \tag{4.12}$$

For sufficiently large n, and for a given ω, whatever decision rule $E \in G$ that one considers,

$$\frac{\text{number of misclassified samples with respect to } E}{n} \geq a - \varepsilon, \tag{4.13}$$

from the properties of the Bayes' classifier. This contradicts Eq. (4.12). Thus $b < a$ is impossible. Hence for any convergent sequence in $U(\omega)$ with limit as b, $b \geq a$. From Theorem 4.1, we have $b \leq a$, since b is an accumulation point. This shows that $b = a$. Since every bounded infinite set has at least one accumulation point, and every accumulation point in $U(\omega)$ here is a, $\lim_{n\to\infty} f_n(\omega) = a$.

Remarks

- It is to be noted that when the class conditional densities and the a priori probabilities are known, the error probability associated with any decision rule whatsoever (including the one provided by the *GA-classifier*) can never be less than a. Here, the claim is that the error rate of the *GA-classifier* (obtained by dividing the number of misclassified data points by the size of the training data) during training is less than or equal to a. As the size of the training set goes to infinity, from Corollary 4.2, the error rate provided by the *GA-classifier* approaches a. Thus the performance of the *GA-classifier* approaches that of the Bayes' classifier as n goes to infinity. This indicates that the boundary generated by the *GA-classifier* approaches the Bayes' boundary if it exists and is unique under limiting conditions. In case the Bayes' boundary is not unique, then the *GA-classifier* will generate any such boundary where its error rate is a.

- The proof is not typical of the *GA-classifier*. It will hold for any other classifier where the criterion is to find a decision boundary that minimizes the number of misclassified points. For example, if simulated annealing is used instead of genetic algorithm, then too the results will hold under limiting conditions.
- Instead of hyperplanes, any other higher-order surface could have been used in *GA-classifier* for obtaining the decision boundaries, provided the number of higher-order surfaces is finite and known, and the Bayes' boundary is indeed provided by such surfaces. It would lead to only minor modifications to the proof presented earlier, with the basic inference still holding good.
- Note that although the class conditional density functions are assumed to be continuous, the above results will still hold good even if the density functions are not continuous.
- The proof assumes that there exist H_{opt} hyperplanes that can model the Bayes' decision boundary. Although theoretically any curved surface can be approximated by a number of hyperplanes, this may become an unrealistic assumption in practice. However, results are included in this chapter to demonstrate that as the size of the data set increases, the hyperplanes provided by the *GA-classifier* approximate the highly curved Bayes' boundary better.
- From Theorem 4.1 and Corollary 4.2, it is evident that for small values of n the result of the *GA-classifier* is such that during training

$$\frac{\text{number of misclassified points}}{n} < a,$$

and also

$$a - \frac{\text{number of misclassified points}}{n}$$

could be significant. As n goes to infinity,

$$\frac{\text{number of misclassified points}}{n} \to a.$$

Thus, while conducting the experiments with known a priori probabilities and densities, we would find that during training

$$\frac{\text{number of misclassified points}}{n} < a,$$

and

$$a - \frac{\text{number of misclassified points}}{n}$$

could be large or small depending on n being small or large, respectively. It is to be noted that

$$\frac{\text{number of misclassified points}}{n} < a$$

is not a complete indication of the performance of the classifier. One should check the performance using test data sets also. Some such results are provided later in this chapter.

The results proved analytically in this section are also verified experimentally in Sect. 4.4 under different situations. Note that in the proof described above it is assumed that H_{opt} is known a priori. However in practice, as mentioned earlier, H_{opt} may not be known. In that case we try to use a conservative value, which may lead to the presence of some redundant hyperplanes in the resultant decision boundary. The next section describes a method of eliminating the redundant hyperplanes, thereby generating the optimum number of hyperplanes H_{GA} of the *GA-classifier*. A relationship of H_{GA} and the optimum hyperplanes H_{opt} yielding the Bayes' boundary is also established in the next section.

4.3 Relationship Between H_{opt} and H_{GA}

Here, we first present a technique for obtaining H_{GA} hyperplanes from the initial overestimation of H. Subsequently, we establish that H_{GA} is equal to H_{opt} when there exists exactly one partition of the feature space that provides the Bayes' error probability a. In case more than one partition can provide the Bayes' error probability, then all we can say is that H_{GA} will be greater than or equal to H_{opt}.

4.3.1 Obtaining H_{GA} from H

A hyperplane is considered to be redundant if its removal has no effect on the recognition score of the classifier for the given training data set. In order to arrive at the optimal number of hyperplanes, one of the ways is to consider first of all, all possible combinations of the H hyperplanes. For each such combination, the first hyperplane is removed and it is checked whether the remaining hyperplanes can successfully classify all the patterns. If so, then this hyperplane is deleted, and the test is repeated for the next hyperplane in the combination.

Obviously, testing all possible combinations results in an algorithm with exponential complexity in H. To avoid this, a branch and bound technique is adopted where the search within a combination is discontinued (before considering all the H hyperplanes) if the number of hyperplanes found to be nonredundant so far, is greater than or equal to the number of hyperplanes declared to be nonredundant by some earlier combination. The complexity may be reduced further by terminating the algorithm if the combination being tested provides a set of $\lceil \log_2 K \rceil$ nonredundant hyperplanes, since this is the minimum number of hyperplanes that is required in order to provide K distinct regions. This method guarantees removal of all the redundant hyperplanes from the decision boundary, thereby yielding exactly H_{GA} hyperplanes.

4.3.2 How H_{GA} Is Related to H_{opt}

Let H_{GA} be the number of hyperplanes (after elimination of redundancy) found by the *GA-classifier* to provide an error rate $= f_n(\omega)$, which is less than or equal to a (the Bayes' error probability) when $n \to \infty$. If H_{opt} is the optimal number of hyperplanes, then obviously H_{GA} cannot be less than H_{opt}. It now needs to be ascertained whether $H_{GA} > H_{opt}$ or $H_{GA} = H_{opt}$. For this the following situations must first be considered:

(a) The number of partitions which provides the Bayes' error probability is exactly one. Since this partition, formed from H_{opt} hyperplanes, provides the Bayes' error probability, which is known to be optimal, hence for $n \to \infty$, the regions provided by the H_{GA} hyperplanes must be exactly the same as the regions provided by the H_{opt} hyperplanes. Thus H_{GA} must be same as H_{opt} for large values of n.

(b) On the contrary, the number of partitions that provide the Bayes' error probability may be greater than one. For example, this will be the case if at least one of the K classes is totally disconnected from the other classes. In these cases, the regions provided by the H_{GA} hyperplanes may not be identical to the ones provided by the H_{opt} hyperplanes. Consequently, H_{GA} can be greater than H_{opt} for such situations, although the classifier still provides an error rate equal to $f_n(\omega)$.

4.3.3 Some Points Related to n and H

In practice we always deal with finite data sets (or finite n). Obviously, in that case, additional hyperplanes, beyond H_{opt}, may be placed appropriately in order to further reduce the number of misclassified points, at the cost of possibly reduced generalizability. These hyperplanes will not be eliminated by the redundancy removal process. However, as n increases, the effect of introducing additional hyperplanes will decrease and also the performance of the classifier in terms of the test data will gradually improve. In the limiting case, for $n \to \infty$, only the optimum number of hyperplanes with a specific arrangement will provide the requisite decision boundary. Any additional hyperplane will obviously be detected as redundant. At the same time, the generalization of the classifier will be optimum.

4.4 Experimental Results

Empirical results provided in this section show that the theoretical findings of Sect. 4.2 indeed hold for the *GA-classifier*. It is also found experimentally that the decision boundary obtained by *GA-classifier* approaches that of Bayes' classifier as n increases. Data sets following triangular and normal distributions are considered, having both linear and nonlinear class boundaries. All the data sets have considerable amounts of overlap. In a part of the experiments, instead of lines, circular segments (in two dimension) are considered as the constituting elements of the decision boundaries. Its objective is to demonstrate the theoretical findings in Sect. 4.2 for higher-order surfaces. The effect of class a priori probability on the recognition score has also been experimentally investigated.

Different situations considered for conducting the experiments are as follows:

  i. The decision boundary is provided by H_{opt} hyperplanes, and H_{opt} is known a priori.
 ii. The decision boundary is provided by H_{opt} higher-order surfaces, and H_{opt} is known a priori.
iii. It is known that the decision boundary can be approximated by H_{opt} hyperplanes but the value of H_{opt} is not known.
 iv. It is known that the decision boundary can be approximated by H_{opt} higher-order surfaces but the value of H_{opt} is not known.
  v. It is known that no finite number of hyperplanes can approximate the decision boundary.
 vi. It is known that no finite number of higher-order surfaces can approximate the decision boundary.
vii. Nothing is known about the given data set. In that case we may try to approximate the boundary by a fixed number of hyperplanes or any other higher-order surfaces.

This section is divided into three parts. A brief description of the data sets is given in the first part. Comparison of the decision boundaries and recognition scores (during training and testing) obtained by the *GA-classifier* (using linear as well as circular surfaces) and the Bayes' classifier for different sizes of the training data is provided in the second part. Finally, the third part demonstrates the variation of the generalization capability of the classifier as a function of the class a priori probability, for two-class problems.

4.4.1 Data Sets

Four data sets are used for empirically verifying the theoretical results described in Sect. 4.2. The details of these data sets along with their sources are mentioned in Appendix B. They are briefly mentioned below for convenience.

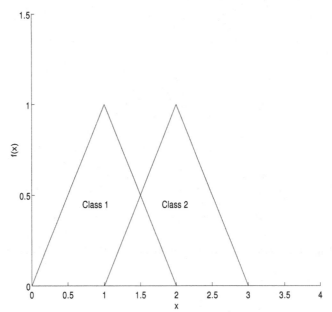

Fig. 4.1. Triangular distribution along X axis for *Data Set 1* having two classes

Data Set 1 is a two-dimensional data set generated using a triangular distribution along the X axis of the form shown in Fig. 4.1 for the two classes, 1 and 2. If P_1 is the a priori probability of class 1 then the Bayes' decision boundary is given by

$$x = 1 + P_1. \tag{4.14}$$

Data Set 2 is a normally distributed data set consisting of two classes. It can be shown that the Bayes' boundary for this data is given by the following equation:

$$a_1 x_1{}^2 + a_2 x_2{}^2 + 2a_3 x_1 x_2 + 2b_1 x_1 + 2b_2 x_2 + c = 0.$$

The data set and the Bayes' decision boundary are also shown in Fig. 4.2.
Data Set 3 consists of nine classes generated using a triangular distribution. All the classes are assumed to have equal a priori probabilities ($= 1/9$). The data set, along with the Bayes' boundary, is shown in Fig. 4.3.
Data Set 4 is a two-class, two-dimensional data set, used specifically for the purpose of investigating the generalization capability of the *GA-classifier* as

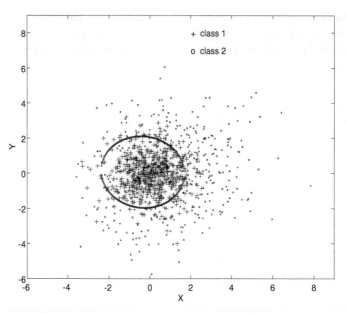

Fig. 4.2. *Data Set 2* for $n = 1000$ along with the Bayes' decision boundary for two classes

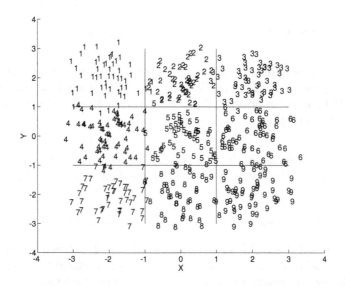

Fig. 4.3. *Data Set 3* for $n = 900$ along with the Bayes' decision boundary for nine classes

a function of the a priori probability of class 1. The data set is normally distributed with the following parameters:

$$\mu_1 = (0.0, 0.0), \ \mu_2 = (1.0, 0.0) \text{ and}$$

$$\Sigma_1 = \Sigma_2 = \Sigma = \begin{pmatrix} 1.0 \ 0.0 \\ 0.0 \ 1.0 \end{pmatrix}.$$

Since $\Sigma_1 = \Sigma_2$, the Bayes' boundary will be linear [155] of the following form:

$$(\mu_2 - \mu_1)^T \Sigma^{-1} X + \frac{1}{2} (\mu_1^T \Sigma^{-1} \mu_1 - \mu_2^T \Sigma^{-1} \mu_2).$$

Note:

- For each data set and for a particular value of n, two different training data sets are generated using different seed values. Training is performed with each data set for five different initial populations. This means that a total of ten runs are performed for a given n. The results of the *GA-classifier* presented here are average values over these ten runs.
- The test data set comprises 1000 points. Testing of the *GA-classifier* is performed for each of the ten training runs. Results shown are the average values over these ten runs.
- Roulette wheel selection strategy with elitism is used. Population size and crossover probability (μ_c) are kept equal to 20 and 0.8, respectively. The mutation probability value (μ_m) varies over the range of [0.015, 0.333], initially having a high value, then gradually decreasing and finally increasing again in the later stages of the algorithm. One hundred iterations of the GA are performed for each μ_m. A total of 1500 iterations is performed. As proved in Sect. 2.4, GA will always provide the optimal string as the number of generations goes to infinity, provided the probability of going from any population to the one containing the optimal string is greater than zero [60]. This proof is irrespective of the population sizes and different probability values. However, appropriate selection of GA parameters is necessary in view of the fact that it has to be terminated after a finite number of iterations.

4.4.2 Learning the Class Boundaries and Performance on Test Data

Utilizing Linear Surfaces

An extensive comparison of the performance of the *GA-classifier* with that of Bayes' classifier is performed for *Data Set 1*, *Data Set 2* and *Data Set 3*. The following cases are considered [37] for the Bayes' classifier:

Case 1: The distributions as well as the class a priori probabilities (P_1 and P_2) are known.

Case 2: The distribution is known, with $P_1 = P_2 = 0.5$.

Case 3: The distribution is known, with $P_1 = n_1/n$ and $P_2 = n_2/n$. Here n_1 and n_2 are the number of points belonging to classes 1 and 2, respectively, $n = n_1 + n_2$.

Case 4: Normal distribution with unequal covariance matrices, while P_1 and P_2 are known.

Case 5: Normal distribution with unequal covariance matrices, and $P_1 = P_2 = 0.5$.

Case 6: Normal distribution with unequal covariance matrices, and $P_1 = n_1/n$ and $P_2 = n_2/n$, where n_1 and n_2 are defined above.

(a) *Data Set 1:* The different values of n considered are 100, 500, 1000, 1500, 2000, 3000 and 4000 and $P_1 = 0.4$. The Bayes' boundary for this data set is a straight line ($x = 1.4$). The training recognition scores of Bayes' classifier and *GA-classifier* are shown in Table 4.1 for $H = 1$ and 3. As expected, for each value of n, the recognition score during training of the *GA-classifier* with $H = 3$ was found to be at least as good as that with $H = 1$. Table 4.2 shows the recognition scores for the test data. Here, it is found that the Bayes' classifier yields a better performance than the *GA-classifier*. Also, *GA-classifier* with $H = 3$ provides a consistently lower score than with $H = 1$. This is expected since a larger H leads to overfitting of the training data, thereby yielding better training performance (Table 4.1), while the generalization capability of the classifier degrades (Table 4.2).

It is seen from Table 4.1 that for all values of n, the recognition scores for *GA-classifier* during training are better than those of Bayes' (all cases). However, this difference in performance gradually decreases for larger values of n. This is demonstrated in Fig. 4.4, which plots α as a function of n, where α = overall training recognition score of the *GA-classifier*−overall recognition score of the Bayes' classifier (case 1). This indicates that as n increases, the decision boundaries provided by the *GA-classifier* gradually approach the Bayes' boundary. This observation is demonstrated in Fig 4.5, which shows the results for $n = 100, 1000, 2000$ and 4000, respectively.

A point to be mentioned here is that although the result of Bayes' classifier (case 1) is the ideal one considering the actual distribution and *a priori* probabilities, interestingly some of the other cases (e.g., case 5 for $n=100$, case 3 for $n=500$) are seen to produce better scores. A reason for this discrepancy may be because of statistical variations. It is also found that results for case 1 and case 4 are similar (having same values for $n=100$, 1000 and 1500, and close values for other values of n). The reason for this is that normal distribution with proper variance has been found to be able to approximate the triangular distribution closely.

(b) *Data Set 2:* Four values of n considered here are $n = 500, 1000, 2000$ and 5000. Figure 4.2 shows the data set for $n = 1000$ and the corresponding Bayes' boundary. (It is found that the Bayes' boundary totally surrounds

Table 4.1. Comparative classwise and overall recognition scores (%) during training for *Data Set 1*

n	Class	*GA-classifier*		Bayes' classifier					
		$H = 1$	$H = 3$	Case 1	Case 2	Case 3	Case 4	Case 5	Case 6
	1	88.77	89.26	81.08	85.94	81.08	82.60	88.77	83.71
100	2	93.24	95.00	91.21	87.88	90.30	89.60	87.18	89.60
	Overall	91.50	92.70	87.00	87.5	87.00	87.00	88.00	87.50
	1	76.48	78.24	80.29	87.75	81.41	83.43	87.51	82.87
500	2	95.69	95.59	89.39	84.73	90.15	87.87	85.38	87.85
	Overall	88.00	88.65	85.70	85.90	86.20	86.10	86.20	86.00
	1	86.91	86.17	82.68	87.64	83.03	84.32	87.40	84.80
1000	2	91.01	92.00	92.93	87.71	92.75	91.84	87.81	91.58
	Overall	89.35	89.67	88.80	87.70	88.85	88.80	87.65	88.50
	1	87.17	87.70	82.26	88.18	82.70	84.38	88.33	84.59
1500	2	90.73	90.45	92.07	88.17	92.10	91.11	87.94	90.54
	Overall	89.08	89.35	88.10	88.16	88.33	88.10	88.10	88.16
	1	80.77	81.00	83.13	88.71	83.25	84.24	87.97	84.24
2000	2	94.81	94.93	91.71	86.77	91.54	91.37	87.52	91.29
	Overall	89.15	89.36	88.25	87.55	88.20	88.50	87.70	88.45
	1	75.96	78.72	80.82	85.59	80.57	82.41	85.34	82.24
3000	2	95.68	94.02	91.86	87.65	91.86	90.75	87.76	90.81
	Overall	87.83	87.90	87.46	86.83	87.36	87.43	86.80	87.40
	1	82.54	82.50	87.73	87.39	81.89	84.35	87.39	83.83
4000	2	91.81	92.08	91.52	86.88	91.93	90.18	86.67	90.59
	Overall	88.22	88.25	88.12	87.07	88.05	87.92	86.95	87.97

Table 4.2. Comparative overall recognition scores (%) during testing for *Data Set 1*

n	*GA-classifier*		Bayes' classifier
	$H=1$	$H = 3$	(Case 1)
100	85.30	85.00	87.90
500	85.90	85.75	87.90
1000	86.20	85.87	87.90
1500	86.21	86.00	87.90
2000	86.55	86.41	87.90
3000	87.10	87.10	87.90
4000	87.32	87.20	87.90

class 1.) Cases 1, 3, 4 and 6 are investigated since only these are relevant in this context. Table 4.3 presents the results corresponding to them during training for $H = 4$ and 6.

For smaller values of n (500 and 1000), it is found, as in the case of *Data Set 1*, that the *GA-classifier* yields a better score than the Bayes' classifier for the training data set (Table 4.3). Figure 4.6 shows the deci-

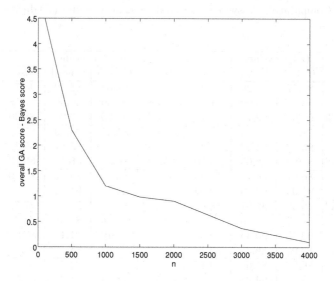

Fig. 4.4. Variation of α (= overall GA score−Bayes' score) with n for *Data Set 1*

Table 4.3. Comparative classwise and overall recognition scores (%) during training for *Data Set 2*

n	Class	*GA-classifier*		Bayes' classifier			
		$H{=}4$	$H{=}6$	Case 1	Case 3	Case 4	Case 6
500	1	89.23	89.23	85.38	86.54	85.77	87.31
	2	62.92	65.83	65.83	64.17	64.17	63.33
	Overall	76.60	78.00	76.00	75.80	75.40	75.80
1000	1	83.92	84.90	83.33	84.12	83.33	84.71
	2	67.96	67.44	66.53	65.92	65.71	65.10
	Overall	76.10	76.35	75.10	75.20	74.70	75.10
2000	1	87.98	87.58	85.27	83.73	84.77	85.22
	2	61.78	63.37	64.77	65.43	65.17	65.18
	Overall	74.85	75.45	75.00	74.87	74.95	74.98
5000	1	85.18	83.04	85.42	85.50	85.74	85.82
	2	65.04	68.39	66.33	66.01	66.09	65.81
	Overall	75.18	75.76	75.94	75.82	75.98	75.88

sion boundary obtained for $n = 1000$ and $H = 4$, along with the Bayes' boundary. Although all the four lines of *GA-classifier* are found to be necessary, and they surround class 1, the boundary formed by them could not approximate the Bayes' boundary well. In spite of this fact, its recognition score during training is relatively larger than that of the Bayes'

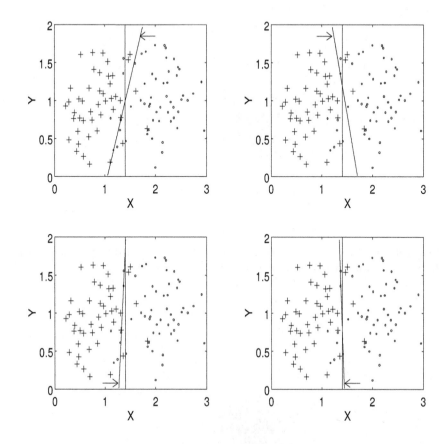

Fig. 4.5. *Data Set 1* for $n = 100$ (*top left*), $n = 1000$ (*top right*), $n = 2000$ (*bottom left*) and $n = 4000$ (*bottom right*). Class 1 is represented by "+" and class 2 by "o". The boundary provided by *GA-classifier* for $H = 1$ (marked with an *arrow*) and the Bayes' decision boundary are shown

classifier. Increasing H to a value 6 improves both the approximation of the boundary and the recognition score during training (Table 4.3).

For $n = 2000$, one out of the four lines is found to be redundant by the *GA-classifier* (Fig. 4.7) and they fail to surround class 1 for the same number (1500) of iterations. The training score is accordingly found to be lower than that of the Bayes' classifier. For $H = 6$ (Fig. 4.8), the recognition score during training exceeds the one obtained by Bayes', and the approximation to the Bayes' boundary improves (although only 5 lines are utilized effectively).

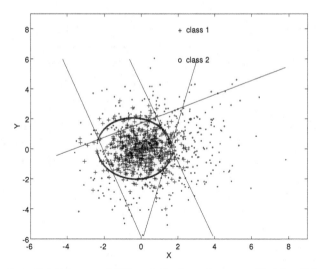

Fig. 4.6. *Data Set 2* for $n = 1000$ and the boundary provided by *GA-classifier* for $H = 4$ along with Bayes' decision boundary (*circular one*)

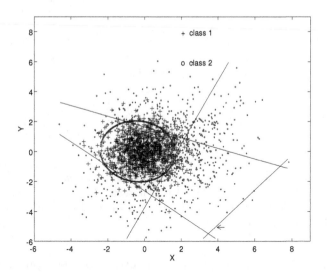

Fig. 4.7. *Data Set 2* for $n = 2000$ and the boundary provided by *GA-classifier* for $H = 4$ along with Bayes' decision boundary (*circular one*). The redundant line is marked with an *arrow*

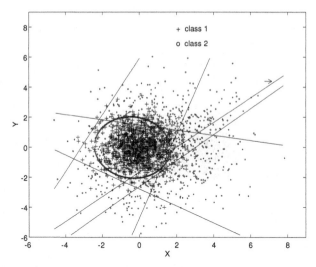

Fig. 4.8. *Data Set 2* for $n = 2000$ and the boundary provided by *GA-classifier* for $H = 6$ along with Bayes' decision boundary (*circular one*). The redundant line is marked with an *arrow*

For a further increase in n to 5000, the training recognition scores (Table 4.3) for both $H = 4$ and 6 are seen to be worse than Bayes' scores for the same number of iterations. However, the approximation to Bayes' boundary is seen to be much improved here as compared to $n = 1000$ and 2000. This is evident from Figs. 4.9 and 4.10, where $H = 6$, as expected, provides better performance than $H = 4$. From Figs. 4.7 and 4.10 it is interesting to note that one line is found to be redundant and at the same time the score is worse than that of Bayes' classifier. This may be attributed to the premature termination of GA.

Table 4.4 shows the recognition scores for *Data Set 2* during testing. Again, it is found that the *GA-classifier* provides poorer scores than Bayes' classifier. Also, *GA-classifier* with $H = 6$ provides better performance than with $H = 4$. This is expected since $H = 6$ can better approximate the complex decision surface.

Table 4.4. Comparative overall recognition scores (%) during testing for *Data Set 2*

n	*GA-classifier*		Bayes' classifier
	$H = 4$	$H = 6$	(case 1)
500	72.75	73.20	74.80
1000	72.92	73.35	74.80
2000	73.05	73.62	74.80
5000	73.34	74.15	74.80

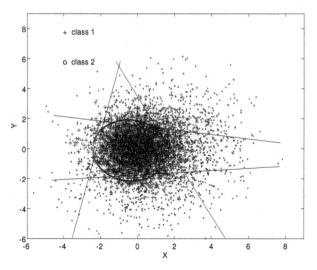

Fig. 4.9. *Data Set 2* for $n = 5000$ and the boundary provided by *GA-classifier* for $H = 4$ along with Bayes' decision boundary (*circular one*)

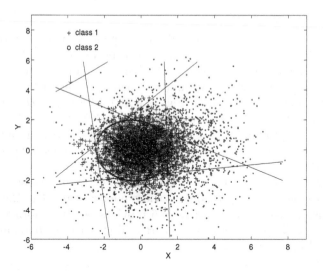

Fig. 4.10. *Data Set 2* for $n = 5000$ and the boundary provided by *GA-classifier* for $H = 6$ along with Bayes' decision boundary (*circular one*). The redundant line is marked with an *arrow*

(c) *Data Set 3:* The nine-class, two-dimensional data set for $n = 1800$ along with the corresponding Bayes' boundary is shown in Fig. 4.3. The experiments on training as well as test data sets have been conducted for $H = 4$ and $n = 450, 900, 1350$ and 1800. Only the training results are presented here. As before, the test results show a superior performance of the Bayes' classifier. These are not included here. Comparison with the Bayes' classifier is made for case 1 only. The results for the overall training recognition scores are shown in Table 4.5. Figure 4.11 shows the boundary obtained using the *GA-classifier* along with the Bayes' decision boundary for $n=1800$, where it is seen that the four lines approximate the Bayes' boundary reasonably well. It is found from Table 4.5 that although for each value of n the training scores of the *GA-classifier* and the Bayes' classifier are comparable, the latter one appears to provide a slightly better performance. One of the reasons may again be the premature termination of the GA. Another factor may be the coding itself which does not allow encoding of the actual Bayes' lines (this may be due to an insufficiency of the precision defined for the perpendicular distance of the lines).

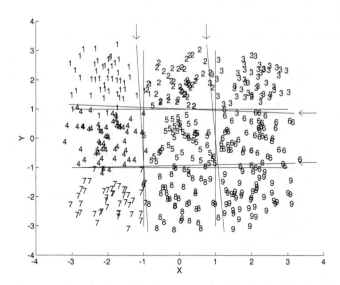

Fig. 4.11. *Data Set 3* and the boundary provided by *GA-classifier* for $n = 1800$ and $H = 4$ (marked with *arrows*) along with Bayes' decision boundary. (Only 900 points are shown.)

Table 4.5. Comparative overall recognition scores (%) during training for *Data Set 3*

n	Overall recognition score	
	GA-classifier (H=4)	Bayes' classifier(case 1)
450	93.56	93.78
900	93.22	93.11
1350	90.08	92.52
1800	92.25	92.50

Utilizing Higher-Order Surfaces

It has been pointed out earlier (Sect. 4.2) that instead of approximating the decision boundary by hyperplanes, we can assume any other higher-order surface with similar effect. This subsection describes the results [37] of utilizing a fixed number of circular segments to constitute the decision surface in two-dimensional space. The method of using circular/hyperspherical segments for modelling the class boundaries is described in Chap. 3.

The results obtained during both training and testing, when C circular segments are used to model the decision boundaries for *Data Set 1* and *Data Set 2*, are shown in Table 4.6. The value of p (which is defined in Sect. 3.5.2 and controls the size of the *outer rectangle*) is assumed to be 8 in order to be able to approximate any arbitrary boundary appropriately.

Table 4.6. Overall recognition scores (%) during training and testing for higher-order surfaces

Data Set	n	No. of Circles C	*GA-classifier* (circular surface)		*GA-classifier* (linear surface $H = C$) from Tables 4.1 and 4.3	
			Training	Testing	Training	Testing
Data Set 1	1000	1	89.40	86.05	89.35	86.20
	2000	1	89.10	86.20	89.15	86.55
	2000	4	74.97	73.10	74.85	73.05
Data Set 2	2000	6	75.55	73.60	75.45	73.62
	5000	6	76.00	74.45	75.76	74.15

Interestingly, the approximation of the decision boundary for *Data Set 1* and *Data Set 2* could be made better using linear and circular segments respectively. This is followed for both data sets. Figure 4.12 shows the GA and Bayes' boundaries obtained for *Data Set 1*, $n = 1000$ and $C = 1$, and *Data Set 2*, $n = 5000$ and $C = 6$ (where one segment is found to be redundant as in Fig. 4.10), respectively.

The test scores for *Data Set 1* with circular segments are found to be poorer compared to those with linear segments. The case is reversed for *Data Set 2*

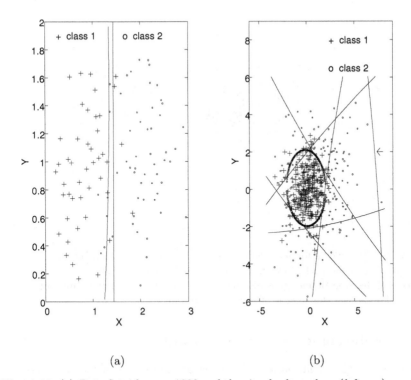

(a)   (b)

Fig. 4.12. (a) *Data Set 1* for $n = 1000$ and the circular boundary (*left one*) provided by *GA-classifier* when circular segments are considered to constitute the decision boundary for C = 1 along with Bayes' decision boundary. (Only 100 data points are plotted for clarity.); (b) *Data Set 2* for $n = 5000$ and the boundary provided by *GA-classifier* when circular segments are considered to constitute the decision boundary for $C = 6$ along with Bayes' decision boundary. The redundant segment is marked with an *arrow* (only 500 data points are plotted for clarity)

(except with $n = 2000$ and $H = 6$, where the two are comparable). This is expected since the decision boundary is linear for *Data Set 1*, while it is circular for *Data Set 2*. Note that all these results are, however, inferior to the Bayes' scores for the test data (Tables 4.2 and 4.4).

Note that for $p = 8$, the circular property of the segments does not appear to be fully exploited (Fig. 4.12(b)) because of large radius. This was done in order to be able to better approximate any type of decision boundary using circular segments only. Figure 4.13 shows another result for $n = 2000$, $p = 0$ and $C = 2$ which corresponds to smaller radius, and hence shows better circular characteristics. Here, the recognition scores over the training and the test data are 75.10% and 74.65%, respectively.

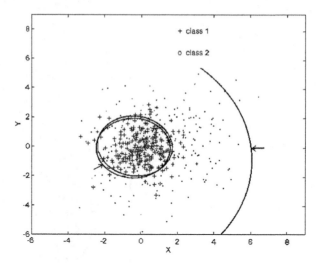

Fig. 4.13. *Data Set 2* for $n = 2000$ and the boundary provided by *GA-classifier* (marked with *arrows*) when circular segments are considered for $C = 2$ and $p = 0$. The Bayes' decision boundary is also shown. (Only 500 data points are plotted for clarity.) The arc on the *right* is obviously redundant

4.4.3 Variation of Recognition Scores with P_1

In this section the variation of the test recognition scores of the *GA-classifier* with P_1 (a priori probability of class 1) is demonstrated, and it is shown that the variation is similar to that of the Bayes' classifier for *Data Set 1*, *Data Set 2* and *Data Set 4*. Figures 4.14 ($H = 1$ and 3), 4.15 ($H = 6$) and 4.16 ($H = 1$) show the results for the three data sets, respectively. Here one training data set having 200 points and two test data sets of 1000 points are taken. Training of the *GA-classifier* is performed for five initial conditions. Subsequently, testing is performed for the two test data sets. The results shown are the average values over the ten runs. Note that the variation of the test recognition scores of the *GA-classifier* with P_1 is similar to that of Bayes' classifier for all the three data sets.

For the convenience of the readers, the above-mentioned variation for Bayes' classifier is discussed theoretically in Appendix C with reference to triangular and normal distribution of data points. It is shown that for triangular distribution the error probability varies symmetrically with P_1 having the maximum value for $P_1 = 0.5$. Similar observations were made in the investigation for *Data Set 1*. Consequently, results are presented for P_1 lying in the range [0,0.5] only. The test scores for the *GA-classifier* with $H=3$ are found to be consistently poorer than those with $H = 1$ for this data set (Fig. 4.14). It is to be mentioned here that recognition of the training data was better for $H = 3$ than for $H = 1$ (Table 4.1). This again supports the observations that

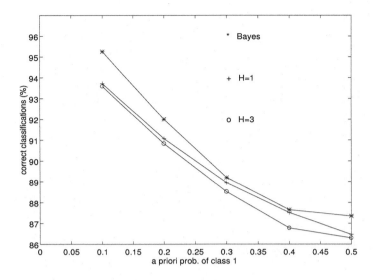

Fig. 4.14. Variation of recognition score of test data with P_1 for *Data Set 1* corresponding to Bayes' classifier and *GA-classifier* ($H = 1$ and 3), $n = 200$

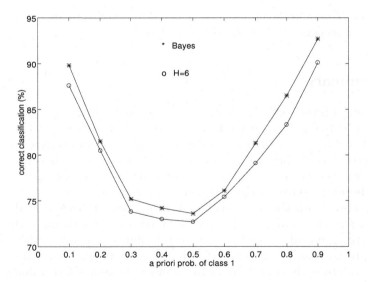

Fig. 4.15. Variation of recognition score of test data with P_1 for *Data Set 2* corresponding to Bayes' classifier and *GA-classifier* ($H = 6$), $n = 200$

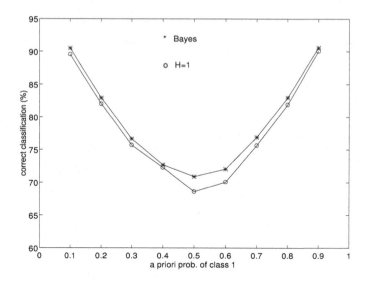

Fig. 4.16. Variation of recognition score of test data with P_1 for *Data Set 4* corresponding to Bayes' classifier and *GA-classifier* ($H = 1$), $n = 200$

increasing H beyond the optimum value leads to better training performance but poorer generalization of the *GA-classifier*.

4.5 Summary

A theoretical investigation is reported here to establish the relationship between the *GA-classifier* and the Bayes' classifier. It is shown that for a sufficiently large number of training data points, and for a sufficiently large number of iterations the performance of the *GA-classifier* approaches that of the Bayes' classifier when the number of hyperplanes (or higher-order surfaces) giving rise to the Bayes' decision regions are known.

It is also shown that the number of hyperplanes provided by the *GA-classifier* for constituting the decision boundary will be optimum if the minimal partition providing the Bayes' error probability is unique. Otherwise, it will be greater than or equal to the optimum value. In either case, the classifier will still be optimal in terms of the number of misclassified training data points.

The experimental results on the different distributions of overlapping data with both linear and nonlinear boundaries show that the decision regions provided by the *GA-classifier* gradually approach the ones provided by the Bayes' classifier as the number of training data points increases. Specifically, it has been found that for the *GA-classifier*

(a) *error rate* $\leq a$ during training,

(b) *error rate* $\geq a$ during testing, and

(c) *error rate* $\rightarrow a$ during training and testing as $n \rightarrow \infty$,

where a is the Bayes' error probability, and *error rate* is defined as

$$error\ rate = \frac{\text{number of misclassified points}}{n}.$$

For *Data Set 2*, where the decision boundaries are nonlinear, the *GA-classifier* using circular segments performs better than the one using linear surfaces (Table 4.6, for *Data Set 2*). On the other hand, for *Data Set 1* where, the decision boundary is actually linear, the *GA-classifier* with lines performs better. In real-life, data sets are often overlapping with nonlinear boundaries. Thus using higher-order surfaces for modelling the decision boundary appears to be better in real-life problems. This also conforms to the findings in Chap. 3, where the approximation of the class boundaries could be made better by using hyperspherical surfaces for *ADS 1* and *Vowel* data.

As far as the generalization capability is concerned, the Bayes' classifier generalizes better than the *GA-classifier* for all the data sets, and also for all values of a priori class probability P_1. It must be noted that when the class a priori probabilities and the probability distributions are known, the Bayes' classifier provides the optimal performance. No other classifier, in that case, can provide a better performance. However, in most practical cases, these quantities are usually unknown. It is in such situations that the *GA-classifier*, which is nonparametric in nature, can be utilized.

As has been mentioned earlier, determination of an appropriate values of H (or, C when higher-order surfaces are utilized) is crucial for the success of the *GA-classifier*. Chap. 5 describes a solution to this problem where the concept of variable length chromosomes is incorporated in the *GA-classifier*.

5

Variable String Lengths in *GA-classifier*

5.1 Introduction

We have seen in the previous chapters that the estimation of a proper value of H is crucial for a good performance of the *GA-classifier*. Since this is difficult to achieve, one may frequently use a conservative value of H while designing the classifier. This first of all leads to the problem of an overdependence of the algorithm on the training data, especially for small sample sizes. In other words, since a large number of hyperplanes can readily and closely fit the classes, this may provide good performance during training but poor generalization capability. Secondly, a large value of H unnecessarily increases the computational effort, and may lead to the presence of redundant hyperplanes in the final decision boundary. (A hyperplane is termed redundant if its removal has no effect on the classification capability of the *GA-classifier*.)

In this chapter, a method has been described to automatically evolve the value of H as a parameter of the problem. For this purpose, the concept of variable length strings in GA has been adopted. Unlike the conventional GA, here the length of a string is not fixed. Crossover and mutation operators are accordingly defined. A factor has been incorporated into the fitness function that rewards a string with smaller number of misclassified samples as well as smaller number of hyperplanes. Let the classifier so designed utilizing the concept of variable string lengths in genetic algorithm (VGA) be called *VGA-classifier* [38].

Issues of minimum misclassification error and minimum number of hyperplanes generated by the *VGA-classifier* are theoretically analyzed under limiting conditions. Moreover, it is proved that for infinitely large training data set, and infinitely many iterations, the error rate provided by the *VGA-classifier* during training will approach the Bayes' error probability. As mentioned in Sect. 1.2.3, the Bayes' classifier is known to provide optimal performance when the class conditional densities and the a priori probabilities are known.

Experimental results on one artificial data, *ADS 1*, and two real-life data sets, *Iris* and *Vowel* (used in Chap. 3), show that the *VGA-classifier* is able to reduce the number of hyperplanes significantly while providing improved generalization capability as compared to the *GA-classifier*. A similar work was described by Srikanth et al. in [448], where the decision boundary was modelled by a variable number of ellipsoids. Not only does this have a higher degree of complexity than hyperplanes, there was no incentive in the fitness function for reducing the number of ellipsoids. Additionally, no theoretical basis was provided in [448] to demonstrate the validity of the methodology. However, the operators in [448] are implemented in the *VGA-classifier* for the purpose of comparison.

There is a clear analogy between the principle of hyperplane fitting in the genetic classifiers and the process of classification in the multilayer perceptron (MLP) with hard limiting neurons. The latter also approximates the class boundaries with piecewise linear surfaces. Thus the *VGA-classifier* can be applied effectively for the determination of MLP architecture and connection weights with hard limiting neurons. Such an algorithm is described in a later part of this chapter along with some experimental results.

Section 5.2 describes briefly some other attempts to use variable string lengths in GA, and discusses the criteria for the application of VGA to the classification problem. A detailed description of *VGA-classifier*, including the modified genetic operators, is then presented in Sect. 5.3, along with suitable examples. Section 5.4 contains the theoretical analysis of *VGA-classifier* in terms of the number of hyperplanes and error probability. The experimental results on different data sets are provided in Sect. 5.5. Section 5.6 describes how the results of *VGA-classifier* can be applied for automatically determining the architecture and the connection weights of an MLP. A comparative performance study is also included. Finally, Sect. 5.7 summarizes this chapter.

5.2 Genetic Algorithm with Variable String Length and the Classification Criteria

The concept of variable string lengths in genetic algorithms has been used earlier in [439] to encode sets of fixed length rules. Messy genetic algorithm [168, 169] also uses the concept of variable string lengths for constructing the chromosomes which may be under- or overspecified. Use of GA with variable string length has been made in [185] for encoding variable number of fixed length blocks in order to construct layers of a neural network, and in [290] for the genetic evolution of the topology and weight distribution of neural networks.

As mentioned in Sect. 5.1, the *GA-classifier* with fixed H, and consequently fixed string length has several limitations like overfitting of the training data and presence of redundant hyperplanes in the decision boundary when a conservative value of H is used. To overcome these limitations, the use of variable length strings representing a variable number of hyperplanes for modelling optimally the decision boundary therefore seems natural and appropriate. This would eliminate the need for fixing the value of H, evolving it adaptively instead, thereby providing an optimal value of H.

It is to be noted that in the process, if we aim at reducing the number of misclassified points only, as was the case for fixed length strings, then the algorithm may try to fit as many hyperplanes as possible for this purpose. This, in turn, would obviously be harmful with respect to the generalization capability of the classifier. Thus the fitness function should be defined such that its maximization ensures primarily the minimization of the number of misclassified samples and also the requisite number of hyperplanes.

While incorporating the concept of variable string lengths, one may note that it is necessary to either modify the existing genetic operators or introduce new ones. In order to utilize the existing operators as much as possible, a new representation scheme involving the consideration of the ternary alphabet set $\{0, 1, \# \}$, where $\#$ represents the "don't care" position, is used. For applying the conventional crossover operator, the two strings, which may now be of unequal lengths, can be made of equal length by appropriately padding one of them with $\#$'s. However, some extra processing steps have to be defined in order to tackle the presence of $\#$'s in the strings. Similarly, the mutation operator needs to be suitably modified such that it has sufficient flexibility to change the string length while retaining the flavour of the conventional operator. (As will be evident in the next section, the genetic operators are defined in such a way that the inclusion of $\#$ in the strings does not affect their binary characteristics for encoding and decoding purposes.) The classifier thus formed using variable string length GA (or VGA) is referred to as the *VGA-classifier*.

Therefore the objective of the *VGA-classifier* is to place an appropriate number of hyperplanes in the feature space such that it, first of all, minimizes the number of misclassified samples and then attempts to reduce the number of hyperplanes. Using variable length strings enables one to check, automatically and efficiently, various decision boundaries consisting of different number of hyperplanes in order to attain the said criterion. Note that the *GA-classifier*, with fixed length strings, does not have this flexibility. The description of such a classifier is given in the next section.

5.3 Description of *VGA-Classifier*

The basic operations of GAs are followed sequentially as shown in Fig. 2.2. However, the operators themselves are suitably defined for this problem.

5.3.1 Chromosome Representation and Population Initialization

The chromosomes are represented by strings of 1, 0 and $\#$ (don't care), encoding the parameters of variable number of hyperplanes. As mentioned in

Sect. 3.4.1, in $\mathbb{R}^N$, N parameters are required for representing one hyperplane. These are $N - 1$ angle variables, $\alpha_1^i, \alpha_2^i, \ldots, \alpha_{N-1}^i$, indicating the orientation of hyperplane i ($i = 1, 2, \ldots, H$, when H hyperplanes are encoded in the chromosome), and one perpendicular distance variable, d^i, indicating its perpendicular distance from the origin. Let H_{max} represent the maximum number of hyperplanes that may be required to model the decision boundary of a given data set. It is specified a priori. Let the angle and perpendicular distance variables be represented by b_1 and b_2, bits respectively.

Then l_H, the number of bits required to represent a hyperplane, and l_{max}, the maximum length that a string can have, are

$$l_H = (N - 1) * b_1 + b_2, \qquad (5.1)$$

$$l_{max} = H_{max} * l_H, \qquad (5.2)$$

respectively. Let string i represent H_i hyperplanes. Then its length l_i is

$$l_i = H_i * l_H.$$

Unlike the conventional *GA-classifier*, the number of hyperplanes encoded in a string may vary, and is therefore stored. The initial population is created in such a way that the first and the second strings encode the parameters of H_{max} and 1 hyperplanes, respectively, to ensure sufficient diversity in the population. For the remaining strings, the number of hyperplanes, H_i, is generated randomly in the range $[1, H_{max}]$, and the l_i bits are initialized randomly to 1's and 0's.

Example 5.1. Let $N = 2$, $H_{max} = 3$, $b_1 = 2$ and $b_2 = 2$, indicating that the features are in two-dimensional space, and a maximum of three hyperplanes are allowed to model the decision boundary. Also, two bits are allowed to encode each of the angle and perpendicular distance variables. Then

$$l_{max} = 3 * (1 * 2 + 2) = 12,$$

i.e., maximum string length in the population will be 12. In the initial population, let strings i and j comprise 2 and 3 hyperplanes, respectively. Or $H_i = 2$, $H_j = 3$. Therefore

$$l_i = 8 \text{ and } l_j = 12.$$

Let the randomly generated strings be

$$string_i = 1\,0\,1\,1 \quad 0\,0\,1\,0$$
$$string_j = 0\,1\,0\,0 \quad 1\,0\,1\,1 \quad 0\,0\,0\,1$$

In the *GA-classifier* (Chap. 3), if $l = 12$, then all the strings must essentially contain 12 bits.

5.3.2 Fitness Computation

As mentioned in Sect. 5.2, the fitness function (which is maximized) is defined in such a way that

(a) a string with smaller value of misclassifications is considered to be fitter than a string with a larger value, irrespective of the number of hyperplanes, i.e., it first of all minimizes the number of misclassified points, and then

(b) among two strings providing the same number of misclassifications, the one with the smaller number of hyperplanes is considered to be fitter.

The number of misclassified points for a string i encoding H_i hyperplanes is found as in Chap. 3. However, it is again explained here for the sake of completeness. Let the H_i hyperplanes provide r_i distinct regions which contain at least one training data point. (Note that although $r_i \leq 2^{H_i}$, in reality it is upper bounded by the size of the training data set.) For each such region and from the training data points that lie in this region, the class of the majority is determined, and the region is considered to represent (or be labelled by) the said class. Points of other classes that lie in this region are considered to be misclassified. The sum of the misclassifications for all the r_i regions constitutes the total misclassification $miss_i$ associated with the string. Accordingly, the fitness of string i may be defined as

$$fit_i = (n - miss_i) - \alpha H_i, \; 1 \leq H_i \leq H_{max}, \qquad (5.3)$$
$$= 0, \qquad\qquad otherwise, \qquad (5.4)$$

where $n =$ size of the training data set and $\alpha = \frac{1}{H_{max}}$. Let us now explain how the first criterion is satisfied. Let two strings i and j have number of misclassifications $miss_i$ and $miss_j$, respectively, and number of hyperplanes encoded in them be H_i and H_j, respectively. Let $miss_i < miss_j$ and $H_i > H_j$. (Note that since the number of misclassified points can only be integers, $miss_j \geq miss_i + 1$.) Then,

$$fit_i = (n - miss_i) - \alpha H_i,$$
$$fit_j = (n - miss_j) - \alpha H_j.$$

The aim now is to prove that $fit_i > fit_j$, or that $fit_i - fit_j > 0$. From the above equations,

$$fit_i - fit_j = miss_j - miss_i - \alpha(H_i - H_j).$$

If $H_j = 0$, then $fit_j = 0$ from (Eq. (5.4)), and therefore $fit_i > fit_j$. When $1 \leq H_j \leq H_{max}$, we have $\alpha(H_i - H_j) < 1$, since $(H_i - H_j) < H_{max}$. Obviously, $miss_j - miss_i \geq 1$. Therefore $fit_i - fit_j > 0$, or, $fit_i > fit_j$.
The second criterion is also fulfilled since $fit_i < fit_j$ when $miss_i = miss_j$ and $H_i > H_j$.

Example 5.2. From Fig. 5.1, for a two-dimensional ($N=2$), two-class problem where $n = 15$, let the two lines ($H_i = 2$) encoded in a chromosome be as shown. There are three corresponding regions, *Region 1*, *Region 2* and *Region 3*. Points represented by 1 and 2 indicate that they belong to classes 1 and 2 in the training data set, respectively. Hence, as mentioned above, *Region 1* and *Region 3* are associated with class 2 (since the majority of points that lie in this region are from class 2). Similarly, *Region 2* is associated with class 1. The total misclassification associated with the string is 3 (i.e., *miss* = 3). Since here $H_i = 2$, for $H_{max} = 3$ (and hence $\alpha = \frac{1}{3}$) and $n = 15$, the fitness of the string is

$$15 - 3 - 1/3 * 2 = 11.333.$$

Note that in the case of fixed length strings, the term αH_i would not have any significance, since its value would be the same for all the strings ($= 1$).

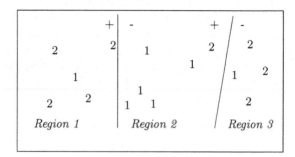

Fig. 5.1. Example of fitness computation

5.3.3 Genetic Operators

Among the operations of selection, crossover and mutation, the selection operation used here may be one of those used in conventional GA, while crossover and mutation need to be newly defined for VGA. These are now described in detail.

Crossover: Two strings, i and j, having lengths l_i and l_j, respectively, are selected from the mating pool. Let $l_i \leq l_j$. Then string i is padded with #'s so as to make the two lengths equal. Conventional crossover is now performed over these two strings with probability μ_c. The following two cases may now arise:

- All the hyperplanes in the offspring are complete. (A hyperplane in a string is called *complete* if all the bits corresponding to it are either defined (i.e., 0's and 1's) or #'s. Otherwise it is incomplete.)

- Some hyperplanes are incomplete (i.e., the bits corresponding to the hyperplanes are a combination of defined and undefined values).

In the second case, let u = number of defined bits (either 0 or 1) and l_H = total number of bits per hyperplane = $(N-1)*b_1 + b_2$ from Eq. (5.1). Then, for each incomplete hyperplane, all the #'s are set to defined bits (either 0 or 1 randomly) with probability $\frac{u}{l_H}$. In case this is not permitted, all the defined bits are set to #. Thus each hyperplane in the string becomes complete. Subsequently, the string is rearranged so that all the #'s are pushed to the end, or, in other words, all the hyperplanes are transposed to the beginning of the strings. The information about the number of hyperplanes in the strings is updated accordingly.

Example 5.3. Let us consider single-point crossover between $string_i$ and $string_j$ described in Example 5.1. They are written again for convenience.

$$string_i = 1\ 0\ \ 1\ 1\ \ \ \ 0\ 0\ \ 1\ 0$$
$$string_j = 0\ 1\ \ 0\ 0\ \ \ \ 1\ 0\ \ 1\ 1\ \ \ \ 0\ 0\ \ 0\ 1$$

First $string_i$ is padded with #'s to make both the strings equal in length. Therefore now,

$$string_i = 1\ 0\ 1\ 1\ \ \ \ 0\ 0\ 1\ 0\ \ \ \ \#\ \#\ \#|\#$$
$$string_j = 0\ 1\ 0\ 0\ \ \ \ 1\ 0\ 1\ 1\ \ \ \ 0\ 0\ 0|1$$

Similar to the standard practice, the crossover point is determined randomly in the range 1 to 12, since $l_j{=}12$. Let the crossover point be 11 (the leftmost bit being 1) as shown above. Hence after crossover the two strings become:

$$offspring_i = 1\ 0\ 1\ 1\ \ \ \ 0\ 0\ 1\ 0\ \ \ \ \#\ \#\ \#|1$$
$$offspring_j = 0\ 1\ 0\ 0\ \ \ \ 1\ 0\ 1\ 1\ \ \ \ 0\ 0\ 0|\#$$

Note that the third hyperplane of both the offspring is incomplete. Considering $offspring_i$ only,

$$u\ \text{(no. of defined bits)} = 1,\quad t\ \text{(total no. of bits for a hyperplane)} = 4.$$

Therefore the #'s in the first three positions in $offspring_i$ are set to defined bits (0 or 1 randomly) with probability $\frac{1}{4}$. Otherwise, the 1 in the last position is set to #. Similarly, probability $\frac{3}{4}$ determines whether the # in position 12 of $offspring_j$ is set to defined bit or not. Otherwise, the defined bits in positions 9 - 11 are set to #. Let the resulting two strings be

$$offspring_i = 1\ 0\ 1\ 1\ \ \ \ 0\ 0\ 1\ 0\ \ \ \ \#\ \#\ \#\ \#$$
$$offspring_j = 0\ 1\ 0\ 0\ \ \ \ 1\ 0\ 1\ 1\ \ \ \ 0\ 0\ 0\ 1$$

There is no need for pushing the #'s in $offspring_i$ to the end of the string, since they are already there. Thus

$$offspring_i = \qquad 1\ 0\ 1\ 1\ \ \ \ 0\ 0\ 1\ 0.$$

Mutation: In order to introduce greater flexibility in the method, the mutation operator is defined in such a way that it can both increase and decrease the string length. For this, the strings are padded with #'s such that the resultant length becomes equal to l_{max}. Now for each defined bit position, it is determined whether conventional mutation [164] can be applied or not with probability μ_m. Otherwise, the position is set to # with probability μ_{m_1}. Each undefined position is set to a defined bit (randomly chosen) according to another mutation probability μ_{m_2}. These are described in Fig. 5.2.

Note that mutation may result in some incomplete hyperplanes, and these are handled in a manner, as done for crossover operation. It may also result in a change of the string length. For example, the operation on the defined bits, i.e., when $k \leq l_i$ in Fig. 5.2, may result in a decrease in the string length, while the operation on #'s, i.e., when $k > l_i$ in the figure, may result in an increase in the string length. Also, mutation may yield strings having all #'s indicating that no hyperplanes are encoded in it. Consequently, this string will have fitness = 0 and will be automatically eliminated during selection.

```
Begin
        l_i = length of string i
        Pad string i with # so that its length becomes l_max
        for k = 1 to l_max do
                Generate rnd, rnd1 and rnd2 randomly in [0,1]
                if k ≤ l_i do /* for defined bits */
                        if rnd < μ_m do /* conventional mutation */
                                Flip bit k of string i
                        else /* try changing to # */
                                if rnd1 < μ_m1 do
                                        Set bit k of string i to #
                                endif
                        endif
                else /* k > l_i i.e., for #'s */
                        if rnd2 < μ_m2 do /* set to defined */
                                Position k of string i is set to 0 or 1 randomly
                        endif
                endif
        endfor
End
```

Fig. 5.2. Mutation operation for string i

Example 5.4. Let us consider *offspring*$_i$ from Example 5.3, which is of length 8. So it is padded with 4 #'s to make the length 12. Or,

$$offspring_i = 1\ 0\ \ 1\ 1\ \ \ 0\ 0\ \ 1\ 0\ \ \ \#\ \#\ \ \#\ \#$$

Applying the different mutation operators described above, let the resultant string be

$$1\ 0\ 1\ 1 \quad 0\ \#\ 1\ \# \quad 0\ 1\ \#\ 0.$$

The #'s in position 6 and 8 in the second hyperplane are set to be defined randomly with probability $\frac{1}{2}$, since there are two defined bits of a total of four positions. Otherwise, the defined bits (positions 5 and 7 in the second hyperplane) are set to #'s. Similarly, the # in position 11 of the third hyperplane is set to a random defined bit with probability $\frac{3}{4}$, since there are three defined bits of a total of four positions. Otherwise, the defined bits (in positions 9, 10 and 12 of the third hyperplane) are set to #'s. Let the resultant string be

$$1\ 0\ 1\ 1 \quad \#\ \#\ \#\ \# \quad 0\ 1\ 1\ 0.$$

The #'s are pushed to the end of the string to result in

$$1\ 0\ 1\ 1 \quad 0\ 1\ 1\ 0 \quad \#\ \#\ \#\ \#$$

or, $1\ 0\ 1\ 1 \quad 0\ 1\ 1\ 0.$

As in conventional GAs, the operations of selection, crossover and mutation are performed here over a number of generations till a user specified termination condition is attained. Elitism is incorporated such that the best string seen up to the current generations is preserved in the population. The best string of the last generation, thus obtained, along with its associated labelling of regions provides the classification boundary of the n training samples. After the design is complete, the task of the classifier is to check, for an unknown pattern, the region in which it lies, and to put the label accordingly.

5.4 Theoretical Study of *VGA-classifier*

This section has two parts. In the first part, a theoretical investigation on the values of *miss* and H of the *VGA-classifier* is provided. The relationship between the *VGA-classifier* and the Bayes' classifier is established in the second part.

5.4.1 Issues of Minimum *miss* and H

In this section we prove that the above-mentioned *VGA-classifier* will provide the minimal misclassification error during training, for infinitely large number of iterations. At the same time it will require minimum number of hyperplanes in doing so.

For proving this, we use the result of [60], where it has been established that for an infinitely large number of iterations, an elitist model of GA will surely provide the optimal string. In order to prove this convergence it was assumed that the probability of going from any string to the optimal one is always greater than zero, and the probability of going from a population containing the optimal string to one not containing the optimal one is zero. Since the mutation operation and elitism of the proposed VGA ensure that both these conditions are met, the result of [60] regarding the convergence to the optimal string is valid for VGA as well.

Let us now consider the fitness function for string i given by Eq. (5.3). Maximization of the fitness function means minimization of

$$miss_i + \alpha H_i = err_i, \quad \text{say}$$

where $\alpha = \frac{1}{H_{max}}$, and err_i is the error function.

Let for any size of the training data set (n), the minimum value of the error function as obtained by the *VGA-classifier* be

$$err_{min} = miss_{opt} + \alpha H_{opt}$$

after it has been executed for infinitely large number of iterations. Following the result in [60], this corresponds to the optimal string. Therefore we may write

$$miss_{opt} + \alpha H_{opt} \leq miss + \alpha H, \quad \forall \, miss, H. \tag{5.5}$$

Theorem 5.1. *For any value of H, $1 \leq H \leq H_{max}$, the minimal number of misclassified points is $miss_{opt}$.*

Proof. The proof is trivial and follows from the definition of the fitness function in Eq. (5.3) and the fact that $miss_{opt} + \alpha H_{opt} \leq miss + \alpha H$, $\forall \, miss$, H (see Eq. (5.5)).

Theorem 5.2. *H_{opt} is the minimal number of hyperplanes required for providing $miss_{opt}$ number of misclassified points.*

Proof. Let the converse be true, i.e., there exists some H', $H' < H_{opt}$, that provides $miss_{opt}$ number of misclassified points. In that case, the corresponding fitness value would be $miss_{opt} + \alpha H'$. Note that now $miss_{opt} + \alpha H_{opt} > miss_{opt} + \alpha H'$. This violates Eq. (5.5). Hence $H' \not< H_{opt}$, and therefore H_{opt} is the minimal number of hyperplanes required for providing $miss_{opt}$ misclassified points.

From Theorems 5.1 and 5.2, it is proved that for any value of n, the *VGA-classifier* provides the minimum number of misclassified points for infinitely large number of iterations, and it requires minimum number of hyperplanes in doing so.

5.4.2 Error Rate

In this part, it is shown that for $n \longrightarrow \infty$, the error rate provided by the *VGA-classifier* during training will approach the Bayes' error probability, a, for infinitely many iterations.

The fitness function of the *VGA-classifier* as defined in Eq. (5.3) is

$$fit = (n - miss) - \alpha H.$$

Dividing the right hand side by n, we get

$$1 - \frac{miss}{n} - \alpha \frac{H}{n}.$$

For $n \longrightarrow \infty$, if the limit exists, then it goes to

$$1 - \lim_{n \longrightarrow \infty} \left[\frac{miss}{n} \right].$$

The term $\frac{miss}{n}$ provides the error rate of the *VGA-classifier*.

It is to be noted that the *VGA-classifier* attempts to minimize $f_{nE}(\omega)$ Eq. (4.11), the average number of misclassified samples, for large n. Assuming that $H_{max} \geq H_{opt}$, we may analyze as in Sect. 4.2, and state the following theorem and corollary:

Theorem 5.3. *For sufficiently large n, $f_n(\omega) \not> a$, (i.e., for sufficiently large n, $f_n(\omega)$ cannot be greater than a) almost everywhere.*

Proof. The proof is the same as that of Theorem 4.1.

Corollary 5.4. $\lim_{n \to \infty} f_n(\omega) = a$ *for $\omega \in B$.*

Proof. The proof is the same as that of corollary to Theorem 4.1.

As for the case with *GA-classifier*, Theorem 5.3 and its corollary indicate that the performance of the *VGA-classifier* approaches that of the Bayes' classifier under limiting conditions. Or in other words, the decision boundary generated by the *VGA-classifier* will approach the Bayes' decision boundary for infinitely large training data set and number of iterations, if the Bayes' boundary is unique. Otherwise the *VGA-classifier* will provide one such boundary for which the error rate is equal to the Bayes' error probability a.

5.5 Experimental Results

In this section the effectiveness of VGA in automatically determining the value of H of the classifier is demonstrated on a number of artificial and real-life data sets. The recognition scores of the *VGA-classifier* are also compared with those of the fixed length *GA-classifier*. A comparison of the present concept of using variable string lengths in GA with another similar approach [448] is provided.

5.5.1 Data Sets

The data sets *ADS 1*, *Vowel* and *Iris* have been used for demonstrating the effectiveness of the *VGA-classifier*. These are described in Appendix B, and were used earlier in Chap. 3.

5.5.2 Results

As in Chap. 3, a fixed population size of 20 is chosen. *Roulette wheel strategy* is used to implement proportional selection. *Single-point crossover* is applied with a fixed crossover probability of 0.8. A variable value of mutation probability μ_m is selected from the range [0.015, 0.333]. Two hundred iterations are performed with each mutation probability value. (Note that the number of hyperplanes constituting the search space for *VGA-classifier* now ranges from 1 to H_{max}. So, unlike the case with fixed H, the number of iterations for each value of μ_m is increased from 100 to 200.) The values of μ_{m_1} and μ_{m_2} mentioned in Sect. 5.3.3 are set to 0.1. The process is executed for a maximum of 3000 iterations. *Elitism* is incorporated as in Chap. 3. Average recognition scores over five different runs of the algorithm are provided. H_{max} is set to 10, so $\alpha = 0.1$. Training sets are the same as used in Chap. 3.

Performance of the *VGA-classifier*

Tables 5.1 and 5.2 show the number of hyperplanes H_{VGA}, for a particular run, as determined automatically by the *VGA-classifier* for modelling the class boundaries of *ADS 1*, *Vowel* and *Iris* data sets when the classifier is trained with 10% and 50% samples, respectively. Two different values of H_{max} are used for this purpose, viz., $H_{max} = 6$ and $H_{max} = 10$. The overall recognition scores obtained during testing of the *VGA-classifier* along with their comparison with those obtained for the fixed length version (i.e., *GA-classifier*) with $H = 6$ and 10 are also shown.

Table 5.1. H_{VGA} and the comparative overall recognition scores (%) during testing (when 10% of the data set is used for training and the remaining 90% for testing)

Data set	VGA-classifier $H_{max} = 10$		Score for GA-classifier $H = 10$	VGA-classifier $H_{max} = 6$		Score for GA-classifier $H = 6$
	H_{VGA}	Score		H_{VGA}	Score	
ADS 1	3	95.62	84.26	4	96.21	93.22
Vowel	6	73.66	69.21	6	71.19	71.99
Iris	2	95.56	76.29	2	95.81	93.33

The results demonstrate that in all the cases, the *VGA-classifier* is able to evolve an appropriate value of H_{VGA} from H_{max}. In addition, its recognition score on the test data set is found, on an average, to be higher than that of the *GA-classifier*. There is only one exception to this for the *Vowel* data when 10% of the samples are used for training (Table 5.1). In this case, $H_{max} = 6$ does not appear to be a high enough value for modelling the decision boundaries of *Vowel* classes with *VGA-classifier*. In fact this is reflected in both the tables,

Table 5.2. H_{VGA} and the comparative overall recognition scores (%) during testing (when 50% of the data set is used for training and the remaining 50% for testing)

Data set	VGA-classifier $H_{max} = 10$		Score for GA-classifier $H = 10$	VGA-classifier $H_{max} = 6$		Score for GA-classifier $H = 6$
	H_{VGA}	Score		H_{VGA}	Score	
ADS 1	4	96.41	95.92	4	96.83	96.05
Vowel	6	78.26	77.77	6	77.11	76.68
Iris	2	97.60	93.33	2	97.67	97.33

where the scores for *VGA-classifier* with $H_{max} = 6$ are less than those with $H_{max} = 10$.

In all the cases where the number of hyperplanes for modelling the class boundaries is less than 6, the scores of *VGA-classifier* with $H_{max} = 6$ are found to be superior to those with $H_{max} = 10$. This is because with $H_{max} = 10$, the search space is larger as compared to that for $H_{max} = 6$, which makes it difficult for the classifier to arrive at the optimum arrangement quickly or within the maximum number of iterations considered here. (Note that it may have been possible to further improve the scores and also reduce the number of hyperplanes, if more iterations of VGA were executed.)

In general, the scores of the *GA-classifier* (fixed length version) with $H = 10$ are seen to be lower than those with $H = 6$ for two reasons: overfitting of the training data and difficulty of searching a larger space. The only exception is with *Vowel* for training with 50% data, where the score for $H = 10$ is larger than that for $H = 6$. This is expected, in view of the overlapping classes of the data set and the larger size of the training data. One must note in this context that the detrimental effect of overfitting on the generalization performance increases with decreasing size of the training data.

As an illustration, the decision boundary obtained by the *VGA-classifier* for *ADS 1* when 10% of the data set is chosen for training is shown in Fig. 5.3.

Comparison With the Method in Srikanth et al. [448]

This section deals with a comparison of the way of incorporating variable length chromosomes as used in the *VGA-classifier* with that described by Srikanth et al. [448]. For this purpose, the operators in [448] are adopted for the same problem of pattern classification using hyperplanes, and the resulting performance is compared to those of the *VGA-classifier* for the three data sets. The method of incorporating variable string lengths in GAs as proposed in [448] is first described below.

The initial population is created randomly such that each string encodes the parameters of only one hyperplane. The fitness of a string is characterized by just the number of training points it classifies correctly, irrespective of

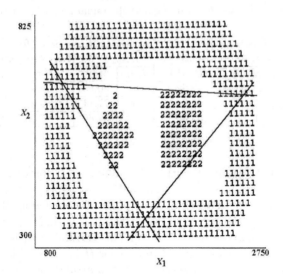

Fig. 5.3. *ADS 1* along with the *VGA* boundary for $H_{max} = 10$

the number of hyperplanes encoded in it. Among the genetic operators, traditional selection and mutation are used. A new form of crossover, called *modulo crossover* is used which keeps the sum of the lengths of the two chromosomes constant both before and after crossover.

Two other operators are used in conjunction with the *modulo crossover* for the purpose of faster recombination and juxtaposition. These are the *insertion* and *deletion* operators. During *insertion*, a portion of the genetic material from one chromosome is inserted at a random *insert-location* in the other chromosome. Conversely, during *deletion*, a portion of a chromosome is deleted to result in a shorter chromosome. Let the resultant *VGA-classifier* obtained using the operators in [448] be referred to as *MVGA-classifier*.

Tables 5.3 and 5.4 show the comparative overall recognition scores during both training and testing of the *VGA-classifier* and *MVGA-classifier* for the above-mentioned three data sets for 10% and 50% training data, respectively. Other parameters are the same as mentioned before. Results shown are the average values taken over five different runs. To keep parity, a maximum of 10 hyperplanes are used in the *MVGA-classifier*. The table also shows the number of hyperplanes, H_{VGA}, generated by the two methods for one particular run. Since the *MVGA-classifier* does not take care of the minimization of the number of hyperplanes while maximizing the fitness function, the resultant H_{VGA} is usually higher than that of *VGA-classifier*.

As is evident from the tables, the performance during training is better for the *MVGA-classifier* than for the *VGA-classifier* for all the data sets. The former, in general, uses more hyperplanes (of which many were found to be redundant [25]), which results in an increase in the execution time. From the training performance, it appears that the operators used in [448] are better able to recombine the subsolution blocks into larger blocks. However, this is seen, in general, to result in comparatively poorer scores during testing. To consider a typical example, in one of the runs with the *Vowel* data set when 10% data was used for training, 10 hyperplanes were used to provide a training recognition score of 97.47%, while the recognition score during testing fell to 68.95%.

It is also found that with increase in the size of the training data, the number of hyperplanes for modelling the class boundaries increase for *MVGA-classifier*. Furthermore, as expected, the performance of all the classifiers is improved with increase in the size of the training data from 10% to 50%.

Table 5.3. Comparative classification performance of *VGA-classifier* for $H_{max}=10$ using two types of variable string lengths (when 10% of the data set is used for training and the remaining 90% for testing)

Data set	VGA-classifier			MVGA-classifier[448]		
	Training score (%)	Test score (%)	H_{VGA}	Training score (%)	Test score (%)	H_{VGA}
ADS 1	100	95.62	3	100	93.16	6
Vowel	80.00	73.66	6	97.36	70.22	9
Iris	100	95.56	2	100	94.98	2

Table 5.4. Comparative classification performance of *VGA-classifier* for $H_{max}=10$ using two types of variable string lengths (when 50% of the data set is used for training and the remaining 50% for testing)

Data set	VGA-classifier			MVGA-classifier[448]		
	Training score (%)	Test score (%)	H_{VGA}	Training score (%)	Test score (%)	H_{VGA}
ADS 1	98.18	96.41	4	100.00	96.01	9
Vowel	79.73	78.26	6	85.48	78.37	9
Iris	100	97.60	2	100.00	94.67	5

5.6 *VGA-classifier* for the Design of a Multilayer Perceptron

It is known that the multilayer perceptron (MLP) [279, 406], with the neurons executing hard limiting nonlinearities, approximates the class boundaries using piecewise linear segments. Thus, a clear analogy exists between the two methodologies, viz., classification based on MLP and the *VGA-classifier*. If the parameters of the hyperplanes produced by the *VGA-classifier* are encoded in the connection weights and threshold values of MLP, then the performance provided by *VGA-classifier* and MLP will be the same. The architecture (including connection weights) of MLP can thus be determined from the results of the *VGA-classifier* [40]. (In turn, this will eliminate the need of training the MLP and hence the learning parameters.)

Based on this analogy, a methodology where the architecture along with the connection weights of MLP, with each neuron executing the hard limiting function, can be determined automatically using the principle of the *VGA-classifier* as described in this section. During the design procedure, it is guaranteed that the number of hidden layers (excluding the input and output layers) of the MLP thus formed will be at most two. The neurons of the first hidden layer are responsible for generation of the equations of hyperplanes. The neurons in the second hidden layer are responsible for generating the regions by performing the AND function, whereas those in the output layer are responsible for producing a combination of different regions by performing the OR function. The algorithm also includes a postprocessing step which removes the redundant neurons, if there are any, in the hidden/output layers. The performance of the resulting MLP is compared with those of its conventional version and some more with different architectures, for the same data sets, viz., *ADS 1*, *Vowel* and *Iris*.

5.6.1 Analogy Between Multilayer Perceptron and *VGA-classifier*

In an MLP with hard limiting neurons, the parameters of the piecewise linear surfaces (that the MLP uses to implicitly approximate the class boundaries) are encoded in the connection weights and threshold biases of the network. Similarly, the *VGA-classifier* also generates decision boundaries by appropriately fitting a number of hyperplanes in the feature space. The parameters are encoded in the chromosomes. Both the methods start from an initial randomly generated state (e.g., the set of initial random weights in MLP, initial population in VGA). Both of them iterate over a number of generations while attempting to decrease the classification error in the process.

The obvious advantage of the GAs over MLP is that the former performs concurrent search for a number of sets of hyperplanes, each representing a different classification in the feature space. On the other hand, the MLP deals with only one such set. Thus it has a greater chance of getting stuck at a local optimum, which the *GA-classifier* can overcome to a large extent. Using the concept of variable string length in a *GA-classifier* further allows it not to assume any fixed value of the number of hyperplanes, while MLP assumes a fixed number of hidden nodes and layers. This results in the problem of overfitting with an associated loss of generalization capability for MLP. In this context one must note that since the *VGA-classifier* has to be terminated after finitely many iterations, and the size of the data set is also finite, it may not always end up with the minimum number of hyperplanes. Consequently, the problem of overfitting exists for *VGA-classifier* also, although it is comparatively reduced.

5.6.2 Deriving the MLP Architecture and the Connection Weights

In this section we describe how the principle of fitting a number of hyperplanes using GA, for approximating the class boundaries, can be exploited in determining the appropriate architecture along with the connection weights of MLP. Since the purpose is to model the equation of hyperplanes, a hard limiting function is used in the neurons of the MLP defined as

$$f(x) = \begin{cases} +1, & \text{if } x \geq 0, \\ -1, & \text{if } x < 0. \end{cases}$$

Terminology

Let us assume that the *VGA-classifier* provides H_{VGA} hyperplanes, designated by

$$\{Hyp_1, Hyp_2, \ldots, Hyp_{H_{VGA}}\},$$

r regions designated by

$$\{R_1, R_2, \ldots, R_r\},$$

and the K classes designated by

$$\{C_1, C_2, \ldots, C_K\}.$$

Note that more than one region may be labelled with a particular class, indicating that $r \geq K$.

Let R^1 be the region representing class C_1, and let it be a union of r_1 regions given by

$$R^1 = R_{j_1^1} \bigcup R_{j_2^1} \bigcup \ldots R_{j_{r_1}^1}, \qquad 1 \leq j_1^1, j_2^1, \ldots, j_{r_1}^1 \leq r.$$

Generalizing the above, let R^i ($i = 1, 2, \ldots, K$) be the region representing class C_i, and let it be a union of r_i regions given by

$$R^i = R_{j_1^i} \bigcup R_{j_2^i} \bigcup \ldots \bigcup R_{j_{r_i}^i}, \qquad 1 \leq j_1^i, j_2^i, \ldots, j_{r_i}^i \leq r.$$

Note that each R^i is disjoint, i.e.,

$$R^i \bigcap R^j = \phi, \qquad i \neq j, \quad i,j = 1,2,\ldots,k.$$

Network Construction Algorithm

The network construction algorithm (NCA) is a four-step process where the number of neurons, their connection weights and the threshold values are determined. It guarantees that the total number of hidden layers (excluding the input and output layers) will be at most two. (In this context, Kolmogorov's Mapping Neural Network Existence Theorem must be mentioned. The theorem states that any continuous function can be implemented exactly by a three-layer, including input and output layers, feed-forward neural network. The proof can be found in [190]. However, nothing has been stated about the selection of connection weights and the neuronal functions.)

The hyperplanes in *VGA-classifier* are represented by Eq. (3.15). The output of the *VGA-classifier* is the parameters of the H_{VGA} hyperplanes. These are obtained as follows:

$$\alpha_1^1, \ \alpha_2^1, \ldots, \ \alpha_{N-1}^1, \ d^1,$$
$$\alpha_1^2, \ \alpha_2^2, \ldots, \ \alpha_{N-1}^2, \ d^2,$$

$$\vdots$$

$$\alpha_1^{H_{VGA}}, \ \alpha_2^{H_{VGA}}, \ldots, \ \alpha_{N-1}^{H_{VGA}}, \ d^{H_{VGA}},$$

where α_is and d are the angles and perpendicular distance values, respectively.

Step 1: Allocate N neurons in the input layer, *layer 0*, where N is the number of input features. The N-dimensional input vector is presented to this layer, where the neurons simply transmit the value in the input links to all the output links.

Step 2: Allocate H_{VGA} neurons in *layer 1*. Each neuron is connected to the N neurons of *layer 0*. Let the equation of the ith hyperplane $(i = 1,2,\ldots,H_{VGA})$ be

$$c_1^i x_1 + c_2^i x_2 + \ldots + c_N^i x_N - d = 0,$$

where from Eq. (3.15) we may write

$$\begin{aligned}
c_N^i &= \cos \alpha_{N-1}^i, \\
c_{N-1}^i &= \cos \alpha_{N-2}^i \sin \alpha_{N-1}^i, \\
c_{N-2}^i &= \cos \alpha_{N-3}^i \sin \alpha_{N-2}^i \sin \alpha_{N-1}^i, \\
&\vdots \\
c_1^i &= \cos \alpha_0^i \sin \alpha_1^i \ldots \sin \alpha_{N-1}^i, \\
&= \sin \alpha_1^i \ldots \sin \alpha_{N-1}^i,
\end{aligned}$$

since $\alpha_0^i = 0$.

Then the corresponding weight on the link to the ith neuron in *layer 1* from the jth neuron in *layer 0* is

$$w_{ij}^1 = c_j^i, \qquad j = 1, 2, \ldots, N,$$

and the threshold is

$$\theta_i^1 = -d^i$$

(since the bias term is added to the weighted sum of the inputs to the neurons).

Step 3: Allocate r neurons in *layer 2* corresponding to the r regions. If the ith region R_i ($i = 1, 2, \ldots, r$) lies on the positive side of the jth hyperplane Hyp_j ($j = 1, 2, \ldots, H_{VGA}$), then the weight on the link between the ith neuron in *layer 2* and the jth neuron in *layer 1* is

$$w_{ij}^2 = +1.$$

Otherwise

$$w_{ij}^2 = -1,$$

and

$$\theta_i^2 = -(H_{VGA} - 0.5).$$

Note that the neurons in this layer effectively serve the AND function, such that the output is high ($+1$) if and only if all the inputs are high ($+1$). Otherwise, the output is low (-1).

Step 4: Allocate K neurons in *layer 3* (output layer), corresponding to the K classes. The task of these neurons is to combine all the distinct regions that actually correspond to a single class. Let the ith class ($i = 1, 2, \ldots, K$) be a combination of r_i regions. That is,

$$R^i = R_{j_1^i} \bigcup R_{j_2^i} \bigcup \ldots \bigcup R_{j_{r_i}^i}.$$

Then the ith neuron of *layer 3*, ($i = 1, 2, \ldots, K$), is connected to neurons $j_1^i, j_2^i \ldots j_{r_i}^i$ of *layer 2* and,

$$w_{ij}^3 = 1, \qquad j \in \{j_1^i, j_2^i \ldots j_{r_i}^i\},$$

whereas

$$w_{ij}^3 = 0, \qquad j \notin \{j_1^i, j_2^i \ldots j_{r_i}^i\},$$

and

$$\theta_i^3 = r_i - 0.5.$$

Note that the neurons in this layer effectively serve the OR function, such that the output is high (+1) if at least one of the inputs is high (+1). Otherwise, the output is low (−1). For any given point, the output of at most one neuron in the output layer, corresponding to the class of the said point, will be high. Also, none of the outputs of the neurons in the output layer will be high if an unknown pattern, lying in a region with unknown classification (i.e., there were no training points in the region) becomes an input to the network.

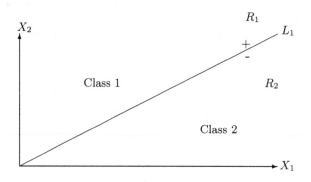

Fig. 5.4. Problem for demonstrating the network construction algorithm

Example 5.5. In order to demonstrate the functioning of the algorithm, let us consider the problem in Fig. 5.4 for $N = 2$ (two-dimensional feature space). Let L_1 be the line resulting from the application of *VGA-classifier* for partitioning the two classes shown. The corresponding angle (considering angles from the normal to the X_2 axis in the anticlockwise direction) and perpendicular distance values are

$$L_1 \rightarrow \alpha_1^1 = 315°, d^1 = 0.$$

In other words, the equation of L_1 is given by

$$x_2 \cos(315) + x_1 \sin(315) = 0,$$

or,

$$\frac{x_2}{\sqrt{2}} - \frac{x_1}{\sqrt{2}} = 0.$$

As can be seen from Fig. 5.4, there are two distinct regions, viz., R_1 and R_2, of which R_1 represents the region for class 1 and R_2 represents the region for class 2. Also,

$$R_1 \rightarrow \text{``positive'' side of } L_1$$
$$R_2 \rightarrow \text{``negative'' of } L_1.$$

Applying NCA, we obtain the following in steps:

Step 1: Two neurons in the input layer, since $N = 2$.

Step 2: One neuron in *layer 1*, since $H_{VGA} = 1$. The connection weights and the threshold are as follows:

$$w_{11}^1 = \cos \alpha_0^1 * \sin \alpha_1^1,$$
$$= -\frac{1}{\sqrt{2}},$$
$$w_{12}^1 = \cos \alpha_1^1,$$
$$= \frac{1}{\sqrt{2}},$$
$$\theta_1^1 = -d^1,$$
$$= 0.0.$$

Step 3: Two neurons in *layer 2*, since there are two distinct regions, $r = 2$. The connection weights and the thresholds are as follows:

$$w_{11}^2 = 1,$$
$$\theta_1^2 = -0.5,$$
$$w_{21}^2 = -1,$$
$$\theta_2^2 = -0.5.$$

Step 4: Two neurons in the output layer, *layer 3*, since there are two classes, $K = 2$. The connection weights and the thresholds are as follows:

$$w_{11}^3 = 1,$$
$$w_{12}^3 = 0,$$
$$\theta_1^2 = 0.5,$$
$$w_{21}^3 = 0,$$
$$w_{22}^3 = 1,$$
$$\theta_2^2 = 0.5.$$

Note that the zero weights effectively mean that the corresponding connections do not exist. The resulting network is shown in Fig. 5.5.

5.6.3 Postprocessing Step

The network obtained from the application of NCA may be further optimized in terms of the links and neurons in the output layer. A neuron in *layer 3* that has an input connection from only one neuron in *layer 2* may be eliminated completely. Mathematically, let for some i, $1 \le i \le K$,

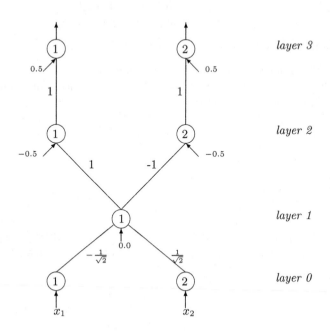

Fig. 5.5. Network for the problem in Fig. 5.4

$$w_{ij}^3 = 1, \quad \text{if } j = j',$$
$$= 0, \quad \text{otherwise.}$$

Then neuron i of *layer 3* is eliminated and is replaced by neuron j' of *layer 2*. Its output then becomes the output of the network. Note that this step produces a network where a neuron in layer i is connected to a neuron in layer $i + 2$.

In the extreme case, when all the neurons in the output layer (*layer 3*) get their inputs from exactly one neuron in *layer 2*, the output layer can be totally eliminated, and *layer 2* becomes the output layer. This reduces the number of layers from three to two. This will be the case when $r = k$, i.e., a class is associated with exactly one region formed by the H_{VGA} hyperplanes.

Applying the postprocessing step to the network obtained in Fig. 5.5, we find that neurons 1 and 2 of *layer 3* have links only from neurons 1 and 2 of *layer 2*, respectively. Consequently, one entire layer may be removed, and this results in a network as shown in Fig. 5.6.

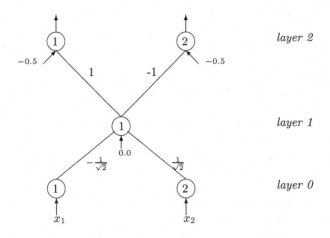

Fig. 5.6. Modified network after postprocessing

5.6.4 Experimental Results

The effectiveness of the network construction algorithm (NCA) is demonstrated on the same data sets used earlier, viz., *ADS 1*, *Vowel* and *Iris*. The MLP is executed using both hard limiters and the sigmoid function in the neurons. The sigmoid function is defined as

$$f(x) = \frac{1}{1 + e^{-x}}.$$

As in Chap. 3, the learning rate and momentum factor are fixed at 0.8 and 0.2, respectively. Online weight update, i.e., update after each training data input, is performed for a maximum of 3000 iterations.

The performance of *VGA-classifier* and consequently that of the MLP derived using NCA (i.e., where the architecture and the connection weights have been determined using NCA) is compared with that of a conventional MLP having the same architecture as provided by NCA, but trained using the back propagation (BP) algorithm with the neurons executing the sigmoid function. For the purpose of comparison, three more typical architectures for the conventional MLP have also been considered which were used in Chap. 3, viz., those having two hidden layers with 5, 10 and 20 nodes in each layer, respectively. Tables 5.5–5.7 summarize the results obtained. The MLP architecture is denoted by "Arch." in the tables.

The number of hyperplanes (H_{VGA}) and regions (r) obtained by the *VGA-classifier* starting from $H_{max} = 10$ are mentioned in columns 2–3. These are used to select the MLP architectures as shown in columns 10–11 and 12–13.

From Table 5.5 corresponding to *ADS 1*, it is found that the MLP trained using BP does not succeed in learning the boundaries of class 2, for all the architectures (columns 4–11). In fact, as seen from Fig. 3.7, class 2 is totally surrounded by class 1. The *VGA-classifier*, on the other hand, is able to place the lines appropriately, thereby yielding a significantly better score both during training and testing (columns 2–3). Consequently, the network derived using NCA (whose performance is the same as that of the *VGA-classifier*) also provides a significantly better score (columns 12–13).

The case is similar for *Vowel* and *Iris* data (Tables 5.6 and 5.7, respectively) where the *VGA-classifier*, and consequently the MLP derived using NCA, provides a superior performance than the MLPs trained with BP. The overall recognition score during testing for *Vowel* is found to increase with the increase in the number of nodes in the hidden layers (columns 5, 7 and 9) since the classes are highly overlapping. For *Iris* data, the reverse is true, indicating a case of overfitting the classes.

Note that the Arch. values of the MLPs mentioned in columns 10–11 and 12–13 of the tables are the ones obtained without the application of the post-processing step. These values are put in order to clearly represent the mapping from *VGA-classifier* to MLP, in terms of the number of hyperplanes and regions, although the postprocessing task could have reduced the size of the network while keeping the performance same. For example, in the case of *Iris* data, the number of hyperplanes and regions are 2 and 3 (columns 2–3), respectively. Keeping analogy with this, the Arch. value in column 12–13 are mentioned to be 4:2:3:3. In practice, after postprocessing, the said values became 4:2:3. Similarly, for *ADS 1* and *Vowel* data, the values after postprocessing were found to be 2:3:4:2 and 3:6:2:6, respectively.

5.7 Summary

The problem of fixing the value of H a priori of the *GA-classifier* is resolved by introducing the concept of variable string lengths in the classification algorithm. New genetic operators are defined to deal with the concept of variable string lengths for formulating the classifier. The fitness function has been defined so that its maximization indicates minimization of the number of mis-classified samples as well as the required number of hyperplanes. It is proved that for infinitely large number of iterations the method is able to arrive at the minimum number of misclassified samples and will need minimum number of hyperplanes for this purpose. It is also established that as the size of the training data set n and the number of iterations go to infinity, the performance of the *VGA-classifier* will approach that of Bayes' classifier.

Experimental evidence for different percentages of training and test data indicates that given a value of H_{max}, the algorithm can not only be able to automatically evolve an appropriate value of H for a given data set, but also to result in improved performance of the classifier. Although it is proved

Table 5.5. Classwise and overall recognition scores (%) for *ADS 1*

Class	VGA-classifier $H_{VGA}=3, r=5$		MLP									
			SIGMOID								Hard Limiter	
			Arch.=2:5:5:2		Arch.=2:10:10:2		Arch.=2:20:20:2		Arch.=2:3:5:2		Arch.=2:3:5:2	
	Trn.	Testing	Trn.	Testing	Trn.	Testing	Trn.	Testing	Trn.	Testing	Trn.	Testing
1	100.00	95.89	100.00	100.00	100.00	100.00	100.00	100.00	100.00	100.00	100.00	95.89
2	100.00	94.31	0.0	0.0	0.0	0.0	0.0	0.0	0.0	0.0	100.00	94.31
Overall	100.00	95.62	83.63	82.47	83.63	82.47	83.63	82.47	83.63	82.47	100.00	95.62

Table 5.6. Classwise and overall recognition scores (%) for *Vowel*

Class	VGA-classifier $H_{VGA}=6, r=7$		MLP										
			SIGMOID								Hard Limiter		
			Arch.=3:5:6		Arch.=3:10:10:6		Arch.=3:20:20:6		Arch.=3:6:7:6		Arch.=3:6:7:6		
	Training	Testing	Training	Testing	Training	Testing	Training	Testing	Training	Testing	Training	Testing	
δ	10.30	8.21	85.71	12.30	87.51	47.69	100.00	47.23	85.71	6.15	10.30	8.21	
a	100.00	91.35	75.00	43.20	100.00	16.04	100.00	27.62	87.50	13.58	100.00	91.35	
i	94.11	84.51	100.00	69.67	100.00	79.35	100.00	80.58	100.00	78.70	94.11	84.51	
u	73.33	66.91	86.67	80.14	100.00	77.94	100.00	83.29	100.00	91.91	73.33	66.91	
e	89.99	85.56	80.00	68.44	94.98	90.37	100.00	87.30	80.00	59.35	89.99	85.56	
o	83.33	75.92	88.89	77.16	94.44	54.93	94.44	51.70	77.78	43.21	83.33	75.92	
Overall	80.00	73.66	87.05	65.26	95.82	67.55	98.82	68.48	88.23	56.36	80.00	73.66	

that the classifier will always provide the minimum number of hyperplanes for infinitely large number of iterations, in practice, one may sometimes find redundancy in the output boundary due to premature termination. These would surely be removed if the algorithm is executed for sufficiently large number of iterations.

Note that since any nonlinear boundary can be approximated by a number of linear surfaces, the *VGA-classifier* with a large value of H_{max} can be applied to data sets having any kind of nonlinear class boundaries. Further, any choice of $H_{max} \geq H_{opt}$ would provide similar performance of the *VGA-classifier* if n and the number of iterations are taken to be sufficiently large. Thus, given a large enough value of H_{max}, the *VGA-classifier* can automatically evolve an appropriate number of hyperplanes to model any kind of decision boundary for large values of n and numbers of iterations. Moreover, being nonparametric in nature, the *VGA-classifier* can be applied to any kind of data set. (In this context we mention about the proof that exists in the literature [155] showing that the k-NN classifier approaches the Bayes' classifier if the values of k and $\frac{k}{n}$ can be made to approach ∞ and 0, respectively, as $n \to \infty$. Several choices of k are therefore possible that satisfy this criterion, with each choice providing a different approximation of the boundary and hence different performance of the k-NN classifier. Similarly, it is known that when the MLP, given an appropriate architecture, is trained using back-propagation, it approximates the Bayes' optimal discriminant function under limiting conditions [402]. However, a proper choice of the architecture is not an easy task, thereby making the performance of MLP vary widely with different architectures.)

Regarding the choice of the number of iterations to be executed before termination and the size of the training data set (n) of the *VGA-classifier*, one may note the following. Since the class boundaries are explicitly generated in the *VGA-classifier*, one indication to terminate the algorithm may be when the boundaries undergo very little or no change over a number of iterations. In this context one may refer to the investigation in [324], where the concept of ϵ-optimal stopping time for GAs is formulated. Estimation of a proper value of n which would ensure that the performance of the *VGA-classifier* is within some factor of the optimal performance is an open issue and requires further research.

The method of using variable string length in the algorithm of Srikanth et al. [448] is also implemented in the *VGA-classifier* for comparison. Since the former method does not include a factor for reducing the number of surfaces, it is found to use more hyperplanes for constituting the decision boundary. This results in better performance during training, mostly at the cost of reduced generalization capability. Additionally, the execution time is also greater since no explicit effort is made to decrease the number of hyperplanes.

A method for automatic determination of MLP architecture and the associated connection weights is described, based on its analogy with the *VGA-classifier* in terms of placement capability of hyperplanes for approximating the class boundaries. The method guarantees that the architecture will involve at most two layers (excluding the input and output layers), with the neurons in the first and second hidden layers being responsible for hyperplane and region generation, and those in the output providing a combination of regions for the classes. Since the principle of *VGA-classifier* is used for developing the network construction algorithm (NCA), it becomes mandatory to consider hard limiting neurons in the derived MLP. Although this makes the network rigid and susceptible to noise and corruption in the data, one may use NCA for providing a possible appropriate structure of conventional MLPs.

Table 5.7. Classwise and overall recognition scores (%) for *Iris*

Class	VGA-classifier $H_{VGA} = 2, \tau = 3$		MLP									
			SIGMOID								Hard Limiter	
			Arch.=4:5:5:3		Arch.=4:10:10:3		Arch.=4:20:20:3		Arch.=4:2:3:3		Arch.=4:2:3:3	
	Training	Testing	Training	Testing	Training	Testing	Training	Testing	Training	Testing	Training	Testing
1	100.00	100.00	100.00	100.00	100.00	98.81	100.00	66.67	100.00	100.00	100.00	100.00
2	100.00	93.33	100.00	66.67	100.00	64.77	100.00	71.11	100.00	77.78	100.00	93.33
3	100.00	93.33	100.00	77.78	100.00	80.51	100.00	97.78	100.00	66.67	100.00	93.33
Overall	100.00	95.56	100.00	81.48	100.00	81.36	100.00	78.51	100.00	81.48	100.00	95.56

6

Chromosome Differentiation in *VGA-classifier*

6.1 Introduction

Chapter 5 dealt with a description of how the concept of variable-length chromosomes in GA (VGA) can be utilized for approximating the class boundaries of a data set using a variable number of hyperplanes. In the present chapter, the effectiveness of incorporating the concept of chromosome differentiation (CD) into two classes, say male (M) and female (F), is first studied. The modified GA is called GA with chromosome differentiation (GACD) [43]. Here two classes of chromosomes exist in the population. Crossover (mating) is allowed only between individuals belonging to these categories. A schema analysis of GACD, vis a vis that of conventional GA (CGA), is provided. It is also shown that in certain situations, the lower bound of the number of instances of a schema sampled by GACD is greater than or equal to that of CGA.

The concept of variable length chromosomes is then integrated with GACD for designing the *VGACD-classifier* [41]. As for the *VGA-classifier* (described in Chap. 5), the fitness function rewards a string with smaller number of misclassified samples as well as smaller number of hyperplanes. The effectiveness of the resultant classifier is established for the task of pixel classification in satellite images.

Section 6.2 describes the GACD in detail, including the motivation for incorporating the concept of chromosome differentiation in GAs. The schema analysis for GACD is provided in Sect. 6.3. A description of Variable-Length GACD classifier (*VGACD-classifier*) is provided in the following section. Application of the *VGACD-classifier* for classifying pixels in a satellite image of a part of the city of Calcutta, and its comparison to other classifiers, is described in Sect. 6.5. Finally, Sect. 6.6 summarizes this chapter.

6.2 GACD: Incorporating Chromosome Differentiation in GA

6.2.1 Motivation

Nature generally differentiates the individuals of a species into more than one class. Sexual differentiation is a typical example. The prevalence of this form of differentiation indicates an associated advantage which appears to be in terms of cooperation between two dissimilar individuals, who can at the same time specialize in their own fields. This cooperation and specialization is expected to give rise to healthier and more fit offspring [164].

In conventional genetic algorithms, since no restriction is placed upon the selection of mating pair for crossover operation, often chromosomes with similar characteristics are mated. Therefore, no significant new information is gained out of this process, and the result is wastage of computational resources. These observations motivated the investigation into the effects of differentiating the chromosomes of a population into two different classes, namely M and F, respectively, thereby giving rise to two separate populations [43]. In addition, the two populations are required to be as dissimilar as possible in order to increase the diversity. Therefore, the populations are initially generated in such a way that the Hamming distance between them is maximized. Crossover is allowed between individuals belonging to the two distinct populations only. (The concept of restricted mating through Hamming distance has also been used in [134, 135].) The other genetic operators are kept the same as in conventional GA.

As crossover is allowed between these two dissimilar groups only, a greater degree of diversity is introduced in the population leading to greater exploration in the search. At the same time conventional selection is performed over the entire population which serves to exploit the information gained so far. Thus GACD is expected to attain a greater balance between exploration and exploitation, which is crucial for any adaptive system, thereby making GACD superior to conventional GA (CGA). The following section describes the GACD in detail [43].

6.2.2 Description of GACD

The basic steps of GA are followed in GACD as well. However, the individual processes are modified that are now discussed in detail.

Population Initialization and Fitness Computation

The structure of a chromosome of GACD is shown in Fig. 6.1. Here the l bits, termed *data bits*, encode the parameters of the problem. The initial two bits, termed the *class bits*, are used to distinguish the chromosomes into two classes,

M and F. If the *class bits* contain either 01 or 10, the corresponding chromosome is called an M chromosome, and if it contains 00, the corresponding chromosome is called an F chromosome. These bits are not allowed to assume the value 11. (This is in analogy with the X and Y chromosomes in natural genetic systems, where XY/YX indicates male, while XX indicates female.)

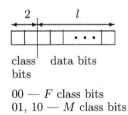

00 — F class bits
01, 10 — M class bits

Fig. 6.1. Structure of a chromosome in GACD

Two separate populations, one containing the M chromosomes (M population) and the other containing the F chromosomes (F population), are maintained over the generations. The sizes of these two populations, P_m and P_f, respectively, may vary. Let $P_m + P_f = P$, where P is fixed (equivalent to the population size of CGA). Initially $P_m = P_f = \frac{P}{2}$. The data bits for each M chromosome are first generated randomly. One of the two *class bits*, chosen randomly, is initialized to 0 and the other to 1. The data bits of the F chromosomes are initially generated in such a way that the Hamming distance between the two populations (in terms of the data bits) is maximum. The Hamming distance between two chromosomes c_1 and c_2, denoted by $h(c_1, c_2)$, is defined as the number of bit positions in which the two chromosomes differ. Hamming distance between two populations, Pop_1 and Pop_2, denoted by $h(Pop_1, Pop_2)$, is defined as follows:

$$h(Pop_1, Pop_2) = \sum_i \sum_j h(c_i, c_j), \quad \forall c_i \in Pop_1, \forall c_j \in Pop_2.$$

The pseudo code of a method of generating the F chromosomes such that the above-mentioned restriction is satisfied while allowing a certain amount of randomness is shown in Fig 6.2. Here $M(i, j)$ and $F(i, j)$ indicate the jth bit of the ith chromosome in the M and F populations, respectively. $check(i, j)$ is an auxiliary data structure used to keep track of the bits of the M chromosome that have been chosen for complementing. The *class bits* of each F chromosome are initialized to 0's.

Only the l data bits of the M and F populations are used to compute the fitness for the chromosomes in a problem specific manner.

Selection

Selection is performed over all the P ($= P_m + P_f$) chromosomes, (i.e., disregarding the class information) using their fitness values. In other words, all

Begin

     $i=0$ to P_m

       $j=0$ to $l-1$

         $check(i,j) = 0$ /* Initialization */

     $i=0$ to P_f

       $j=0$ to $l-1$

         repeat

           $k = random(P_m)$ /* returns an integer in
                   the range of 0 to $P_m - 1$ */

          until $(check(k,j)=0)$ /* $M(k,j)$ not chosen
                 before */

         $check(k,j)=1$ /* $M(k,j)$ now chosen.
              Not to be chosen again */

         $F(i,j)= complement(M(k,j))$

end

Fig. 6.2. Algorithm for initializing the F population from the initial M population in GACD

the chromosomes compete with one another for survival. The selected chromosomes are placed in the mating pool.

Crossover, Mutation and Elitism

Crossover is applied with probability μ_c between an M and an F parent chromosome. Each parent contributes one class bit to the offspring. Since the F parent can only contribute a 0 (its class bits being 00), the class of the child is primarily determined by the M parent which can contribute a 1 (yielding an M child) or a 0 (yielding an F child), depending upon the bit position (among the two class bits) of the M parent chosen. This process is performed for both the offspring whereby either two M or two F or one M and one F offspring will be generated.

  Crossover is carried on until (a) there are no chromosomes in the mating pool, or (b) there are no M (or, F) chromosomes in the mating pool. In the former case the crossover process terminates. In the latter case, the remaining F (or M) chromosomes are mated with the best M (or, F) chromosome. Note that if at the start of the crossover process, it is found that the mating pool contains chromosomes of only one class, then the crossover process is discontinued.

  Bit-by-bit mutation is performed over the data bits only with probability μ_m. The *class bits* are not mutated. Elitism is incorporated by preserving the best chromosome, among both the M and F chromosomes, seen till the current generation in a location outside the population. A detailed description of GACD is available in [43] where its superiority over GA is demonstrated for a large number of function optimization and pattern classification problems.

6.3 Schema Theorem for GACD

In this section the schema theorem for CGA [164], as described in Sect. 2.3.6, is modified appropriately to incorporate the ideas of the GACD algorithm. In general the terms corresponding to M and F populations are denoted by subscripts m and f, respectively, while those with no subscript denote that these are applicable over both the populations. The terminology used for developing the schema theorem for GACD are first mentioned here.

6.3.1 Terminology

P: the total population size which is assumed to be constant
$P_m(t)$: the M population size at time t
$P_f(t)$: the F population size at time t
$\overline{f}$: the average fitness of the entire population
h: a schema
$\overline{f_h}$: the average fitness of instances of schema h over the entire population
$\overline{f_m}$: the average fitness of the M population
$\overline{f_f}$: the average fitness of the F population
l: the length of a string
$m(h,t)$: number of instances of schema h in the population at time t
$m_m(h,t)$: number of instances of schema h in the M population at time t
$m_f(h,t)$: number of instances of schema h in the F population at time t
$\delta(h)$: the defining length of the schema h
$O(h)$: the order of the schema h
μ_c: the probability of crossover
μ_m: the probability of mutation

Superscripts s and c with any of the above-mentioned symbols indicate the corresponding values after selection and crossover, respectively. It is to be noted that the following equalities will hold for GACD for any value of t.

$$P = P_m(t) + P_f(t), \tag{6.1}$$

$$\overline{f} = \frac{\overline{f_m} * P_m + \overline{f_f} * P_f}{P}, \tag{6.2}$$

$$m(h,t) = m_m(h,t) + m_f(h,t). \tag{6.3}$$

6.3.2 Analysis of GACD

The effects of each operation, *selection, crossover* and *mutation* are considered separately.

Selection: Proportional selection is performed over the entire population. Hence, similar to the discussion in Sect. 2.3.6, the number of instances of the schema h after selection will be given by

$$m^s(h, t+1) = m(h, t) * \frac{\overline{f_h}}{\overline{f}}. \tag{6.4}$$

The number of instances of the schema h that will be present in the M and F populations, respectively, must obviously be proportional to the fraction present in the two populations before selection takes place. Or in other words,

$$m^s{}_m(h, t+1) = m^s(h, t+1) * \frac{m_m(h, t)}{m(h, t)}. \tag{6.5}$$

Similarly,

$$m^s{}_f(h, t+1) = m^s(h, t+1) * \frac{m_f(h, t)}{m(h, t)}. \tag{6.6}$$

Crossover: To analyze the effect of crossover (assuming single-point crossover) on the instances of the schema h, its probability of disruption is first calculated. Instances of the schema that are members of the M population are considered first. The analysis for the F population is analogous. For the present, let us assume that an instance of the schema from the M population, if not disrupted by crossover, is placed in the M population again.

Schema h will most likely be disrupted due to crossover if all the following conditions hold.

(a) Crossover occurs (with probability μ_c).

(b) Crossover site falls within the first and the last defining positions (with probability $\frac{\delta(h)}{l-1}$).

(c) Crossover occurs with an instance of some schema h^* in the female population such that h^* is not contained in h (with probability $1 - \frac{m^s{}_f(h, t+1)}{P^s_f(t+1)}$).

(Note that if h^* is contained in h, then crossover can never disrupt h, i.e., schema h will survive in both the offspring. Schema h^*, on the other hand, may not survive crossover at all.)

Taking the above-mentioned three conditions into account, the probability of disruption of h in one instance of the schema may be written as

$$\mu_c * \frac{\delta(h)}{l-1} * \left(1 - \frac{m^s{}_f(h, t+1)}{P^s_f(t+1)}\right). \tag{6.7}$$

Hence the probability of survival of one instance of the schema in the M population is given by

$$1 - \mu_c * \frac{\delta(h)}{l-1} * \left(1 - \frac{m^s{}_f(h, t+1)}{P^s_f(t+1)}\right). \tag{6.8}$$

Consequently, considering $m^s{}_m(h, t+1)$ instances (after selection), after crossover we get

$$m_m^c(h, t+1) \geq m_m^s(h, t+1) \left(1 - \mu_c * \frac{\delta(h)}{l-1} * \left[1 - \frac{m_f^s(h, t+1)}{P_f^s(t+1)}\right]\right).$$

$$(6.9)$$

The greater than sign comes because even after disruptive crossover, the schema h may survive. For example, let h and h^* be as follows:

$$h = \# 1 \# 1 \# | \# 0 \# \# \#$$
$$h^* = \# 0 \# 0 \# | 0 \ 0 \# \# \#.$$

Let the crossover site be as shown. Then after crossover the offspring are

$$child \ 1 = \# 1 \# 1 \# \ 0 \ 0 \# \# \#,$$
$$child \ 2 = \# 0 \# 0 \# \# 0 \# \# \#.$$

Here $child$ 1 is an instance of h, i.e., h survives possibly disruptive crossover. Other than this, the schema h may be generated due to crossover between two other strings.

Similarly the number of instances of h that will survive crossover in the F population is given by the relation

$$m^c{}_f(h, t+1) \geq m^s{}_f(h, t+1) \left(1 - \mu_c * \frac{\delta(h)}{l-1} * \left[1 - \frac{m^s{}_m(h, t+1)}{P_m^s(t+1)}\right]\right).$$

$$(6.10)$$

It had previously been assumed that if an instance of h is present in the M (or F) population, and if h is not disrupted due to crossover, then it survives in the M (or, F) population. In reality the situation may not be so. Let P_1 be the probability that h survives in the M population, when it is originally present in the M population. Hence $(1 - P_1)$ is the probability that h goes to the F population after crossover. Similarly, let P_2 and $(1 - P_2)$ be the probabilities that h survives in the M and F populations, respectively, when it is originally present in the F population. Thus the modified equation for schema survival due to crossover is:

$$m_m^{*c}(h, t+1) = P_1 \times m_m^c(h, t+1) + P_2 \times m_f^c(h, t+1).$$

The second term is introduced on considering the instances of h that are present in the F population, which survive crossover but are placed in the M population. Similarly,

$$m_f^{*c}(h, t+1) = (1 - P_2) \times m_f^c(h, t+1) + (1 - P_1) \times m_m^c(h, t+1).$$

Therefore the number of instances of h present in the entire population after crossover is

$$m^c(h, t+1) = P_1 \times m_m^c(h, t+1) + P_2 \times m_f^c(h, t+1),$$
$$+(1 - P_2) \times m_f^c(h, t+1) + (1 - P_1) \times m_m^c(h, t+1),$$
$$= m_m^c(h, t+1) + m_f^c(h, t+1).$$

Or,

$$m^c(h, t+1) \geq m_m^s(h, t+1) \left\{ 1 - \frac{\mu_c \delta(h)}{l-1} \left[1 - \frac{m_f^s(h,t+1)}{P_f^s(t+1)} \right] \right\}$$
$$+ m_f^s(h, t+1) \left\{ 1 - \frac{\mu_c \delta(h)}{l-1} \left[1 - \frac{m_m^s(h,t+1)}{P_m^s(t+1)} \right] \right\}. \quad (6.11)$$

Using Eq. (6.5) and (6.6), the right-hand side of inequality Eq. (6.11) may be written as

$$\frac{m^s(h, t+1)}{m(h, t)} \left\{ m_m(h, t) + m_f(h, t) - \frac{\mu_c \delta(h)}{l-1} \left[m_m(h, t) \left(1 - \frac{m_f^s(h, t+1)}{P_f^s(t+1)} \right) \right. \right.$$
$$\left. \left. + m_f(h, t) \left(1 - \frac{m_m^s(h, t+1)}{P_m^s(t+1)} \right) \right] \right\}.$$

$$= m^s(h, t+1) \left(1 - \frac{\mu_c \delta(h)}{(l-1)} \left[1 - \left\{ \frac{m_m(h, t) m_f^s(h, t+1)}{P_f^s(t+1) m(h, t)} \right. \right. \right.$$
$$\left. \left. \left. + \frac{m_f(h, t) m_m^s(h, t+1)}{P_m^s(t+1) m(h, t)} \right\} \right] \right).$$

Let the term in the curly brackets be denoted by α. Thus

$$m_{GACD}^c(h, t+1) \geq m^s(h, t+1) \left(1 - \mu_c \frac{\delta(h)}{l-1} [1 - \alpha] \right). \quad (6.12)$$

In this context a slight modification of the schema theorem, available in [164] and described in Sect. 2.3.6, is made which provides a better lower bound of the number of instances of h that survive after selection and crossover. An instance of schema h may be disrupted due to crossover *iff* it is crossed with an instance of another schema h^* such that h^* is not contained in h and the other conditions for disruptive crossover hold. Accounting for this detail, the disruption probability is recalculated as

$$\mu_c \frac{\delta(h)}{l-1} \left(1 - \frac{m^s(h, t+1)}{P} \right).$$

Hence after selection and crossover

$$m_{CGA}^c(h, t+1) \geq m^s(h, t+1) \left\{ 1 - \mu_c \frac{\delta(h)}{l-1} \left[1 - \frac{m^s(h, t+1)}{P} \right] \right\}. \quad (6.13)$$

Let the term $\frac{m^s(h,t+1)}{p}$ be denoted by β. Thus

$$m_{CGA}^c(h, t+1) \geq m^s(h, t+1) \left\{ 1 - \mu_c \frac{\delta(h)}{l-1} [1 - \beta] \right\}.$$

Mutation: Since the conventional bit-by-bit mutation is applied on the strings with a probability μ_m, the probability of disruption of one bit of the schema is μ_m. Probability of its survival is $1 - \mu_m$. Hence the probability of survival of the schema is $(1 - \mu_m)^{O(h)}$. Thus the number of instances of the schema h that are present in the population at time $t+1$ (after selection, crossover and mutation) is given by

$$m_{GACD}(h, t+1) \geq m^s(h, t+1) \left\{ 1 - \frac{\mu_c \delta(h)}{(l-1)}(1-\alpha) \right\} \left\{ (1 - \mu_m)^{O(h)} \right\}.$$

Approximating the right hand side, the inequality may be written as

$$m_{GACD}(h, t+1) \geq m^s(h, t+1) \left\{ 1 - \frac{\mu_c \delta(h)}{l-1}(1-\alpha) - \mu_m O(h) \right\}. \quad (6.14)$$

Similarly, the equation for CGA is given by

$$m_{CGA}(h, t+1) \geq m^s(h, t+1) \left\{ 1 - \frac{\mu_c \delta(h)}{l-1}(1-\beta) - \mu_m O(h) \right\}. \quad (6.15)$$

In order to compare $m_{GACD}(h, t+1)$ and $m_{CGA}(h, t+1)$, the following cases are considered.

Case i: Let $m_m(h, t) = m_f(h, t) = m_1$. In that case $m_m^s(h, t+1) = m_f^s(h, t+1) = m_2$. Note that $m(h, t) = 2m_1$, $m^s(h, t+1) = 2m_2$ and $\beta = \frac{2m_2}{p}$. Then

$$\alpha = \frac{1}{2m_1} \left(\frac{m_1 m_2}{P_f^s(t+1)} + \frac{m_1 m_2}{P_m^s(t+1)} \right),$$

$$= \frac{m_2}{2} \left(\frac{P}{P_f^s(t+1) P_m^s(t+1)} \right),$$

$$= \beta \left[\frac{P^2}{4 P_f^s(t+1) P_m^s(t+1)} \right].$$

The minimum value of the term in square brackets is 1 when $P_m^s(t+1) = P_f^s(t+1)$. Hence $\alpha \geq \beta$. This in turn indicates that $lower\_bound(m_{GACD}(h, t+1)) \geq lower\_bound(m_{CGA}(h, t+1))$, i.e., the lower bound of the number of instances of some schema h sampled by GACD is better than that of CGA.

Case ii: Let $m_m(h, t) \neq m_f(h, t)$. Let $m_m(h, t) = \gamma m_f(h, t)$ where $\gamma \neq 1$. Then $m_m^s(h, t+1) = \gamma m_f^s(h, t+1)$. Note that $m(h, t) = m_f(h, t)(1+\gamma)$, $m^s(h, t+1) = m_f^s(h, t+1)(1+\gamma)$ and $\beta = \frac{m_f^s(h, t+1)(1+\gamma)}{P}$. Thus,

$$\alpha = \frac{1}{m_f(h, t)(1+\gamma)} \left(\frac{\gamma m_f(h, t) m_f^s(h, t+1)}{P_f^s(t+1)} + \frac{m_f(h, t) \gamma m_f^s(h, t+1)}{P_m^s(t+1)} \right),$$

$$= \frac{m_f^s(h, t+1)\gamma}{(1+\gamma)} \left(\frac{P}{P_f^s(t+1)P_m^s(t+1)} \right),$$

$$= \frac{m_f^s(h, t+1)(1+\gamma)}{P} \frac{\gamma}{(1+\gamma)^2} \frac{P^2}{P_f^s(t+1)P_m^s(t+1)},$$

$$= \beta \frac{\gamma}{(1+\gamma)^2} \frac{P^2}{P_f^s(t+1)P_m^s(t+1)}.$$

In this case $\alpha \geq \beta$, if the following holds:

$$\frac{\gamma}{(1+\gamma)^2} \frac{P^2}{P_f^s(t+1)P_m^s(t+1)} \geq 1.$$

Or,

$$\frac{P}{\sqrt{P_f^s(t+1)P_m^s(t+1)}} \geq \frac{1+\gamma}{\sqrt{\gamma}}. \tag{6.16}$$

Since the above-mentioned condition cannot be always ensured, it cannot be concluded that $lower\_bound(m_{GACD}(h, t+1)) \geq lower\_bound(m_{CGA}(h, t+1))$. (Note also that both the functions $\frac{(1+\gamma)}{\sqrt{\gamma}}$ and $\frac{P}{\sqrt{P_f^s(t+1)P_m^s(t+1)}}$ have minimum value 2.)

In order to experimentally compare the values of m_{CGA} and m_{GACD}, an optimization problem is considered. Let $f(x) = x^2$ be the function to be optimized. A population size of 30 (initially 15 male and female strings are considered for GACD) and string length of 10 are taken. $\mu_c = 0.8$, and $\mu_m = 0.01$. The variation of the number of instances of four schemata with different characteristics is presented over the first five generations in Figs. 6.3(a)–(d). It is found that the growth rates for schemata with high fitness values are greater for GACD as compared to CGA (see Figs. 6.3(a)–(c)). At the same time the decay rate for schema with low fitness value is also greater for GACD (see Fig. 6.3(d)).

6.4 *VGACD-classifier*: Incorporation of Chromosome Differentiation in *VGA-classifier*

In *VGACD-classifier*, since the length of a chromosome may vary, the *data bits* can take values from $\{1, 0, \#\}$, where $\#$ denotes "don't care". The *data bits* encode the parameters of H_i hyperplanes, where $1 \leq H_i \leq H_{max}$. The details regarding the encoding of a hyperplane are described in Chap. 5.

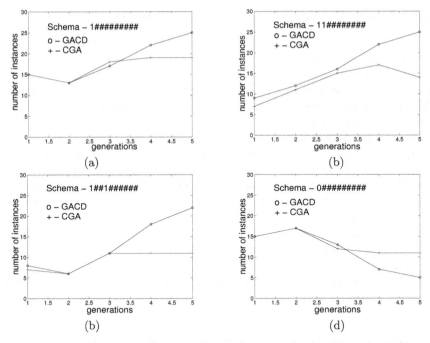

Fig. 6.3. Variation of the number of instances of a schema with the number of generations for the function $f(x) = x^2$: (a) schema 1#########; (b) schema 11########; (c) schema 1##1######; and (d) schema 0#########

6.4.1 Population Initialization

The M population is first generated in such a way that the first two chromosomes encode the parameters of 1 and H_{max} hyperplanes, respectively. The remaining chromosomes encode the parameters of H_i hyperplanes where $1 \leq H_i \leq H_{max}$. For all the M chromosomes, one of the two *class bits*, chosen randomly, is initialized to 0 and the other to 1. All the chromosomes are then made of the maximum length by padding with an appropriate number of #'s. The F population is initialized using the algorithm described in Fig. 6.2. However, a minor modification needs to be made since the *data bits* of the M chromosomes may now contain #'s, in addition to the normal 0's and 1's. The complement of # is defined to be #. Or, in other words $complement(\#) = \#$.

Once the F chromosomes are initialized, some of the hyperplanes may be incomplete. (As mentioned in Sect. 5.3.3, a hyperplane in a string is called *complete* if all the bits corresponding to it are either defined (i.e., 0's and 1's) or #'s. Otherwise it is incomplete.) These are tackled in a manner described for the case of crossover in *VGA-classifier* (Sect. 5.3.3). The information about the number of hyperplanes in the F chromosomes is updated accordingly.

6.4.2 Fitness Computation and Genetic Operators

The fitness of a chromosome is computed as explained in Sect. 5.3.2. The *class bits* are not taken into consideration when computing the fitness value. The genetic operators, crossover and mutation, are as follows:

Crossover: Two strings, i and j, having lengths l_i and l_j are selected from the M and F populations, respectively. Thereafter, crossover is performed between them only on the *data bits* (i.e., by ignoring the *class bits*), as described in Sect. 5.3.3. This results in two offspring. Subsequently, each parent contributes one class bit to the offspring. Since the F parent can only contribute a 0 (its class bits being 00), the class of the child is primarily determined by the M parent which can contribute a 1 (yielding an M child) or a 0 (yielding an F child) depending upon the bit position (among the two class bits) of the M parent chosen. This process is performed for both the offspring whereby either two M or two F or one M and one F offspring will be generated. These are put in the respective populations.

Mutation:

As for crossover, only the *data bits*, and not the *class bits*, are considered for mutation. This is performed in a manner as described in Sect. 5.3.3.

6.5 Pixel Classification of Remotely Sensed Image

6.5.1 Relevance of GA

Classification of pixels for partitioning different landcover regions is an important problem in the realm of satellite imagery. Satellite images usually have a large number of classes with overlapping and nonlinear class boundaries. Figure 6.4 shows, as a typical example, the complexity in scatter plot of 932 points belonging to 7 classes which are taken from the *SPOT* image of a part of the city of Calcutta. Therefore, for appropriate modelling of such nonlinear and overlapping class boundaries, the utility of an efficient search technique is evident. Moreover, it is desirable that the search technique does not need to assume any particular distribution of the data set and/or class a priori probabilities. Therefore, the application of *VGACD-classifier* for pixel classification in remote sensing imagery appears to be natural and appropriate [41]. A comparative study of the *VGACD-classifier*, *VGA-classifier*, and the classifiers based on the k-NN rule and the Bayes' maximum likelihood ratio is provided in this regard. Several quantitative measures [398] and visual quality of the classified image are considered for this purpose. Results are reported for a *SPOT* image of a part of the city of Calcutta.

6.5.2 Experimental Results

The 512×512 *SPOT* image of a part of the city of Calcutta is available in three bands. It is described in detail in Appendix B. The image in the near infrared

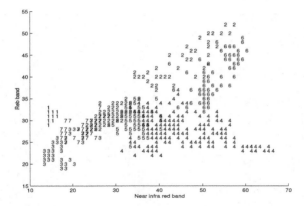

Fig. 6.4. Scatter plot for a training set of *SPOT image* of Calcutta containing seven classes (1,...,7)

band is shown in Fig. 6.5. The design set comprises 932 points belonging to 7 classes that are extracted from the above image. (A two-dimensional scatter plot of the training data is shown in Fig. 6.4.) The 7 classes are *turbid water* (TW), *pond water* (PW), *concrete* (Concr.), *vegetation* (Veg), *habitation* (Hab), *open space* (OS) and *roads* (including bridges) (B/R). Different percentages (30%, 50% and 80%) of these 932 points are used for training, while the remaining are used for the purpose of testing. The classifier trained using 80% of the data set is then utilized for classifying the pixels in the 512×512 image. The experimental parameters were kept the same as in Sect. 5.5.2.

The comparative performance of the genetic classifiers, *VGACD-classifier* and *VGA-classifier*, and the Bayes' maximum likelihood classifier (capable of handling overlapping classes) assuming normal distribution of the data set for each class with different covariance matrices and a priori probabilities for the classes, and the k-NN classifier (well-known for generating highly nonlinear boundaries through piecewise linear segments) with k=$\sqrt{n}$ is shown in Table 6.1. It presents the results for different percentages of training data, and $H_{max} = 15$ for the genetic classifiers. (Note that the choice of H_{max}, as long as it is sufficiently high, is not crucial for the performance of the variable string length genetic classifiers.) Comparison is made in terms of two measures, the percentage recognition scores and kappa values [398]. While the percentage recognition score measures the overall percentage of the points correctly classified, the kappa value measures the relationship of beyond chance agreement to expected disagreement. The latter is explained more in detail in Sect. 7.7. Higher values of kappa indicate better performance. The classwise recognition scores for the different classifiers when perc = 80% is shown in Table 6.2. Table 6.3 presents, as an example, the confusion matrix obtained for the 80% training data corresponding to the Bayes' classifier.

As seen from Table 6.1, the performance of the *VGACD-classifier*, is always better than that of the *VGA-classifier*, irrespective of the percentage

Fig. 6.5. *SPOT* image of Calcutta in the near infrared band

Table 6.1. Comparative results in terms of recognition scores "recog" (%) and kappa values (%). Here "perc" indicates percentage of data used for training

perc	*VGACD-classifier*		*VGA-classifier*		Bayes'		k-NN	
	recog	kappa	recog	kappa	recog	kappa	recog	kappa
30	83.38	81.56	80.94	76.56	82.16	78.38	82.77	78.94
50	85.01	82.16	84.36	81.07	85.86	82.85	88.01	85.34
80	88.89	87.07	84.12	80.62	86.24	83.28	89.41	87.13

of data used for training. This indicates that incorporation of the concept of chromosome differentiation leads to an improvement in performance of the variable string length GA classifier. Overall, the performance of the *VGACD-classifier* is found to be better than or comparable to that of the k-NN rule for 30% and 80% training data, while for 50% training data the k-NN rule outperformed all other classifiers.

From the classwise scores shown in Table 6.2, it is found that the *VGACD-classifier* recognizes the different classes consistently with a higher degree of accuracy. On the contrary, the other classifiers can recognize some classes

Table 6.2. Comparative classwise recognition scores (%) for perc = 80

Class	Recognition Scores			
	VGACD-classifier	*VGA-classifier*	Bayes'	k-NN
TW	100.00	100.00	100.00	100.00
PW	93.93	89.55	100.00	96.97
Concr	80.83	91.42	91.42	92.53
Veg	92.29	85.31	88.45	94.21
Hab	70.73	43.75	49.55	62.50
OS	94.73	87.27	89.47	94.74
B/R	72.72	21.35	36.36	45.45

Table 6.3. Confusion Matrix for training data of *SPOT* image of Calcutta obtained using Bayes' maximum likelihood classifier

		Recognized as						
		TW	PW	Concr	Veg	Hab	OS	B/R
	TW	100	0	0	0	0	0	0
	PW	0	115	11	0	2	8	12
	Concr	0	11	116	0	0	0	4
Actual	Veg	0	0	0	173	24	3	0
	Hab	0	7	0	10	42	0	1
	OS	0	5	0	5	0	64	0
	B/R	0	11	8	0	0	0	22

very nicely, while for some other classes their scores are much poorer. For example, the Bayes' maximum likelihood classifier provides 100% accuracy for the classes TW and PW, but its scores for classes B/R and Hab. are only 36.36% and 49.55%, respectively.

Figure 6.6 demonstrates the variation of the the number of points misclassified by the best chromosome with the number of generations for the *VGACD-classifier* and *VGA-classifier* (when *perc* = 80). As is seen from Fig. 6.6, both the classifiers consistently reduce the number of misclassified points. The best value is obtained just after the 1700 and 1900 iterations for *VGACD-classifier* and *VGA-classifier*, respectively. Incorporating the concept of chromosome differentiation therefore helps in faster convergence of the algorithm, since any given value of the number of misclassified points is achieved earlier by the *VGACD-classifier* than the *VGA-classifier* (if at all).

Figure 6.7 shows the full Calcutta image classified using the *VGACD-classifier*. As can be seen, most of the landmarks in Calcutta have been properly classified. For the purpose of comparison, Figs. 6.8(a)–(f) provide the results obtained by the different classifiers (including results for k-NN rule with k = 1 and 3) for partitioning the 512 × 512 image, by zooming a characteristic portion of the image containing the *race course* (a triangular structure). Here 80% of the design set is used for training.

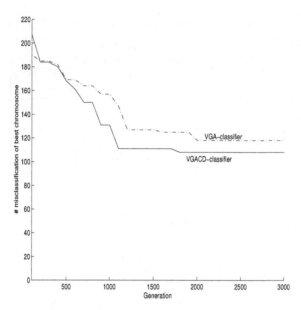

Fig. 6.6. Variation of the number of points misclassified by the best chromosome with generations for *VGACD-classifier* and *VGA-classifier*

As seen from the figures, although all the classifiers (with the exception of k=1 for k-NN rule) are able to identify the *race course*, only the *VGACD-classifier* and the *VGA-classifier* are able to identify a triangular lighter outline (which is an open space, corresponding to the tracks) within the *race course* properly. The performance of k-NN rule is found to gradually improve with the value of k, being the best for k=$\sqrt{n}$. On inspection of the full classified images it was found that the Bayes' maximum likelihood classifier tends to over estimate the roads in the image. On the other hand, the *VGA-classifier* tends to confuse between the classes bridges and roads (B/R) and pond water (PW). It was revealed on investigation that a large amount of overlap exists between the classes Concr and B/R on one hand (in fact, the latter class was extracted from the former) and PW and B/R on the other hand. This is also evident from Table 6.3, where 27% and 20% of the points belonging to the class B/R went to class PW and Concr, respectively. These problems were not evident for the case of the *VGACD-classifier*, thereby indicating the significant superiority of the former.

6.6 Summary

The concept of chromosome differentiation has been incorporated in GAs resulting in a modified GA called GACD. Its schema theorem is provided where it is found that in certain situations, the lower bound of the number of in-

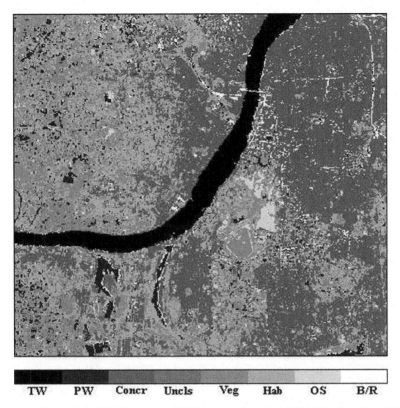

| TW | PW | Concr | Uncls | Veg | Hab | OS | B/R |

Fig. 6.7. Classified *SPOT* image of Calcutta using the *VGACD-classifier* (H_{max} =15, final value of H=13)

stances of a schema sampled by GACD is greater than or equal to that of CGA. The concepts of variable string length and chromosome differentiation in genetic algorithms have been integrated for the development of a nonparametric *VGACD-classifier*. The superiority of the *VGACD-classifier*, as compared to the *VGA-classifier*, Bayes' maximum likelihood classifier, and k-NN rule, is established for a *SPOT* image of Calcutta.

An interesting analogy of the concept of chromosome differentiation can be found in the sexual differentiation found in nature. The class bits in VGACD are chosen in tune with the way the X and Y chromosomes help to distinguish the two sexes. Because the initial M and F populations are generated so that they are at a maximum Hamming distance from each other, and crossover is restricted only between individuals of these two classes, VGACD appears to be able to strike a greater balance between exploration and exploitation of the search space. This is in contrast to its asexual version. It is because of this fact that the former is consistently seen to outperform the latter.

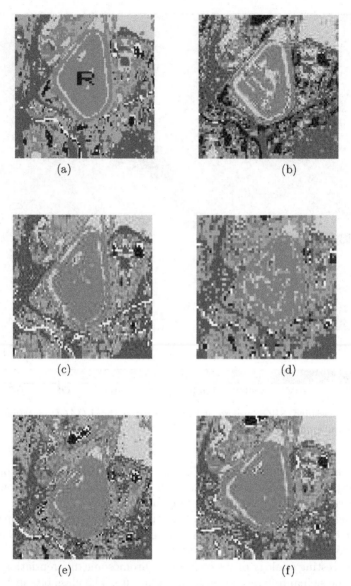

Fig. 6.8. Classified *SPOT* image of Calcutta (zooming the *race course*, represented by R on the first figure, only) using (a) *VGACD-classifier*, H_{max} =15, final value of H=13, (b) *VGA-classifier*, H_{max} =15, final value of H=10, (c) Bayes' maximum likelihood classifier (d) k-NN rule, k = 1, (e) k-NN rule, k = 3, (f) k-NN rule, k = $\sqrt{n}$. Training set = 80% of design set.

The genetic classifiers take significantly large amounts of time during training. However, the time taken during testing is very small. On the contrary, the k-NN rule (with k=$\sqrt{n}$) takes significant amount of time for testing, while for the Bayes' classifier both the training and testing times are quite small. As an illustration, the *VGA-classifier* took 515.76 seconds during training on a DEC-Alpha machine (when 3000 iterations were executed). Note that the problem is compounded by the fact that no appropriate criterion for terminating GAs is available in the literature. The k-NN rule took 659.90 seconds when it was tested on the full *SPOT* image of Calcutta whereas for the *VGA-classifier* and the Bayes' maximum likelihood classifier these values were 3.54 seconds and 2.06 seconds, respectively.

7

Multiobjective *VGA-classifier* and Quantitative Indices

7.1 Introduction

Some classifiers using variable string length genetic algorithms were described in Chaps. 5 and 6. The *VGA-classifier* evolves an appropriate number of hyperplanes, which minimizes first the number of misclassified points *miss*, and then the number of hyperplanes H, though the former is given primary/more importance. The superiority of the genetic classifier over the conventional Bayes' maximum likelihood classifier and the k-NN rule, as well as some neural network-based classifiers, was demonstrated for a wide variety of data sets.

It may be noted that as a result of the combination of the two objective criteria, *miss* and H, into a single-objective function in the *VGA-classifier*, a situation may result where a large number of hyperplanes may be utilized for reducing *miss* by a small fraction. Consequently, although such a classifier would provide good abstraction of the training data, its generalization capability is likely to be poor.

This chapter describes a multiobjective GA-based classifier for pattern recognition that deals with these problems [42]. The ability of GAs and other evolutionary algorithms for handling complex problems, involving features such as discontinuities, multimodality, disjoint feasible spaces and noisy function evaluations, as well as their population-based nature, makes them possibly well suited for optimizing multiple objectives.

Three multiobjective optimization (MOO) techniques, *NSGA-II* [122] and Pareto Archived Evolutionary Strategies (*PAES*) [246], and a new constrained elitist multiobjective GA (*CEMOGA*) [42], are used for designing the classifier. Unlike *NSGA-II*, *CEMOGA* utilizes some domain-specific knowledge in the process that helps in increasing the diversity of the system, thereby leading to a larger exploration of the search space. The classifiers simultaneously optimize three objective criteria, namely, *miss*, H and *ClassAccuracy*, described in detail in Sect. 7.4.2. This ensures that overfitting/overlearning is avoided, while classes of smaller size are not ignored during training. Since the MOO algorithms will provide a set of solutions after training, the concepts of validation set (in addition to training and test sets) and validation functional are introduced for selecting one solution from the set. For comparing the performance of different MOO techniques, two quantitative measures, viz., *purity* and *minimal spacing*, are defined. These two measures, along with the number of hyperplanes required, time required for training, percentage of correct classification, user's accuracy and kappa, are used for comparing the performance of different classifiers.

The chapter is organized as follows: Section 7.2 describes the different concepts and issues in optimizing multiple objectives. The relevance of using the MOO technique for the classifier design task is explained in Sect. 7.3. A detailed description of the multiobjective classifiers, and their validation and testing steps are provided in Sects. 7.4 and 7.5, respectively. Some indices for comparing the relative performance of the MOO algorithms are described in Sect. 7.6. Section 7.7 provides the experimental results along with a comparative study. Finally, Sect. 7.8 summarizes the chapter.

7.2 Multiobjective Optimization

Many real-world problems involve multiple measures of performance, or objectives, which should be optimized simultaneously. Optimizing multiple objectives involves finding a solution, or possibly a set of solutions, which would provide the values of all the objective functions acceptable to the designer [94, 96]. A preliminary discussion on some of the basic issues in multiobjective optimization has been presented in Sect. 2.6. For the sake of completeness, some of it is again mentioned here. A general minimization problem of k objectives can be mathematically stated as:

$$\text{Minimize } \overline{f}(\overline{x}) = [f_1(\overline{x}), f_2(\overline{x}), \ldots, f_k(\overline{x})]^T, \tag{7.1}$$

$$\text{subject to:}$$

$$g_i(x) \leq 0, \qquad i = 1, 2, \cdots, m, \tag{7.2}$$
$$h_i(x) = 0, \qquad i = 1, 2, \cdots, p,$$

where there are m inequality constraints and p equality constraints. The MOO problem then reduces to finding $\overline{x}$ such that $\overline{f}(\overline{x})$ is optimized [95]. A suitable solution to such problems involving conflicting objectives should offer *acceptable*, though possibly suboptimal in the single-objective sense, performance in all objective dimensions, where acceptable is a problem-dependent and ultimately subjective concept. In MOO, a solution x_i is said to dominate another solution x_j if both the following conditions are true [121]:

- The solution x_i is no worse than x_j in all objectives.
- The solution x_i is strictly better than x_j in at least one objective.

A decision vector $x_i \in U$, where U is the universe, is said to be *Pareto-optimal* if and only if there exists no x_j, $x_j \in U$, such that x_i is dominated by x_j. Solution x_i is said to be *nondominated*. The set of all such nondominated solutions, in the global sense, is called the *Pareto-optimal set*. In general, MOO problems tend to achieve a family of alternatives which must be considered equivalent in the absence of information concerning the relevance of each objective relative to the others. Some of the important issues in MOO are designing a suite of test problems with varying degrees of difficulty [124] and defining a set of appropriate quantitative measures for comparing the performance of different MOO strategies [520].

Evolutionary algorithms have been used widely for solving multiobjective problems because of their population-based nature and their ability to come out of local optima. Some recent and popularly used MOO algorithms based on these approaches are NSGA-II [122], strength Pareto evolutionary algorithms (SPEA) [521, 522] and Pareto archived evolutionary strategies (PAES), which are based on evolutionary strategies [247, 246]. The next section describes how the characteristics of multiobjective evolutionary algorithms may be exploited for extending the hyperplane-based *VGA-classifier* so as to incorporate simultaneous optimization of more than one objective function. For this purpose, NSGA-II [122] and PAES [246], and a constrained elitist multiobjective GA (CEMOGA) [42] are considered.

7.3 Relevance of Multiobjective Optimization

As already mentioned, the *VGA-classifier* described in Chap. 5 uses a single criterion, which gives primary importance to *miss*, and thereafter to H, for optimization during training. The preference given to *miss* over H is user defined and ad hoc in nature. It could result in overfitting of the data, when a large number of hyperplanes could be used for providing maximum accuracy during training, resulting in reduced generalizability. Thus, optimizing *miss* and H separately and simultaneously could be beneficial for the classifier design task.

Although *miss* provided a measure of the overall classification accuracy, it did not incorporate any classwise information. Therefore, a new term called *ClassAccuracy*, defined as the product of the classwise recognition scores, is taken as the third objective for optimization. It may be noted that minimizing *miss* does not automatically lead to maximizing *ClassAccuracy*, and vice versa. For example, consider a data set having two classes, with 100 points in class 1 and 10 points in class 2. Assume that in one situation, 50% of both the classes are correctly recognized. Hence, *ClassAccuracy* is 0.25 and *miss* = 55. In another situation, let 60% and 40% of the classes be recognized, respectively. Here the *ClassAccuracy* becomes 0.24, while *miss* is 46. It is evident that although in the first case *ClassAccuracy* is more, in the second *miss* is smaller. The question as to which one is better is therefore seen to be problem dependent, and cannot be resolved without any domain knowledge.

Since any combination of the three objectives would again, in general, be ad hoc in nature, it is imperative to adopt a MOO strategy. An important consequence of this is that the decision maker will be provided with a set of possible alternative solutions, as well as intermediate ones, which the decision maker may subsequently refine. The next section describes how the principles of multiobjective GA for can be used for extending the design of the genetic classifiers described in the earlier chapters. Note that the objectives to be optimized simultaneously are *miss* (the number of misclassified points), *ClassAccuracy* (the product of the rates of correct classification of each class) and *H* (the number of hyperplanes *H* required to model the class boundaries). The purpose is to minimize *miss* and *H*, and to maximize *ClassAccuracy*. As mentioned earlier, in the *VGA-classifier* [38], only one objective was taken as the optimizing criterion.

7.4 Multiobjective GA-Based Classifier

In this section we describe different multiobjective classifiers based on CE-MOGA, NSGA-II and PAES. Note that the earlier versions of multiobjective GA methods, e.g., NSGA and NPGA, are not chosen since these are essentially nonelitist in nature. NSGA-II has been chosen for the purpose of comparison since the operations of the proposed *CEMOGA-classifier* are based on it. PAES has been chosen since it is based on evolutionary strategies, another widely used evolutionary algorithm.

7.4.1 Chromosome Representation

As earlier, the chromosomes are represented by strings of 1, 0 and # (don't care), encoding the parameters of variable number of hyperplanes. If angle and distance bits are represented by b_1 and b_2 bits, respectively, then the length, l_i, of the ith chromosome encoding H_i ($1 \leq H_i \leq H_{Max}$, where H_{Max} represents the maximum number of hyperplanes, specified a priori) hyperplanes is

$$l_i = H_i * [(N - 1) * b_1 + b_2], \qquad i = 1, \cdots, P, \qquad (7.3)$$

where N is the number of dimensions, and P is the population size. Note that if the number of classes is represented by K, then the minimum number of hyperplanes, H_{Min}, required to generate those many regions is:

$$H_{Min} = \log_2 K. \qquad (7.4)$$

7.4.2 Fitness Computation

The fitness of a chromosome is assessed on the basis of *miss*, *H* and *ClassAccuracy*.

Computing miss:

The number of misclassified points for a string i encoding H_i hyperplanes is found as in Sect. 5.3.2.

Computing ClassAccuracy:

For a data set having K classes, the fraction of the points in class j ($j = 1, 2, \ldots, K$) correctly classified, $ClassAccuracy_j$, is defined as:

$$ClassAccuracy_j = \frac{n_{j_c}}{n_j}, \tag{7.5}$$

where n_j and n_{j_c} represent the number of points and the number of correctly classified points of class j, respectively. Overall class accuracy of a chromosome, considering all the classes, is calculated as

$$ClassAccuracy = \Pi_{j=1}^{K} ClassAccuracy_j. \tag{7.6}$$

Calculating H:

If the length of a chromosome is l_i, then the number of hyperplanes encoded in the chromosome is calculated as follows:

$$H_i = \frac{l_i}{(N-1) * b_1 + b_2}. \tag{7.7}$$

Any chromosome with $H = 0$ is assigned $ClassAccuracy = 0$ so that it will be automatically eliminated during selection.

7.4.3 Selection

In MOO, the selection procedure differs significantly from the corresponding single-objective version. This section describes, in brief, the selection scheme used in the *CEMOGA-classifier* and *NSGAII-classifier* [122] since this operation of *CEMOGA-classifier* is the same as that of *NSGAII-classifier*. Details of the selection scheme may be found in [121, 122]. The *PAES-classifier* is described separately in Sect. 7.4.7.

The chromosomes in a population are first sorted based on their *domination* status using the procedure *Nondominated sort* [121, 122]. Here the chromosomes that are not dominated by any other member of the population form the first front and are put in rank 1. These chromosomes are then removed from consideration, and the chromosomes which thereafter become nondominated are assigned rank 2. The process is repeated until all the chromosomes have been assigned a rank. The chromosomes are then put in a sorted order according to their ranks. The overall complexity of the *Nondominated sort* algorithm described in detail in [121] is $O(mp^2)$, where m is the number of objectives and P is the number of chromosomes in the population. The first front represents a *nondominated set* with respect to the current population, since none of the solutions in this front is dominated by any other solution in the population.

After applying the *Nondominated sort* algorithm the chromosomes are assigned a *crowding distance* and selection is performed using the *crowded tournament selection*. The *crowding distance* is computed as the sum of the difference of the objectives values of the solutions preceding and following the current solution (corresponding to the chromosome under consideration) for each objective [121, 122]. This provides a measure of the density of the solutions surrounding a particular point in the population.

In *crowded tournament selection* [122], a pair of chromosomes is selected randomly, and the one with the lower rank is selected. If the ranks of the chromosomes are equal, then the chromosome with a larger crowding distance is chosen. The large crowding distance ensures that the solutions are spread along the *Pareto-optimal front*.

7.4.4 Crossover

The crossover operation is modified to be applicable to two strings of (possibly) different lengths. The lengths of the two are first made equal by padding the smaller chromosome with #'s. Conventional single-point crossover is then performed on them with a probability μ_c. The resultant chromosome may have some *complete* and some *incomplete* hyperplanes. As mentioned in Chap. 5, a hyperplane is complete if all the bits representing it are either defined (i.e., 0's and 1's) or undefined (i.e., #'s); it is invalid otherwise (i.e, when it consists of a combination of defined and undefined values).

Suppose a hyperplane is incomplete, and let u be the number of defined bits in that hyperplane. Then, all its # bits are set to defined bits with a probability u/l_H, where l_H is the number of bits required to encode one hyperplane. In case this is not permitted, all the defined bits are set to #'s. After each hyperplane becomes complete, all the #'s are pushed to the end of the chromosome so that we have all the hyperplanes in the beginning of the chromosome.

7.4.5 Mutation

As in Chap. 5, the mutation operation is also modified to tackle variable-length strings. The length of a chromosome is first made equal to the maximum chromosome length by appropriately padding with #'s. Then for each defined bit position, mutation is carried out with *mutation probability* μ_m. μ_m is varied with the number of iterations such that it goes though a complete cycle of high values, then low values, then high followed by low values once again. Note that when μ_m is high, greater exploration of the search space results, while exploitation is given more importance for low values of μ_m.

By alternating this cycle, we intend to provide a better balance between exploration and exploitation of the search space; a crucial factor in the good performance of GAs. The variation of μ_m with the number of generations is depicted in Fig. 7.1 [42].

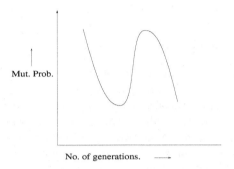

Fig. 7.1. Mutation probability graph

If mutation cannot be carried out on a defined bit, it is set to # with a probability μ_{m1}. Each undefined bit is set to a defined bit (chosen randomly) with probability μ_{m2}. The resultant chromosome may again have some *invalid* hyperplanes which are handled in the same way as in crossover. Note that mutation operation on a defined bit may result in a decrease in the chromosome length, while the operation on # positions may result in an increase in the chromosome length. Here, the values of μ_{m1} and μ_{m2} are fixed and are kept equal so that the probability of increasing and decreasing the string length is the same.

7.4.6 Incorporating Elitism

In this section we describe how elitism is incorporated in the *NSGAII-classifier* and *CEMOGA-classifier*. As for selection, these are again mentioned together since they are somewhat similar, though the procedure in *CEMOGA-classifier* is so designed such that some domain-specific constraints are incorporated.

Elitism in NSGAII-classifier

Here the parent population and the child population, obtained using selection, crossover and mutation on the parent population, are first combined together to form a combined generation of size $2P$. The combined population is sorted and the best P chromosomes are selected from this combined generation to constitute the new parent generation. Chromosomes from rank 1 get selected first, followed by rank-2 chromosomes, and so on till the total number of chromosomes exceeds P. Thereafter the last front chromosomes are removed, and they are selected one by one on the basis of their crowding distance, so that the total number of chromosomes in the new parent generation is P. This new parent generation again undergoes crowded tournament selection, crossover and mutation to form the child generation. This process is repeated for a maximum number of generations, fixed a priori. Details are available in [122].

Elitism in CEMOGA-classifier

As in the *NSGAII-classifier*, an initial population of size P is created. This is also called the *combined generation* for the sake of convenience. Note that, although in the initial stage the size of the *combined generation* is equal to P, for subsequent generations this will be equal to $P + P_C$, where P_C is the size of the *constrained generation*, described below. Now two sets of operations are invoked on the same *combined generation*. First, crowded tournament selection [122] is used to select P chromosomes from the *combined generation*, after which crossover and mutation operators are applied to form a generation of size P, which is called the *current generation*. Simultaneously, a *constrained generation* is formed from the *combined generation* by selecting chromosomes which satisfy the following constraints [42]:

(a) H is greater than H_{Min} (Eq. 7.4).
(b) *miss* is less than $\alpha_{m1} * S$, for the first θ_1 generations and less than $\alpha_{m2} * S$ for later generations, where S is the size of the data sets and $\alpha_{m1}, \alpha_{m1} < 1$.
(c) *ClassAccuracy* is greater than zero for the first θ_2 generations, and greater than α_{c1} for the later generations, where $0 < \alpha_{c1} < 1$.

Note that the size of *constrained generation*, P_C, may vary, depending upon the number of chromosomes that satisfy the constraints, but is restricted to $\delta * N$, where $\delta > 0$. If the number of chromosomes that satisfy the constraints exceeds $\delta * N$, then in a process analogous to the one described for Sect. 7.4.6, the chromosomes from the first front onwards, and which satisfy the constraints, are added to the *constrained generation* one after the other till the size of the *constrained generation* exceeds the limit. Thereafter, the last front chromosomes are removed from the *constrained generation*, and crowded tournament selection is applied to select the appropriate number of chromosomes from this front. We call this process *Crowded ConstrainedGen Formation*. The *current generation* and the *constrained generation* are combined to form the *combined generation*, which will now be of size $P + P_C$. The entire process is continued for a specified number of generations.

Note that in contrast to the elitism procedure for *NSGAII-classifier*, where better ranked chromosomes are retained, even though they may not satisfy the practical constraints of the problem, this is not the case in *CEMOGA-classifier*. Here, the *constrained generation* is formed in such a way that even lower ranked chromosomes have a chance of surviving, provided they satisfy the constraints. This, in turn, leads to a better diversity, and hence larger exploration of the search space, which is likely to result in improved performance. To illustrate the difference in operation of NSGA-II and CEMOGA, consider the following example.

Example 7.1. Let us consider only the constraint on the number of hyperplanes by requiring that $H \geq 4$, i.e., $H_{Min} = 4$ (the other constraints are ignored for the sake of simplicity). Suppose it is required to select three chromosomes from a set of six chromosomes shown below.
Chr 1: $H = 6$, $miss = 15$, $ClassAccuracy = 0.2$,
Chr 2: $H = 8$, $miss = 9$, $ClassAccuracy = 0.3$,
Chr 3: $H = 3$, $miss = 20$, $ClassAccuracy = 0.1$,
Chr 4: $H = 7$, $miss = 16$, $ClassAccuracy = 0.15$,
Chr 5: $H = 8$, $miss = 12$, $ClassAccuracy = 0.25$,
Chr 6: $H = 5$, $miss = 22$, $ClassAccuracy = 0.08$.
Note that Chr 1, Chr 2, and Chr 3 are rank-1 chromosomes, while the other three are rank-2 chromosomes. In *NSGAII-classifier* the first three rank-1 chromosomes will be selected, although Chr 3 does not satisfy the constraint that $H \geq 4$. Note that although this solution uses a smaller number of hyperplanes, it is not a good solution since these many hyperplanes are simply insufficient for the given data set. This practical constraint is enforced in the proposed elitist strategy since, as desired, it will select Chr 4 (which is of a lower rank) over Chr 3. Fig. 7.2 illustrates the *CEMOGA* procedure.

7.4.7 *PAES-classifier*: The Classifier Based on Pareto Archived Evolution Strategy

PAES-classifier utilizes the Pareto-archived evolution strategy (*PAES*) [246] as the underlying MOO technique. PAES uses a $(1 + 1)$ evolutionary strategy (ES) [423]. This algorithm performs a local search by using a small change operator to move from one solution to a nearby solution. At any generation t, a parent solution (P_t), offspring solution (C_t) and a number of best solutions called *Archive* are maintained. The offspring C_t is obtained by applying the mutation operator on a copy of P_t. P_t and C_t are then compared for domination, and the following situations may arise:

(a) If P_t dominates C_t, C_t is not accepted and parent P_t is mutated to get a new solution.
(b) If C_t dominates P_t, C_t is accepted as parent of next generation and it replaces P_t in the archive. The domination status of all members of the archive, with respect to the newly introduced member C_t is checked, and all dominated members are deleted.
(c) If P_t and C_t are nondominating with respect to each other, then C_t is compared with all the members of the archive, and three possibilities may arise:
    (1) C_t is dominated by a member of archive. Then C_t is rejected.

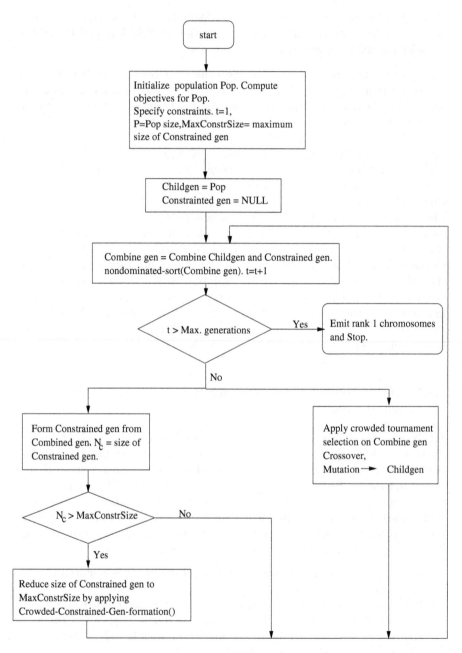

Fig. 7.2. Flowchart of CEMOGA-classifier

(2) C_t dominates some members of the archive. Then the dominated members are deleted and C_t is accepted as parent for the next generation.

(3) C_t and all members of archive are nondominating with respect to each other. Then C_t is added to archive. In case the archive overflows, a member residing in the most dense region (described below) is deleted.

To decide who qualifies as parent (as well as to decide a victim for deletion when the archive overflows), the density of solutions in the neighborhood is checked, and the one residing in the least crowded area qualifies as parent. (In the case of selecting a victim for deletion, one residing in the most crowded area is selected.) For determining density, each objective space is divided into 2^d equal divisions, where d is a user-defined *depth* parameter. Therefore, the entire search space is divided into $(2^d)^k$ unique, equal-sized k-dimensional hypercubes, where k is the number of objectives. The number of solutions in each hypercube is counted. This is used as a measure of the density of the solutions in the archive.

7.5 Validation and Testing

In contrast to the single-objective *VGA-classifier*, which provides only a single final solution, the multiobjective classifiers described here provide a set of solutions. These are the chromosomes in the first front (or rank 1) of the final population for the *NSGAII-classifier* and *CEMOGA-classifier*, while for the *PAES-classifier* these are found in the archive. A validation function therefore needs to be defined to select a chromosome which is best suited for the problem at hand, and which can be subsequently used for designing the final classifier. Accordingly, the data set has to be divided into three parts: training, validation and test sets. The validation function, applied on the validation set, is defined as follows:

$$Validity = 0.5 * ClassAccuracy + 0.3 * \left(\frac{V - miss}{V}\right)^K - 0.2 * \left(\frac{H}{H_{Max}}\right)^K,$$
(7.8)

where V and K are the size of validation set and the number of classes, respectively. From Eq. 7.5, it is found that for individual classes, the value of $ClassAccuracy_j$ can range from 0 to 1. The overall $ClassAccuracy$ is a product of the K such $ClassAccuracy_j$ values corresponding to K classes. Therefore, in order to combine the three objectives into a single validity function, the terms corresponding to $miss$ and H are so defined that these values also range from 0 to 1, and thereafter these are raised to the power of K. The coefficients reflect the relative importance of the three objectives. It may be noted that these values may be changed depending on the problem domain and other user-specific considerations. Moreover, the validity function may be defined in some other form as well.

The classifiers designed by the chromosomes of the first front of the last generation are subjected to the validation phase. The validation function (*Validity*) is then computed corresponding to each classifier. The classifier providing the highest value of *Validity* is taken as the solution.

The final classifier thus obtained from the validation phase is then used for testing. For each unknown pattern in the test set, the region in which it lies is first determined. The pattern is then labelled by the class associated with that region.

7.6 Indices for Comparing MO Solutions

In general, the solutions obtained using any MOO algorithm should be as close as possible to the Pareto-optimal front. Moreover, the solutions should be distributed uniformly over the solution space. In other words, a maximal set of solutions over the Pareto-optimal front is desired which must cover the entire gamut of the front, and also be uniformly distributed over it. Thus the different criteria in this regard are:

- There must be as many nondominated solutions as possible.
- The solutions must be as close to the Pareto-optimal front as possible.
- The solutions must be as diverse as possible.
- The solutions must be as distributed over the nondominated front as possible.

Some measures of comparing the nondominated solutions provided by different MOO strategies are discussed in this section.

7.6.1 Measures Based on Position of Nondominated Front

One of the ways of evaluating the performance of MOO algorithms is based on the position of the nondominated front (the surface formed by the nondominated solutions). Some such measures are the error ratio [472] and set coverage metric [519]. The error ratio represents the fraction of the solutions that is not able to reach the Pareto-optimal front. In this way, it measures the distance of the obtained nondominated front from the Pareto-optimal front. The error ratio is defined as the ratio of the number of nondominated solutions that do not belong to a set of Pareto-optimal front to the total cardinality of the nondominated set. The first measure, error ratio, assumes that a knowledge of the Pareto-optimal set is available, an information that may not be readily available in real-life problems. The other measure, set coverage, $C(\mathcal{A}, \mathcal{B})$, between two sets of solution vectors $\mathcal{A}$ and $\mathcal{B}$, calculates the proportion of solutions in $\mathcal{B}$ that are dominated by solutions in $\mathcal{A}$. This measure essentially computes the relative goodness of only two MOO strategies. Another measure, called *purity*, is now described that can be used to compare the solutions obtained by not only two but several MOO strategies (which is in contrast to the set coverage measure), while not assuming any knowledge about the Pareto-optimal front (which is in contrast to the error ratio) [42, 206].

Suppose there are τ, $\tau \geq 2$, MOO strategies applied to a problem. Let $r_i = |R_1^i|$, $i = 1, 2, \ldots, \tau$ be the number of rank-1 solutions obtained from each MOO strategy. Compute the union of all these solutions as $R^* = \bigcup_{i=1}^{\tau} \{R_1^i\}$. Thereafter a ranking procedure is applied on R^* to obtain the new rank-1 solutions, called R_1^*. Let r_i^* be the number of rank-1 solutions which are present in R_1^*. That is,

$$r_i^* = |\{\gamma | \gamma \in R_1^i, \text{ and } \gamma \in R_1^*\}|. \tag{7.9}$$

Then the purity measure for the ith MOO strategy, P_i, is defined as

$$P_i = \frac{r_i^*}{r_i}, \qquad i = 1, 2, \ldots, \tau. \tag{7.10}$$

Note that the purity value may lie between $[0, 1]$, where a value nearer to 1 indicates a better performance.

7.6.2 Measures Based on Diversity of the Solutions

Schott [421] suggested a measure called spacing which reflects the uniformity of the distribution of the solutions over the nondominated front. Spacing (S) between solutions is calculated as

$$S = \sqrt{\frac{1}{|Q|} \sum_{i=1}^{|Q|} (d_i - d)^2}, \tag{7.11}$$

where $d_i = \min_{k \in Q \text{ and } k \neq i} \sum_{m=1}^{M} |f_m^i - f_m^k|$ and f_m^i (or f_m^k) is the mth objective value of the ith (or kth) solution in the final nondominated solution set Q. d is the mean value of all the d_i's. Note that a value of S nearer to 0 is desirable since it indicates that the solutions are uniformly distributed over the Pareto-optimal front. This diversity measure can be calculated for solution sets where there are more than two solutions in the set. Figure 7.3 demonstrates a common scenario when this measure is expected to fail. In Fig. 7.3(a), since point a has b as its closest neighbor and vice versa, the value of S will be low, indicating wrongly a uniform spread over the Pareto-optimal front. This measure is unable to indicate that a large gap exists between b and c. Note that with this assumption the value of S will be also low in Fig. 7.3(b). It is obvious that the situation in Fig. 7.3(b) is preferable, since it shows a more uniform distribution over the Pareto-optimal front. However, the measure S is unable to indicate this. In order to overcome this limitation, a modified measure called *minimal spacing*, S_m, is described in [42]. The essence of this measure is to consider the distance from a solution to its nearest neighbor which has not already been considered. S_m is calculated among

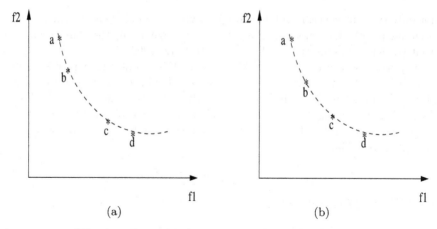

Fig. 7.3. Example of a set of nondominated solutions

the solutions in the nondominated set as follows: Initially consider all solutions as unmarked. Take a solution and call it the seed. Mark the seed. Start computing the nearest distance from the last marked solution (initially the seed). Each time while calculating the minimum distance between two solutions, consider only those solutions which are still unmarked. Once a solution is included in the distance computation process, make it as marked. Continue this process until all solutions are considered, and keep track of the sum of the distances.

Repeat the above process by considering each solution as seed and find out the overall minimum sum of distances. The sequence of solutions and the corresponding distance values are used for the computation of S_m, where again Eq. 7.11 is used with $|Q|$ replaced by $|Q|-1$ (since here we have $|Q|-1$ distances). As for S, a lower value of S_m indicates better performance of the corresponding MOO technique. Note that with this definition, the value of S_m in Fig. 7.3(b) will be lower than that in Fig. 7.3(a), indicating that the former solution set is better. In this regard, it may be mentioned that the range of values of the different objectives often varies widely. Consequently, they are generally normalized while computing the distances. In other words, when computing the d_i's, the term $|f_m^i - f_m^k|$ is divided by $|F_m^{max} - F_m^{min}|$ in order to normalize it, where F_m^{max} and F_m^{min} are the maximum and minimum values, respectively, of the mth objective.

7.7 Experimental Results

The experimental results are reported for three real-life data sets, viz., *Vowel*, *Iris* and *Mango*. *Vowel* is a three-dimensional, six-class data set having 871 points. *Iris* is a four-dimensional, three-class data set having 150 points.

Mango is an 18-dimensional, three-class data set having 166 points. These data sets are described in detail in Appendix B.

7.7.1 Parameter Values

The parameter setting for *CEMOGA-classifier* and *NSGAII-classifier* are as follows:

- population size = 30;
- crossover probability = 0.85;
- conventional mutation probability varied in the range [0.45, 0.002];
- $\mu_{m1} = \mu_{m2} = 0.95$;
- $b_1 = 8$; $b_2 = 16$;
- maximum number of generations = 3000;
- $H_{max} = 20$;
- $\theta_1 = \theta_2 = 50$ (see Sect. 7.4.6);
- $\alpha_{m1} = 0.50$; $\alpha_{m2} = 0.35$; $\alpha_{c1} = 0.05$.

For the *PAES-classifier* (which operates on only one string at a time)

- maximum number of generations = 60,000; archive size = 30; depth parameter = 3.

7.7.2 Comparison of Classification Performance

Here, a comparison of the performance of all the classifiers is provided for the three data sets with respect to the percentage recognition scores, user's accuracy and kappa [97, 398]. The nondominated solutions obtained by the three multiobjective classifiers are compared in the following section, using the measures *purity* and *minimal spacing*, defined in Sect. 7.6.

Table 7.1. Training results on various data sets

	VGA		NSGAII		PAES		CEMOGA	
	H	Score%	H	Score%	H	Score%	H	Score%
Vowel	8	81.61	8	75.86	13	70.11	7	76.43
Iris	3	97.8	2	100	6	100	2	100
Mango	2	93.93	3	78.79	6	81.82	6	84.84

Comparison with Respect to Percentage Recognition Score

Tables 7.1 and 7.2 show the comparative results of the four classifiers on different data sets during training and testing, respectively. As can be seen from Table 7.1, except for *Mango* data, the *CEMOGA-classifier* uses a smaller

Table 7.2. Test results on various data sets

	VGA		NSGAII		PAES		CEMOGA	
	ClassAcc.	Score%	ClassAcc.	Score%	ClassAcc.	Score%	ClassAcc.	Score%
Vowel	0.07	68.51	0.06	74.48	0.07	69.19	0.15	79.08
Iris	0.76	93.33	0.83	94.67	0.86	94.67	0.89	96.00
Mango	0.00	55.42	0.00	60.24	0.00	63.85	0.28	66.26

number of hyperplanes than the other classifiers for approximating the class boundaries of the remaining two data sets [42]. The use of a smaller number of hyperplanes indicates the good performance of *CEMOGA-classifier* since this is likely to result in an improvement of its generalization capability. This is supported by the results provided in Table 7.2, where it is seen that the percentage recognition scores of the *CEMOGA-classifier* is better than those of the other classifiers for all the data sets.

It is interesting to note that although the use of a smaller number of hyperplanes by the *CEMOGA-classifier*, in general, resulted in best performance during testing (Table 7.2), this may not be necessarily the case during training (Table 7.1). In general, using a large number of hyperplanes for approximating the class boundaries is likely to lead to overfitting of the data, thereby leading to better performance during training (better abstraction), but poorer generalization capability. This observation is reflected in Table 7.1 where it is found that the training score of the *CEMOGA-classifier* is, in general, not the best in comparison to the other classifiers for the different data sets. Only for *Iris* is its training performance best (where some of the other classifiers are also found to provide the same score). These results were picked up as the solution to the classification problem from a set of nondominated solutions because they provided better values of the validity function during the validation phase. In fact, this is one of the major issues for developing a multiobjective classifier, since in a single-objective classifier (e.g., *VGA-classifier*), because of the combination of the different criteria into one, one is stuck with a single best solution (with respect to the combined criteria). This reduces the flexibility by eliminating the possibility of looking at different solutions along the trade-off (Pareto-optimal) surface. Moreover, retaining a set of nondominated solutions in the multiobjective classifiers also increases the diversity in the process, since now only a single best chromosome is not able to take over the entire or a large portion of the population, as often happens in the single-objective case.

The only departure with regard to the number of hyperplanes is observed for *Mango*, where *CEMOGA-classifier* uses six hyperplanes, as against two and three used by *VGA-classifier* and *NSGAII-classifier*, respectively. However, it can be seen from Table 7.2 that both the latter two classifiers actually recognize only two of the three classes, and consequently their *ClassAccuracy* values are zero. In contrast, the *CEMOGA-classifier* uses six hyperplanes to recognize all the three classes to a reasonable extent, thereby providing a *ClassAccuracy* value of 0.28. Tables 7.3–7.6, which show the confusion matrix for all the four classifiers during testing, also demonstrate this observation. As can be seen from the tables, only *CEMOGA-classifier* is able to recognize, to a large extent, all the three classes. In contrast, the *VGA-classifier* fails to recognize class 2, while the *NSGAII-classifier* and *PAES-classifier* fail to recognize class 3. In fact, in both these cases it is found that there are no patterns that are either correctly or incorrectly classified to the respective classes (as evident from the confusion matrices where the respective columns have all 0's). Moreover, it was revealed under investigation that for *PAES-classifier*, one point from class 2 was left unclassified (this may happen since this point may lie in a region from where there was no training point, and hence this region was not labelled). This is reflected in Table 7.5 where the sum of the corresponding second row equals 23, while for the other classifiers this value equals 24. Since the *CEMOGA-classifier* uses six hyperplanes to distinguish all the three classes, it has better generalization capability. In fact, this is the characteristic for which the corresponding solution consequently got selected during the validation phase.

Table 7.3. Confusion matrix during testing for Mango with *VGA-classifier*

Actual	Computed Class			
Class	C1	C2	C3	Score%
C1	38	0	3	92.68
C2	9	0	15	0.00
C3	10	0	8	44.44

Table 7.4. Confusion matrix during testing for Mango with *NSGAII-classifier*

Actual	Computed Class			
Class	C1	C2	C3	Score%
C1	34	7	0	82.93
C2	8	16	0	66.67
C3	6	12	0	0.00

Comparison with Respect to User's Accuracy and Kappa

Here we provide comparative results on two more statistical measures, namely user's accuracy and kappa [97, 398].

Table 7.5. Confusion matrix during testing for Mango with *PAES-classifier*

Actual	Computed Class			
Class	C1	C2	C3	Score%
C1	37	4	0	90.24
C2	8	15	0	65.21
C3	8	10	0	0.00

Table 7.6. Confusion matrix during testing for Mango with *CEMOGA-classifier*

Actual	Computed Class			
Class	C1	C2	C3	Score%
C1	37	1	3	90.24
C2	10	12	2	50.00
C3	10	2	6	33.33

User's accuracy: User accuracy (U_a) [398] is a measure of confidence that a classifier attributes to a region as belonging to a class. In other words, it is a measure of purity associated with a region specified by a classifier. It is defined as

$$U_a = n_{i_c}/n_i^p, \qquad (7.12)$$

where n_{i_c} is defined earlier in Eq. (7.5) and n_i^p is the number of points classified into class i.

Kappa: The coefficient of agreement, *kappa* [398] measures the relationship of beyond chance agreement to expected disagreement. It uses all the cells of the confusion matrix. The estimate of kappa is the proportion of the agreement after chance agreement is removed from consideration rather than just the diagonal cells as is done while computing the recognition scores and user's accuracy. The estimate of kappa for class i, κ_i, is defined as

$$\kappa_i = \frac{n * n_{i_c} - n_i * n_i^p}{n * n_i^p - n_i * n_i^p}. \qquad (7.13)$$

The numerator and denominator of the overall kappa are obtained by summing the respective numerators and denominators of κ_i separately over all classes i.

Table 7.7 shows the overall kappa values for all the classifiers for different data sets. As is evident from the table, the *CEMOGA-classifier* consistently provides the best kappa values for all the data sets, once again highlighting its superior performance. In order to indicate the classwise user's accuracy and kappa values, we have chosen the *Vowel* data set as an example, since it has been used extensively in other experiments with the genetic classifiers. The results are shown in Table 7.8. Both the user's accuracy and kappa values are, in general, relatively poorer for class 1 (δ) for all the classifiers. The reason for this is evident from Fig. 3.8, where it is found that this class is largely overlapped with its neighboring classes.

Table 7.7. Overall kappa values during testing

Data Set	VGA-classifier	NSGAII-classifier	PAES-classifier	CEMOGA-classifier
Vowel	0.636	0.664	0.635	0.731
Iris	0.859	0.919	0.919	0.939
Mango	0.248	0.328	0.367	0.418

Table 7.8. User accuracy and kappa measure for *Vowel* data during testing

Class	VGA		NSGAII		PAES		CEMOGA	
	Users' Acc	kappa	Users' Acc	kappa	Users' Acc	kappa	Users' Acc	kappa
1	0.472	0.423	0.653	0.599	0.375	0.315	0.535	0.492
2	0.741	0.709	0.607	0.561	0.760	0.733	0.689	0.653
3	0.859	0.823	0.701	0.627	0.983	0.979	0.948	0.935
4	0.651	0.575	0.945	0.933	0.730	0.674	0.723	0.664
5	0.667	0.561	0.793	0.723	0.674	0.563	0.840	0.786
6	0.841	0.806	0.659	0.583	0.662	0.587	0.763	0.710

Comparison with Respect to Nondominated Solutions

In this section, we compare the performance of the three MOO technique based classifiers on the basis of the quality of nondominated solutions provided by them. This is measured with two indices, viz., *purity* of solutions and their *minimal spacing*, S_m, which are described in Sect. 7.6.

Table 7.9 demonstrates the number of nondominated solutions and the corresponding purity values obtained by the three multiobjective genetic classifiers for the three data sets. Interestingly, in all the cases, *CEMOGA-classifier* attains a purity value of 1, indicating that all the nondominated solutions obtained by it are indeed nondominated even when the solutions of all the three algorithms are combined. On the other hand, almost none of the solutions obtained using *PAES-classifier* were found to be actually nondominated in the combined scenario, while *NSGAII-classifier* lies some where in between. This shows that the latter two classifiers are unable to come up with an effective set of solutions. In this regard it may be noted that *CEMOGA-classifier*, in general, finds a smaller number of better nondominated solutions than the other two algorithms.

For the purpose of demonstration, the nondominated solutions obtained using these three algorithms are shown in Fig. 7.4 for *Vowel* data. Here "*",

Table 7.9. Purity measure on various data sets

| | NSGAII | | PAES | | CEMOGA | |
	# solutions	Purity	# solutions	Purity	# solutions	Purity
Vowel	15	0.33	15	0.00	7	1.00
Iris	2	1.00	4	0.00	1	1.00
Mango	2	1.00	3	0.33	4	1.00

"o" and "+" denote the nondominated solutions provided by the *CEMOGA-classifier*, *PAES-classifier* and *NSGAII-classifier*, respectively. Note that when the solutions of all the three classifiers are combined and they are re-ranked it is found that all the *CEMOGA-classifier* solutions actually belong to rank 1, none of the *PAES-classifier* solutions belong to rank 1, while five of the *NSGAII-classifier* solutions (encircled in the figure as ⊕ for convenience) belong to rank 1. The inability of the *PAES-classifier* in evolving a good set of solutions may be due to its strictly local nature. This precludes situations where a good solution may be obtained after crossing a barrier of bad solutions, something that may arise frequently in real-life problems.

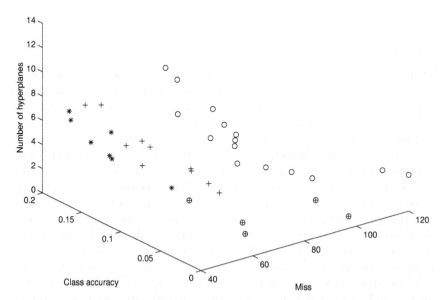

Fig. 7.4. Nondominated solutions obtained by *CEMOGA-classifier* (*), *PAES-classifier* (o) and *NSGAII-classifier* (+). Actual rank 1 solutions for *NSGAII-classifier* are encircled

Table 7.10 gives the values of minimal spacing (S_m) obtained by the three methods. This is reported for both the cases: one when all the respective solutions are considered, and second when only actual rank-1 solutions (obtained after combination of the solutions) are considered. The result in the second category is denoted by S_m (R1) in the table. Interestingly, the two values are exactly the same for *CEMOGA-classifier*, for all the data sets, since all the solutions obtained by this classifier actually belong to rank 1, as indicated by the purity values (Table 7.9). As can be noted from Table 7.10, in several cases (denoted by "-") it was not possible (or meaningful) to calculate S_m, since the number of solutions was less than or equal to two. For *Vowel*, where it was possible to compute this value for two classifiers, *CEMOGA-classifier* provided significantly better performance (i.e., smaller value). Note that for *PAES-classifier*, it was not possible to compute S_m (R1) values for any of the data sets, since in all the cases the purity values are 0.

Table 7.10. Minimal spacing (S_m) measure on various data sets

	NSGAII		PAES		CEMOGA	
	S_m	S_m (R1)	S_m	S_m (R1)	S_m	S_m (R1)
Vowel	0.088	0.117	0.084	-	0.042	0.042
Iris	-	-	0.474	-	-	-
Mango	-	-	0.089	-	0.056	0.056

7.8 Summary

In this chapter, multiobjective variable string length genetic algorithms have been used for extending the design of the *VGA-classifier* described earlier. Three objective criteria, *miss* (total number of misclassified samples), *H* (number of hyperplanes) and *ClassAccuracy* (product of the classwise correct recognition rates), are used as the optimizing criteria. While the first two criteria need to be minimized, the third one is to be maximized simultaneously. Three multiobjective optimization techniques, viz., NSGA-II, PAES and CEMOGA, have been used for developing the classifiers. Unlike NSGA-II, CEMOGA uses some domain-specific constraints in the process, which helps in making the search more efficient. Since the multiobjective classifiers, in general, provide a set of nondominated solutions, a validation phase is used after training in order to select one of the chromosomes (which provided the largest value of the validation function) to be used for testing.

Comparison of the performance of *CEMOGA-classifier*, *NSGAII-classifier* and *PAES-classifier* is made for three real-life data sets, with respect to both the number of required hyperplanes and classification performance. It is found that, in general, the *CEMOGA-classifier* is able to approximate the class boundaries using a smaller number of hyperplanes, thereby providing superior generalization capability. The only point of departure was for the *Mango* data, where although the remaining classifiers required smaller number of hyperplanes as compared to the *CEMOGA-classifier*, they were unable to recognize one of the three classes. In contrast, only the *CEMOGA-classifier* was able to recognize all the three classes, and thus provided superior results during testing.

The nondominated solutions obtained by the three MOO algorithms, NSGA-II, PAES and CEMOGA, are also compared with respect to the position of the corresponding fronts as well as the diversity of the solutions. For the former, a measure, called *purity*, has been used that computes the fraction of solutions that remain nondominated when the solutions resulting from several MOO techniques are combined. In order to measure the diversity, or spread, of the solutions on the Pareto-optimal front, an index called *minimal spacing* has been adopted. This is an improved version of an existing measure called *spacing*. Interestingly, it is found that the CEMOGA consistently provides a *purity* value of 1 for all the data sets, NSGA-II does so for only two of the six data sets, while PAES is unable to do so even once. With respect to *minimal spacing*, in general, CEMOGA again outperforms the other MOO strategies.

8

Genetic Algorithms in Clustering

8.1 Introduction

The earlier chapters dealt with the application of GAs for designing several nonparametric classifiers by approximating the class boundaries of a given data set using a number of hyperplanes. These require the availability of a set of training data points, each associated with a class label. When the data available are unlabelled, the classification problem is referred to as *unsupervised classification*, or *clustering*.

The purpose of any clustering technique is to evolve a $K \times n$ partition matrix $U(X)$ of a data set X ($X = \{\mathbf{x}_1, \mathbf{x}_2, \ldots, \mathbf{x}_n\}$) in $I\!\!R^N$, representing its partitioning into a number, say K, of clusters ($C_1, C_2, \ldots, C_K$). The partition matrix $U(X)$ may be represented as $U = [u_{kj}]$, $k = 1, \ldots, K$ and $j = 1, \ldots, n$, where u_{kj} is the membership of pattern $\mathbf{x}_j$ to cluster C_k. Clustering techniques fall into two broad classes, partitional and hierarchical. K-means and single linkage [212, 461] are widely used techniques in these two domains, respectively. Clustering methods are also often categorized as crisp and fuzzy. In crisp clustering, a point belongs to exactly one cluster, while in fuzzy clustering, a point can belong to more than one cluster with varying degrees of membership.

The two fundamental questions that need to be addressed in any typical clustering system are: (i) How many clusters are actually present in the data, and (ii) how real or good is the clustering itself. That is, whatever may be the clustering method, one has to determine the number of clusters and also the goodness or validity of the clusters formed. The measure of validity of the clusters should be such that it will be able to impose an ordering of the clusters in terms of its goodness. In other words, if $U_1, U_2, \ldots, U_m$ be m partition matrices of X, and the corresponding values of a validity measure be $V_1, V_2, \ldots, V_m$, then $V_{k1} >= V_{k2} >= \ldots >= V_{km}$ will indicate that $U_{k1} \uparrow U_{k2} \uparrow \ldots \uparrow U_{km}$, for some permutation $k1, k2, \ldots, km$ of $\{1, 2, \ldots, m\}$. Here "$U_i \uparrow U_j$" indicates that partitioning represented by U_i is better than that represented by U_j.

It may be noted that the different clustering methods, in general, try to optimize some measure of goodness of a clustering solution either explicitly or implicitly. For example, the K-means algorithm performs a local optimization of the J measure that is defined later. Moreover, often the number of clusters in a data set is not known. The clustering problem then reduces to one of searching for an appropriate number of suitable partitions such that some goodness measure is optimized. It may be noted that searching the exhaustive set of all possible partitions of the data is prohibitively exhaustive since the search space is huge and complex, with numerous local optima. Consequently, heuristic search methods are often employed in this domain, and these often get stuck at local optima. Several researchers have also been actively investigating the effectiveness of GAs for clustering, since GAs are well-known for solving such complex search and optimization problems.

In the present chapter, a few such genetically guided clustering techniques have been described in detail where the issues of local optima and determining the optimal clusters are addressed [27, 28, 31, 34, 298, 299, 300, 339]. These include crisp clustering into a fixed number of clusters, crisp clustering into a variable number of clusters and fuzzy clustering into a variable number of clusters.

Section 8.2 provides the basic concepts and definitions in clustering. Some commonly used clustering techniques are described in Sect. 8.3. Section 8.4 describes how GAs can be applied to solve the problem of crisp clustering into a fixed number of clusters. Sections 8.5 and 8.6 deal with applications of GAs for clustering into a variable number of clusters, both crisp and fuzzy, respectively. Finally, Sect. 8.7 summarizes this chapter.

8.2 Basic Concepts and Preliminary Definitions

Clustering in N-dimensional Euclidean space $I\!\!R^N$ is the process of partitioning a given set of n points into a number, say K, of groups (or, clusters) based on some similarity/dissimilarity metric. Let the set of n points $\{x_1, x_2, \ldots, x_n\}$ be represented by the set X, and the K clusters be represented by $C_1, C_2, \ldots, C_K$. A partitioning of the points can be represented as follows:

$$
\begin{aligned}
n > \sum_{j=1}^{n} u_{kj} &\geq 1, \quad \text{for } k = 1, \ldots, K, \\
\sum_{k=1}^{K} u_{kj} &= 1, \quad \text{for } j = 1, \ldots, n \text{ and,} \\
\sum_{k=1}^{K} \sum_{j=1}^{n} u_{kj} &= n.
\end{aligned} \tag{8.1}
$$

For crisp clustering, $u_{kj} = 1$ if point x_j belongs to cluster C_k, and 0 otherwise. For fuzzy clustering, u_{kj}, $0 \leq u_{kj} \leq 1$, denotes the degree of belongingness of the point x_j to cluster C_k.

Clustering techniques may be hierarchical and nonhierarchical. A well-known method in the nonhierarchical category is the K-means algorithm [461], which is based on the minimization of the mean squared error of the clusters. From a statistical point of view, the data can be considered to be a mixture of different distributions. Estimating the parameters of these distributions is another approach to clustering, and the Expectation Maximization algorithm [69] is a widely used technique in this class. Other methods in the nonhierarchical category include some graph-theoretic methods like the minimal spanning tree-based clustering [461]. Single linkage, complete linkage and average linkage are some popular methods of hierarchical clustering approaches. In these methods, a valid clustering is obtained at each level of the algorithm, which proceeds in a hierarchy that can be either top-down or bottom-up. Good reviews of clustering methods are available in [214, 404].

The K-means algorithm is known to be sensitive to outliers, since such points can significantly affect the computation of the centroids, and hence the resultant partitioning. K-medoid attempts to alleviate this problem by using the medoid, the most centrally located object, as the representative of the cluster. Partitioning around medoid (PAM) [234] was one of the earliest K-medoid algorithms introduced. PAM finds K clusters by first finding a representative object for each cluster, the medoid. The algorithm then repeatedly tries to make a better choice of medoids by analyzing all possible pairs of objects such that one object is a medoid and the other is not. PAM is computationally quite inefficient for large data sets and a large number of clusters. The Clustering LARge Applications (CLARA) algorithm was proposed by the same authors [234] to tackle this problem. CLARA is based on data sampling, where only a small portion of the real data is chosen as a representative of the data, and medoids are chosen from this sample using PAM. CLARA draws multiple samples and outputs the best clustering from these samples. As expected, CLARA can deal with larger data sets than PAM. However, if the best set of medoids is never chosen in any of the data samples, CLARA will never find the best clustering. Ng and Han [327] proposed the CLARANS algorithm, which tries to mix both PAM and CLARA by searching only the subset of the data set. However, unlike CLARA, CLARANS does not confine itself to any sample at any given time, but draws it randomly at each step of the algorithm. Based upon CLARANS, two spatial data mining algorithms, the spatial dominant approach, SD(CLARANS), and the nonspatial dominant approach, NSD(CLARANS), were developed. In order to make CLARANS applicable to large data sets, use of efficient spatial access methods, such as R*-tree, was proposed [138]. CLARANS had a limitation that it could provide good clustering only when the clusters were mostly equisized and convex. DBSCAN [137], another popularly used density clustering technique that was proposed by Ester et al., could handle nonconvex and nonuniformly-sized clusters. Balanced Iterative Reducing and Clustering using Hierarchies (BIRCH), proposed by Zhang et al. [515], is another algorithm for clustering large data sets. It uses two concepts, the clustering feature and the clustering feature tree, to summarize cluster representations which help the method achieve good speed and scalability in large databases. Discussion on several other clustering algorithms may be found in [183].

8.3 Clustering Algorithms

As already mentioned, clustering techniques can be categorized in several ways: crisp/fuzzy and nonhierarchical (partitional)/hierarchical. In this section, three well-known techniques, K-means, single linkage and fuzzy c-means, are described. (It is to be noted that we have used the names K-means, where K denotes the number of clusters, and fuzzy c-means, where c denotes the number of clusters, on purpose, since the algorithms are well-known by these names.)

8.3.1 K-Means Clustering Algorithm

The K-means algorithm [212, 461] is an iterative scheme that evolves K crisp, compact and hyperspherical clusters in the data such that a measure

$$J = \sum_{j=1}^{n} \sum_{k=1}^{K} u_{kj} ||\mathbf{x}_j - \mathbf{z}_k||^2 \tag{8.2}$$

is minimized. Here, u_{kj} is equal to 1 if the jth point belongs to cluster k, and 0 otherwise; $\mathbf{z}_k$ denotes the center of cluster k, and $\mathbf{x}_j$ denotes the jth point of the data. In K-means, the K cluster centers are usually initialized to K randomly chosen points from the data, which is then partitioned based on the minimum squared distance criterion. The cluster centers are subsequently updated to the mean of the points belonging to them. This process of partitioning followed by updating is repeated until one of the following becomes true: (i) the cluster centers do not change, (ii) the J value becomes smaller than a small threshold or (iii) maximum number of iterations have been executed. The different steps of the K-means algorithm are given below in detail.

Step 1:
  Choose K initial cluster centers $\mathbf{z}_1, \mathbf{z}_2, \ldots, \mathbf{z}_K$ randomly from the n points $\{\mathbf{x}_1, \mathbf{x}_2, \ldots, \mathbf{x}_n\}$.

Step 2:
  Assign point $\mathbf{x}_i$, $i = 1, 2, \ldots, n$ to cluster C_j, $j \in \{1, 2, \ldots, K\}$ iff

$$||\mathbf{x}_i - \mathbf{z}_j|| < ||\mathbf{x}_i - \mathbf{z}_p||, \quad p = 1, 2, \ldots, K, \text{ and } j \neq p. \tag{8.3}$$

  Ties are resolved arbitrarily.

Step 3:
  Compute new cluster centers $\mathbf{z}_1^*, \mathbf{z}_2^*, \ldots, \mathbf{z}_K^*$ as follows:

$$\mathbf{z}_i^* = \frac{1}{n_i} \Sigma_{\mathbf{x}_j \in C_i} \mathbf{x}_j, \quad i = 1, 2, \ldots, K, \tag{8.4}$$

where n_i is the number of elements belonging to cluster C_i.

Step 4:
   If $\mathbf{z}_i^* = \mathbf{z}_i$, $i = 1, 2, \ldots, K$ then terminate. Otherwise continue from step 2.

In case the process does not terminate at step 4 normally, then it is executed for a maximum fixed number of iterations. The rules for update of the cluster centers are obtained by differentiating J with respect to the centers, and equating the differential to zero. This analysis is as follows. The essential purpose here is to minimize J as defined in Eq. (8.2). Equating its derivative with respect to $\mathbf{z}_k$ to zero, we get

$$\frac{dJ}{d\mathbf{z}_k} = 2 \sum_{j=1}^{n} u_{kj}(\mathbf{x}_j - \mathbf{z}_k)(-1) = 0, \quad k = 1, 2, \ldots, K. \tag{8.5}$$

Or,

$$\sum_{j=1}^{n} u_{kj}\mathbf{x}_j - \mathbf{z}_k \sum_{j=1}^{n} u_{kj} = 0. \tag{8.6}$$

Or,

$$\mathbf{z}_k = \frac{\sum_{j=1}^{n} u_{kj}\mathbf{x}_j}{\sum_{j=1}^{n} u_{kj}}. \tag{8.7}$$

Note that in crisp clustering $\sum_{j=1}^{n} u_{kj}$ is nothing but the number of elements belonging to cluster k, i.e., n_k. Hence the update equation boils down to the one defined in Eq. (8.4). In order to ensure that this indeed provides a minima of J, the second derivative, $\frac{d^2 J}{d\mathbf{z}_k^2}$ is computed. This is given by

$$\frac{d^2 J}{d\mathbf{z}_k^2} = 2 \sum_{j=1}^{n} u_{kj}. \tag{8.8}$$

Note that the r.h.s of the above equation is positive, indicating that using the update equation will indeed result in the minimum value of J.
   It has been shown in [425] that K-means algorithm may converge to values that are not optimal. Also global solutions of large problems cannot be found with a reasonable amount of computation effort [441]. Moreover, this technique assumes that the clusters are hyperspherical in shape and more or less equal in size. It is because of these factors that several approximate methods are developed to solve the underlying optimization problem. One such method using GAs is described later on in this chapter.

8.3.2 Single-Linkage Clustering Algorithm

The single-linkage clustering scheme is a noniterative method based on a local connectivity criterion [212]. Instead of an object data set X, single linkage processes sets of n^2 numerical relationships, say $\{r_{jk}\}$, between pairs of objects represented by the data. The value r_{jk} represents the extent to which object j and k are related in the sense of some binary relation ρ. It starts by considering each point in a cluster of its own. The single-linkage algorithm computes the distance between two clusters C_i and C_j as

$$\delta_{SL}(C_i, C_j) = \underbrace{\min_{x \in C_i, y \in C_j}}\ \{d(x, y)\}, \tag{8.9}$$

where $d(x, y)$ is some distance measure defined between objects x and y. Based on these distances, it merges the two closest clusters, replacing them by the merged cluster. The distance of the remaining clusters from the merged one is recomputed as above. The process continues until a single cluster, comprising all the points, is formed. The advantages of this algorithm are that (i) it is independent of the shape of the cluster, and (ii) it works for any kind of attributes, both categorical and numeric, as long as a similarity of the data objects can be defined. However, the disadvantages of this method are its computational complexity and its inability to handle overlapping classes.

8.3.3 Fuzzy c-Means Clustering Algorithm

Fuzzy c-means is a widely used technique that uses the principles of fuzzy sets to evolve a fuzzy partition matrix for a given data set. The set of all $c \times n$, where c is equal to the number of clusters, nondegenerate constrained fuzzy partition matrices, denoted by M_{fcn}, is defined as

$$M_{fcn} = \{U \in R^{c \times n}\ \mid\ \textstyle\sum_{i=1}^{c} u_{ik} = 1,\quad \sum_{k=1}^{n} u_{ik} > 0,\ \forall i\ \text{and} \\ u_{ik} \in [0, 1]; 1 \le i \le c; 1 \le k \le n\}. \tag{8.10}$$

The minimizing criterion used to define good clusters for fuzzy c-means partitions is the FCM function, defined as

$$J_\mu(U, Z) = \sum_{i=1}^{c} \sum_{k=1}^{n} (u_{ik})^\mu D^2(\mathbf{z}_i, \mathbf{x}_k). \tag{8.11}$$

Here $U \in M_{fcn}$ is a fuzzy partition matrix; $\mu \in [1, \infty]$ is the weighting exponent on each fuzzy membership; $Z = [\mathbf{z}_1, \ldots, \mathbf{z}_c]$ represents c cluster centers; $\mathbf{z}_i \in I\!\!R^N$; and $D(\mathbf{z}_i, \mathbf{x}_k)$ is the distance of $\mathbf{x}_k$ from the ith cluster center. The fuzzy c-means theorem [57] states that if $S_{ik} > 0$, for all i and k, then (U, Z) may minimize J_μ, only if when $\mu > 1$

$$u_{ik} = \frac{1}{\sum_{j=1}^{c} \left(\frac{S_{ik}}{S_{jk}}\right)^{\frac{2}{\mu-1}}}, \quad \text{for}\ \ 1 \le i \le c,\ \ 1 \le k \le n, \tag{8.12}$$

and

$$\mathbf{z}_i = \frac{\sum_{k=1}^{n}(u_{ik})^{\mu}\mathbf{x}_k}{\sum_{k=1}^{n}(u_{ik})^{\mu}}, \quad 1 \leq i \leq c. \tag{8.13}$$

A common strategy for generating the approximate solutions of the minimization problem in Eq. (8.11) is by iterating through Eqs. (8.12) and (8.13) (also known as the Picard iteration technique). A detailed description of the FCM algorithm may be found in [57].

Note that in fuzzy clustering, although the final output is generally a crisp clustering, the users are free to still utilize the information contained in the partition matrix. The FCM algorithm shares the problems of the K-means algorithm in that it also gets stuck at local optima depending on the choice of the initial clusters, and requires the number of clusters to be specified a priori.

The following sections describe how the search capability of GAs can be utilized for clustering a given data set. The problem of crisp clustering into fixed number of clusters is tackled first, followed by that of clustering into a variable number of clusters. Thereafter, the genetic fuzzy clustering technique is described in detail.

8.4 Clustering Using GAs: Fixed Number of Crisp Clusters

An application of GA for clustering, feature selection and classification is reported in [465]. An early application of GAs for clustering using an ordered representation was reported in [63]. Murthy and Chowdhury [325] have used GAs for finding optimal clusters, without the need for searching all possible clusters. They used a point-based encoding of the chromosomes as illustrated in the example below:

Example 8.1. Consider a set of 8 points, $\mathbf{x}_1, \mathbf{x}_2, \ldots, \mathbf{x}_8$, to be clustered into 3 groups. Then l_i, the length of chromosome i, $= 8$. Let a chromosome be 23231132. This indicates a grouping of
Cluster 1 $\rightarrow (\mathbf{x}_5, \mathbf{x}_6)$,
Cluster 2 $\rightarrow (\mathbf{x}_1, \mathbf{x}_3, \mathbf{x}_8)$,
Cluster 3 $\rightarrow (\mathbf{x}_2, \mathbf{x}_4, \mathbf{x}_7)$.

The K-means measure Eq. (8.2) was used as the fitness function in [325]. The experimental results show that the GA-based scheme may improve the final output of the K-means algorithm, where an improvement is possible.

In [298] it is pointed out that for the encoding used in [325], it may take too long for the algorithm to converge even for moderately large data sets, given that the length of the chromosome is equal to the size of the data. This is demonstrated experimentally. Consequently a centroid-based real encoding of the chromosomes is proposed [298]. This technique, referred to as *GA-clustering*, is now described in detail below.

8.4.1 Encoding Strategy

Each string is a sequence of real numbers representing the K cluster centers. For an N-dimensional space, the length of a chromosome is $N * K$, where the first N positions (or genes) represent the N dimensions of the first cluster center, the next N positions represent those of the second cluster center, and so on. As an illustration, consider the following example.

Example 8.2. Let $N = 2$ and $K = 3$, i.e., the space is two dimensional and the number of clusters being considered is three. Then the chromosome

$$51.6 \; 72.3 \quad 18.3 \; 15.7 \quad 29.1 \; 32.2$$

represents the three cluster centers (51.6,72.3), (18.3,15.7) and (29.1,32.2). Note that each real number in the chromosome is an indivisible gene.

8.4.2 Population Initialization

The K cluster centers encoded in each chromosome are initialized to K randomly chosen points from the data set. This process is repeated for each of the P chromosomes in the population, where P is the size of the population.

8.4.3 Fitness Computation

The fitness computation process consists of two phases. In the first phase, the clusters are formed according to the centers encoded in the chromosome under consideration. This is done by assigning each point $\mathbf{x}_i$, $i = 1, 2, \ldots, n$, to one of the clusters C_j with center $\mathbf{z}_j$ such that

$$\|\mathbf{x}_i - \mathbf{z}_j\| \leq \|\mathbf{x}_i - \mathbf{z}_p\|, \quad p = 1, 2, \ldots, K, \text{ and } p \neq j.$$

All ties are resolved arbitrarily. After the clustering is done, the cluster centers encoded in the chromosome are replaced by the mean points of the respective clusters. In other words, for cluster C_i, the new center $\mathbf{z}_i^*$ is computed as

$$\mathbf{z}_i^* = \frac{1}{n_i} \Sigma_{\mathbf{x}_j \in C_i} \mathbf{x}_j, \quad i = 1, 2, \ldots, K.$$

These $\mathbf{z}_i^*$'s now replace the previous $\mathbf{z}_i$'s in the chromosome. As an illustration, consider the following example.

Example 8.3. The first cluster center in the chromosome considered in Example 8.2 is (51.6, 72.3). With (51.6, 72.3) as center, let the resulting cluster contain two more points, viz., (50.0, 70.0) and (52.0, 74.0) besides itself, i.e., (51.6, 72.3). Hence the newly computed cluster center becomes $\left(\frac{50.0+52.0+51.6}{3}, \frac{70.0+74.0+72.3}{3}\right) = (51.2, 72.1)$. The new cluster center (51.2, 72.1) now replaces the previous value of (51.6, 72.3) in the chromosome.

Subsequently, the clustering metric J is computed as in Eq. (8.2). The clustering metric of the clusters associated with a chromosome has to be minimized for this problem. Therefore, the fitness function is defined as $f = \frac{1}{J}$. Maximization of f indicates minimization of J. Note that there are several other ways in which the fitness function could have been defined.

8.4.4 Genetic Operators

Roulette wheel selection that implements the proportional selection strategy is adopted. Single-point crossover with a fixed crossover probability of μ_c is used.

Each chromosome undergoes mutation with a fixed probability μ_m. Since floating point representation is considered, a different form of mutation is used. A number δ in the range $[0, 1]$ is generated with uniform distribution. If the value at a gene position is v, after mutation it becomes

$$v \pm 2 * \delta * v.$$

If $v = 0$, then after mutation it becomes $\pm 2 * \delta$. The "+" or "−" sign occurs with equal probability. Other forms like

$$v \pm (\delta + \epsilon) * v,$$

where $0 < \epsilon < 1$ may also be used for this purpose.

The processes of fitness computation, selection, crossover and mutation are executed for a maximum number of iterations. The best string seen up to the last generation provides the solution to the clustering problem. Elitism has been implemented at each generation by preserving the best string seen up to that generation in a location outside the population. Thus on termination, this location contains the centers of the final clusters.

8.4.5 Experimental Results

The experimental results comparing the *GA-clustering* algorithm with the K-means technique are provided for two artificial data sets *Data_2_2* and *Data_3_2*, and two real-life data sets *Iris* and *Cancer*. The data sets are described in detail in Appendix B and are briefly mentioned here for convenience.
Data_2_2: is a two-dimensional data set where the number of clusters is two. It has ten points.
Data_3_2: is a two-dimensional data set where the number of clusters is three. It has 76 points. Figure 8.1 shows the data set.
Iris: This is a 4-dimensional data with 150 points in 3 classes.
Cancer: This is a 9-dimensional data with 683 points in 2 classes.

GA-clustering is implemented with the following parameters: $\mu_c = 0.8$ and $\mu_m = 0.001$. The population size P is taken to be 10 for *Data_2_2*, since it

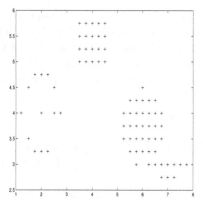

Fig. 8.1. *Data_3_2*

is a very simple data set, while it is taken to be 100 for *Data_3_2*, *Iris* and *Cancer*. Note that it is shown in [8, 441] that if exhaustive enumeration is used to solve a clustering problem with n points and K clusters, then one is required to evaluate

$$\frac{1}{K}\Sigma_{j=1}^{K}(-1)^{K-j}j^n$$

partitions. For a data set of size 10 with 2 clusters, this value is $2^9 - 1$ (= 511), while that of size 50 with 2 clusters is $2^{49} - 1$ (i.e., of the order of 10^{15}).

For K-means algorithm a maximum of 1000 iterations is fixed in case it does not terminate normally. However, in all the experiments the K-means algorithm terminated much before 1000 iterations.

The results of implementation of the K-means algorithm and *GA-clustering* algorithm are shown in Tables 8.1–8.3 for *Data_2_2*, *Data_3_2*, *Iris* and *Cancer*, respectively. Both the algorithms were executed for 100 simulations. For the purpose of demonstration, results of five different initial configurations of the K-means algorithm and five different initial populations of the *GA-clustering* algorithm are shown in the tables.

For *Data_2_2* (Table 8.1) it is found that the *GA-clustering* algorithm provides the optimal value of 2.225498 in all the runs. The K-means algorithm also attains this value most of the times (87% of the total runs). However, in the other cases, it gets stuck at a value of 5.383132. Similarly for *Data_3_2* (Table 8.2), *GA-clustering* attains the best value of 51.013294 in all the runs. K-means, on the other hand, attains this value in 51% of the total runs, while in other runs it gets stuck at different suboptimal values.

For *Iris* (Table 8.3), the *GA-clustering* algorithm again attains the best value of 97.10077 in all the runs. The K-means algorithm, on the other hand, fails to attain this value in any of its runs. The best that the K-means algorithm could achieve was 97.204574, which it reached 60% of the times. For *Cancer Data* (Table 8.4), the K-means algorithm attains the best value of 19323.173817 in 20 out of 100 runs (i.e., 20% of the runs), while in other

Table 8.1. J value obtained by K-means and *GA-clustering* algorithms for five different runs for *Data_2_2* when K=2

Run	K-means	GA-cluster
1	5.383132	2.225498
2	2.225498	2.225498
3	2.225498	2.225498
4	5.383132	2.225498
5	2.225498	2.225498

Table 8.2. J value obtained by K-means and *GA-clustering* algorithms for five different runs for *Data_3_2* when K=3

Run	K-means	GA-cluster
1	51.013294	51.013294
2	64.646739	51.013294
3	67.166768	51.013294
4	51.013294	51.013294
5	64.725676	51.013294

Table 8.3. J value obtained by K-means and *GA-clustering* algorithms for five different runs for *Iris* when K=3

Run	K-means	GA-cluster
1	97.224869	97.10077
2	97.204574	97.10077
3	122.946353	97.10077
4	124.022373	97.10077
5	97.204574	97.10077

Table 8.4. J value obtained by K-means and *GA-clustering* algorithms for five different runs for *Cancer* when K=2

Run	K-means	GA-cluster
1	19323.204900	19323.173817
2	19323.204900	19323.173817
3	19323.204900	19323.173817
4	19323.173817	19323.173817
5	19323.204900	19323.173817

cases it gets stuck at a value of 19323.204900. The *GA-clustering* algorithm, in contrast, attains the best value in all the 100 runs.

8.5 Clustering Using GAs: Variable Number of Crisp Clusters

In most real-life situations the number of clusters in a data set is not known beforehand. The real challenge is to be able to automatically evolve a proper value of the number of clusters and to provide the appropriate clustering under this circumstance. Some attempts in this regard can be found in [392, 461]. The ISODATA algorithm [461] uses a combination of splitting, merging and deleting clusters to adjust the number of cluster centers. Each of these operations depends on several user supplied parameters, which are often very difficult to estimate a priori. Ravi and Gowda [392] used a distributed genetic algorithm based on the ISODATA technique for clustering symbolic objects. However this method also suffers from the same limitations as present in the ISODATA clustering technique. Sarkar et al. [413] have proposed a clustering algorithm using evolutionary programming-based approach, when the clusters are crisp and spherical. The algorithm automatically determines the number and the center of the clusters, and is not critically dependent on the choice of the initial cluster.

In [31], a variable string length genetic algorithm (VGA) [168], with real encoding of the cluster centers in the chromosome [298], is used for automatic clustering. This method is subsequently referred to as *VGA-clustering*. Since it is evident that the K-means-like fitness function, J, cannot be used in this situation (J decreases as the number of clusters is increased), several cluster validity indices, viz., Davies–Bouldin index [110], Dunn's index [132], two of its generalized versions [58], and a newly developed validity index are utilized for computing the fitness of the chromosomes. (Several other indices are available in [310, 500], while a review on this topic is provided in [180].) The results provide a comparison of these indices in terms of their utility in determining the appropriate clustering of the data. The different steps of the algorithm are now described in detail below.

8.5.1 Encoding Strategy and Population Initialization

In *VGA-clustering*, the chromosomes are made up of real numbers (representing the coordinates of the centers). If chromosome i encodes the centers of K_i clusters in N-dimensional space, $K_i \geq 2$, then its length l_i is taken to be $N * K_i$.

Each string i in the population initially encodes the centers of a number, K_i, of clusters, where K_i is given by $K_i = rand() \bmod K^*$. Here, $rand()$ is a function returning an integer, and K^* is a soft estimate of the upper bound of the number of clusters. Note that K^* is used only for the generation of the initial population. The actual number of clusters in the data set is not related to K^*, and may be any number greater than, equal to or less than K^*. The K_i centers encoded in a chromosome are randomly selected points from the data set.

8.5.2 Fitness Computation

For each chromosome, the centers encoded in it are first extracted, and then a partitioning is obtained by assigning the points to a cluster corresponding to the closest center. The cluster centers encoded in the chromosome are then replaced by the centroids of the corresponding clusters. Considering the afore-mentioned partition and the number of clusters, the value of a cluster validity index is computed. The fitness of a chromosome is then defined as a function of the corresponding cluster validity index. The different cluster validity indices that have been utilized for computing the fitness of the chromosomes are described in Sect. 8.5.4.

8.5.3 Genetic Operators

The roulette wheel strategy is used for implementing the proportional selection scheme. For the purpose of crossover, the cluster centers are considered to be indivisible, i.e., the crossover points can only lie in between two clusters centers. The crossover operator, applied stochastically with probability μ_c, must ensure that information exchange takes place in such a way that both the offspring encodes the centers of at least two clusters. For this, the operator is defined as follows:

Let parents P_1 and P_2 encode K_1 and K_2 cluster centers, respectively. C_1, the crossover point in P_1, is generated as $C_1 = rand() \bmod K_1$. As before, $rand()$ is a function that returns an integer. Let C_2 be the crossover point in P_2, and it may vary in between $[LB(C_2), UB(C_2)]$, where $LB()$ and $UB()$ indicate the lower and upper bounds of the range of C_2, respectively. $LB(C_2)$ and $UB(C_2)$ are given by

$$LB(C_2) = \min[2, \max[0, 2 - (K_1 - C_1)]], \tag{8.14}$$

$$\text{and } UB(C_2) = [K_2 - \max[0, 2 - C_1)]]. \tag{8.15}$$

Therefore C_2 is given by

$$C_2 = LB(C_2) + rand() \bmod (UB(C_2) - LB(C_2)).$$

Each chromosome undergoes mutation with a fixed probability μ_m. Since floating point representation is considered, the following mutation is used. A number δ in the range $[0, 1]$ is generated with uniform distribution. If the value at a gene position is v, after mutation it becomes $(1 \pm 2 * \delta) * v$, when $v \neq 0$, and $\pm 2 * \delta$, when $v = 0$. The "+" or "−" sign occurs with equal probability. Note that other ways of implementing floating point mutations may be considered.

8.5.4 Some Cluster Validity Indices

This section contains a description of several cluster validity indices that have been used for computing the fitness of the chromosomes in the *VGA-clustering* scheme.

Davies–Bouldin Index

This index is a function of the ratio of the sum of *within-cluster scatter* to *between-cluster separation* [110]. The scatter within the ith cluster is computed as

$$S_{i,q} = \left(\frac{1}{|C_i|} \sum_{\mathbf{x} \in C_i} \{||\mathbf{x} - \mathbf{z}_i||_2^q\} \right)^{1/q}, \tag{8.16}$$

and the distance between cluster C_i and C_j is defined as

$$d_{ij,t} = ||\mathbf{z}_i - \mathbf{z}_j||_t. \tag{8.17}$$

$S_{i,q}$ is the qth root of the qth moment of the $|C_i|$ points in cluster C_i with respect to their mean $\mathbf{z}_i$, and is a measure of the dispersion of the points in the cluster. Specifically, $S_{i,1}$, used here, is the average Euclidean distance of the vectors in class i from the centroid of class i. $d_{ij,t}$ is the Minkowski distance of order t between the centroids $\mathbf{z}_i$ and $\mathbf{z}_j$ that characterize clusters C_i and C_j. Subsequently we compute

$$R_{i,qt} = \max_{j, j \neq i} \left\{ \frac{S_{i,q} + S_{j,q}}{d_{ij,t}} \right\}. \tag{8.18}$$

The Davies–Bouldin (DB) index is then defined as

$$DB = \frac{1}{K} \sum_{i=1}^{K} R_{i,qt}. \tag{8.19}$$

The objective is to minimize the DB index for achieving proper clustering. The fitness of chromosome j is defined as $\frac{1}{DB_j}$, where DB_j is the Davies–Bouldin index computed for this chromosome. Note that maximization of the fitness function will ensure minimization of the DB index.

Dunn's Index

Let S and T be two nonempty subsets of $\mathbb{R}^N$. Then the diameter $\triangle$ of S and set distance δ between S and T are

$$\triangle(S) = \max_{\mathbf{x}, \mathbf{y} \in S} \{d(\mathbf{x}, \mathbf{y})\},$$

$$\text{and} \quad \delta(S,T) = \min_{\mathbf{x} \in S, \mathbf{y} \in T} \{d(\mathbf{x}, \mathbf{y})\},$$

where $d(\mathbf{x}, \mathbf{y})$ is the distance between points $\mathbf{x}$ and $\mathbf{y}$. For any partition, Dunn defined the following index [132]:

$$\nu_D = \min_{1 \le i \le K} \left\{ \min_{1 \le j \le K, j \ne i} \left\{ \frac{\delta(C_i, C_j)}{\max_{1 \le k \le K} \{\Delta(C_k)\}} \right\} \right\}. \tag{8.20}$$

Larger values of ν_D correspond to good clusters, and the number of clusters that maximizes ν_D is taken as the optimal number of clusters.

Generalized Dunn's Index

The generalized Dunn's index was developed in [58] after demonstrating the sensitivity of the original index ν_D, given by Eq. (8.20), to changes in cluster structure, since not all of the data was involved in the computation of the index. Let δ_i be any positive, semidefinite, symmetric set distance function and Δ_j be any positive, semidefinite diameter function. Then the generalized Dunn's index, ν_{GD}, is defined as

$$\nu_{GD} = \min_{1 \le s \le K} \left\{ \min_{1 \le t \le K, t \ne s} \left\{ \frac{\delta_i(C_s, C_t)}{\max_{1 \le k \le K} \{\Delta_j(C_k)\}} \right\} \right\}. \tag{8.21}$$

Five set distance functions and three diameter functions are defined in [58]. Of these, two combinations, δ_3 and Δ_3 (which was found to be most useful for cluster validation), and δ_5 and Δ_3 are used here. The three measures, viz., δ_3, δ_5 and Δ_3, are defined as follows:

$$\Delta_3(S) = 2 \left(\frac{\sum_{\mathbf{x} \in S} d(\mathbf{x}, \mathbf{z}_S)}{|S|} \right), \tag{8.22}$$

$$\delta_3(S,T) = \frac{1}{|S||T|} \sum_{\mathbf{x} \in S, \mathbf{y} \in T} d(\mathbf{x}, \mathbf{y}), \text{ and} \tag{8.23}$$

$$\delta_5(S,T) = \frac{1}{|S| + |T|} \left(\sum_{\mathbf{x} \in S} d(\mathbf{x}, \mathbf{z}_T) + \sum_{\mathbf{y} \in T} d(\mathbf{y}, \mathbf{z}_S) \right). \tag{8.24}$$

Here $\mathbf{z}_S = \frac{1}{|S|} \sum_{\mathbf{x} \in S} \mathbf{x}$, and $\mathbf{z}_T = \frac{1}{|T|} \sum_{\mathbf{y} \in T} \mathbf{y}$, where S and T are two clusters. The two generalized Dunn's indices ν_{33} and ν_{53} are generated by replacing Δ_i with Δ_3 (in the denominator) and δ_i with δ_3 and δ_5, respectively (in the numerator), in Eq. (8.21).

$\mathcal{I}$-index

The $\mathcal{I}$-index proposed recently [31] is defined as follows:

$$\mathcal{I}(K) = \left(\frac{1}{K} \times \frac{E_1}{E_K} \times D_K \right)^p, \tag{8.25}$$

where K is the number of clusters and p is any real number greater than or equal to 1. Note that p controls the contrast between different cluster configurations. Here,

$$E_K = \sum_{k=1}^{K} \sum_{j=1}^{n} u_{kj} ||\mathbf{x}_j - \mathbf{z}_k||, \tag{8.26}$$

$$\text{and} \quad D_K = \max_{i,j=1}^{K} ||\mathbf{z}_i - \mathbf{z}_j||. \tag{8.27}$$

n is the total number of points in the data set, $U(X) = [u_{kj}]_{K \times n}$ is a partition matrix for the data and $\mathbf{z}_k$ is the center of the kth cluster.

As can be seen from Eq. (8.25), $\mathcal{I}$-index is a composition of three factors, namely, $\frac{1}{K}$, $\frac{E_1}{E_K}$ and D_K. The first factor decreases linearly as K increases. Therefore, this factor will try to reduce $\mathcal{I}$ as K is increased. The second factor consists of the ratio of E_1, which is constant for a given data set, and E_K, which decreases with increase in K. Hence, because of this term, $\mathcal{I}$ increases as E_K decreases. This, in turn, indicates that the formation of more numbers of clusters, which are compact in nature, would be encouraged. Note that although the choice of E_1 does not affect the performance of $\mathcal{I}(K)$, it is used as some sort of normalizing factor in order to avoid extremely low values of the index. Finally, the third factor (which measures the maximum separation between two clusters) will increase with the value of K. However, note that this value is upper-bounded by the maximum separation between two points in the data set. Thus the three factors are found to compete with and balance each other critically. While the first factor will try to decrease K, the second and third factors will try to increase K (thereby encouraging formation of compact and well-separated clusters).

8.5.5 Experimental Results

To experimentally demonstrate the effectiveness of *VGA-clustering*, three artificial and two real-life data sets are considered. The artificial data sets chosen are *AD_5_2*, *AD_10_2* and *AD_4_3*. Among these, the first two data sets are in two dimensions with five and ten clusters, respectively, and the last one is in three dimensions with four clusters. Figures 8.2–8.4 show the three data sets. Note that the artificial data sets are generated in such a way that they present different degrees of difficulty for clustering. For example, although *AD_4_3* (Fig. 8.4) appears to be visually simple, the presence of noisy points in it (situated exactly between two distinct clusters) can mislead the clustering technique, and/or the validity indices. On the other hand, AD_5_2 can be seen to be highly overlapped, and AD_10_2 is overlapping with large number of clusters. As earlier, the two real-life data sets considered are *Iris* and *Cancer*. All these data sets are described in detail in Appendix B.

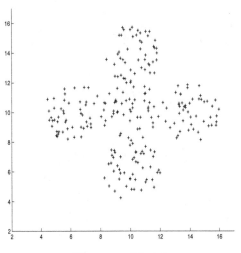

Fig. 8.2. AD_5_2

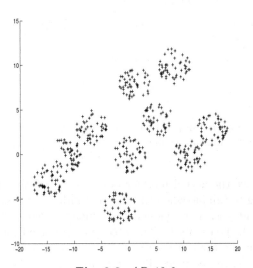

Fig. 8.3. AD_10_2

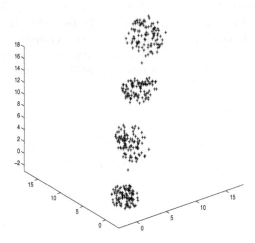

Fig. 8.4. AD_4_3

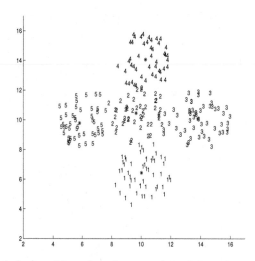

Fig. 8.5. AD_5_2 clustered into five clusters when DB and ν_D index are used for computing fitness; "*" indicate centers

Table 8.5 shows the actual number of clusters, and that provided by the genetic clustering technique when the different indices are used for computing the chromosome fitness. As is evident, for $\mathcal{I}$-index the correct number of clusters is found for all the data sets. The DB index provided the correct number of clusters in five out the seven cases, while ν_D, ν_{33} and ν_{53} do so in three, four and four cases, respectively. This indicates the significant superiority of $\mathcal{I}$-index vis à vis the other ones. The clusterings obtained using the DB, Dunn and Generalized Dunn's indices are shown in Figs. 8.5–8.11 for the artificial

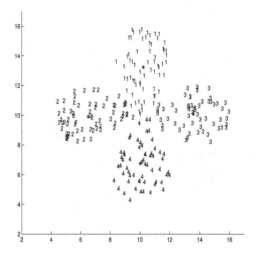

Fig. 8.6. AD_5_2 clustered into 4 clusters when ν_{33} and ν_{53} are used for computing fitness; "*" indicate centers

Table 8.5. Actual and computed number of clusters for the data sets

Data Set	Actual no. clusters	No. clusters with VGA clustering using				
		DB	ν_D	ν_{33}	ν_{53}	$\mathcal{I}(K)$
AD_5_2	5	5	5	4	4	5
AD_10_2	10	8	8	2	2	10
AD_4_3	4	4	2	4	4	4
Iris	3	2	2	2	2	3
Cancer	2	2	2	2	2	2

data sets. Although the $\mathcal{I}$-index provides the correct clusters for all the data sets, its results are demonstrated pictorially for only AD_10_2 and *Iris* (projected on two dimensions), since none of the remaining indices provide the correct clusters for these two data sets.

As can be seen from Fig. 8.5 for AD_5_2 data, while the DB and Dunn's index correctly identify five clusters in the data, the two generalized versions of the Dunn's index identify, incorrectly, four clusters to be appropriate (see Fig. 8.6).

AD_10_2 is generated with ten physical clusters, some of which are compact and separated, while others are overlapping. The clusterings obtained when DB and Dunn's indices are used are shown in Fig. 8.7 when both of them provided eight clusters (resulting from incorrectly merging some physically separate clusters to yield cluster numbers 2 and 5 in the figure). Interestingly, and quite unexpectedly, both ν_{33} and ν_{53} provided two clusters, which are shown in Fig. 8.8. Since this indicates a gross deviation from the actual sce-

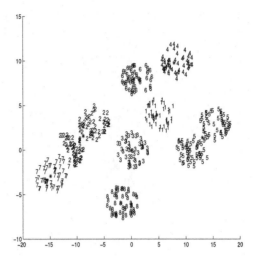

Fig. 8.7. *AD_10_2* clustered into eight clusters when DB and ν_D indices are used for computing fitness; "*" indicate centers

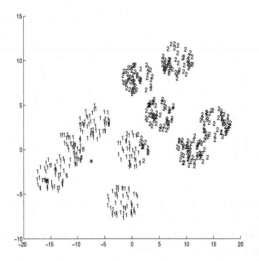

Fig. 8.8. *AD_10_2* clustered into two clusters when ν_{33} and ν_{53} indices are used for computing fitness; "*" indicate centers.

nario, we investigated the values of the indices for different partitionings of *AD_10_2*, including the one that we know to be correct, i.e., with ten clusters. It was found that the values of ν_{33} were 0.986103, 1.421806 and 1.3888610 for the clusterings in Figs. 8.7 (into eight clusters), 8.8 (into two clusters), and the correct partitioning into ten clusters, respectively. Similarly, the values of ν_{53} were 0.963560, 1.367821 and 1.347380, respectively, for the three cases. Thus, among the partitionings that were investigated, the values of ν_{33} and

ν_{53} actually got maximized for the case of two clusters (i.e, corresponding to Fig. 8.8). This indicates that it is not due to the limitation of the *VGA-clustering* technique that two clusters were obtained; on the contrary, this is due to a limitation of the index definition. As is evident from Table 8.5, only $\mathcal{I}$-index, in conjunction with the *VGA-clustering* algorithm, could indicate the correct number of clusters for *AD_10_2* (as also for the other data sets). The corresponding clustered result is shown in Fig. 8.9.

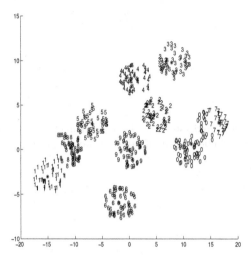

Fig. 8.9. *AD_10_2* clustered into 10 Clusters when $\mathcal{I}$ is used for computing fitness; "*" indicate centers

Figure 8.10 shows that the genetic clustering with DB index, ν_{33} and ν_{53} can correctly evolve the appropriate partitioning for *AD_4_3*, while the original Dunn's index ν_D fails in this regard (Fig. 8.11). On investigation it was found that ν_D for two clusters is larger than that with four clusters, and hence the former is preferred over the latter. Note that this problem arises since all the points are not taken into consideration when computing ν_D by Eq. (8.20). This limitation is removed in the two generalized versions of the Dunn's index.

For *Iris*, the *VGA-clustering* provides two clusters for all the validity indices, except for $\mathcal{I}(K)$. Although it is known that the data has three physical classes, two of them have a significant amount of overlap. Thus many automatic clustering methods reported in the literature have often provided two clusters for this data [58, 253]. Figure 8.12 shows for the purpose of demonstration the *Iris* data set partitioned into two clusters when DB index is used for computing the fitness. In contrast, the *VGA-clustering* in conjunction with the $\mathcal{I}$-index is again able to indicate the correct number of clusters, even when the classes are quite overlapping as for *Iris*. The corresponding clustered data

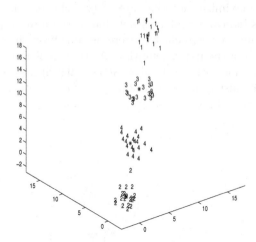

Fig. 8.10. AD_4_3 clustered into four clusters when DB, ν_{33} and ν_{53} are used for computing fitness; "*" indicate centers (only 20 points per class are plotted for the sake of clarity)

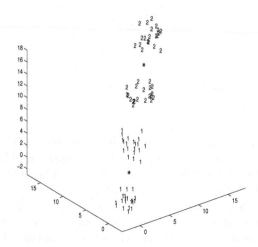

Fig. 8.11. AD_4_3 clustered into two clusters when ν_D is used for computing fitness; "*" indicate centers (only 20 points per class are plotted for the sake of clarity)

is shown in Fig. 8.13. As seen from Table 8.5, the *VGA-clustering* produces two clusters in all the cases for *Cancer* data.

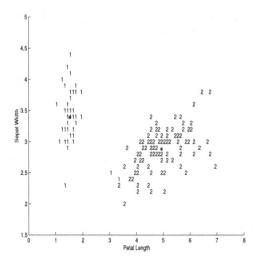

Fig. 8.12. *Iris* data clustered into two clusters when *DB* index is used for computing fitness; "*" indicate centers

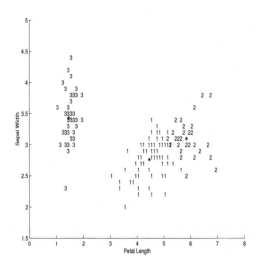

Fig. 8.13. *Iris* data clustered into three clusters when $\mathcal{I}$ is used for computing fitness; "*" indicate centers

Figure 8.14 shows the variation of the number of clusters (of the best chromosome) deemed to be optimal when *DB* index is used, with the number of generations while Fig. 8.15 shows the variation of the value of $\frac{1}{DB\ index}$ of the best chromosome with the number of generations for AD_10_2. The figures show that although the final number of clusters (8) is attained by the *VGA-clustering* with DB index in generation 7, the appropriate partitioning keeps on evolving till around generation 30.

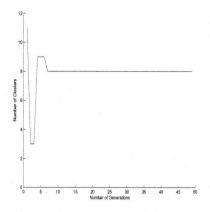

Fig. 8.14. Variation of the number of clusters with generations for AD_10_2 using DB index

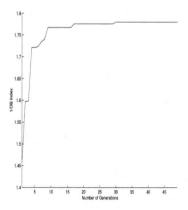

Fig. 8.15. Variation of the value of $\frac{1}{DB\ index}$ with generations for AD_10_2

8.6 Clustering Using GAs: Variable Number of Fuzzy Clusters

In this section, we describe the use of VGAs for automatically evolving the near-optimal $K \times n$ nondegenerate fuzzy partition matrix U^* [299]. The resulting algorithm is referred to as *FVGA-clustering*. For the purpose of optimization, the Xie–Beni cluster validity index $(XB(U, Z, X))$ is used. In other words, the best partition matrix U^* is the one such that

$$U^* \in \mathcal{U} \text{ and } XB(U^*, Z^*, X) = \min_{U_i \in \mathcal{U}} XB(U_i, Z_i, X), \qquad (8.28)$$

where Z^* represents the set of cluster centers corresponding to U^*. Here both the number of clusters as well as the appropriate fuzzy clustering of the data are evolved simultaneously using the search capability of genetic algorithms. The chromosome representation and other genetic operators used are the same as those described in the previous section. The fitness computation and the experimental results are detailed below.

8.6.1 Fitness Computation

The Xie–Beni (XB) index is defined as a function of the ratio of the total variation σ to the minimum separation sep of the clusters. Here σ and sep can be written as

$$\sigma(U, Z; X) = \sum_{i=1}^{c} \sum_{k=1}^{n} u_{ik}^2 D^2(\mathbf{z}_i, \mathbf{x}_k), \qquad (8.29)$$

and

$$sep(Z) = \min_{i \neq j}\{||\mathbf{z}_i - \mathbf{z}_j||^2\}, \qquad (8.30)$$

where $||.||$ is the Euclidean norm, and $D(\mathbf{z}_i, \mathbf{x}_k)$, as mentioned earlier, is the distance between the pattern $\mathbf{x}_k$ and the cluster center $\mathbf{z}_i$. The XB index is then written as

$$XB(U, Z; X) = \frac{\sigma(U, Z; X)}{n \ sep(Z)} = \frac{\sum_{i=1}^{c}(\sum_{k=1}^{n} u_{ik}^2 D^2(\mathbf{z}_i, \mathbf{x}_k))}{n(\min_{i \neq j}\{||\mathbf{z}_i - \mathbf{z}_j||^2\})}. \qquad (8.31)$$

Note that when the partitioning is compact and good, value of σ should be low while sep should be high, thereby yielding lower values of the XB index. The objective is therefore to minimize the XB index for achieving proper clustering.

Given a chromosome, the centers encoded in it are first extracted. These are used to compute the fuzzy partition matrix using Eq. (8.12) followed by the calculation of the corresponding XB index using Eq. (8.31). The fitness function for a chromosome is then defined as $\frac{1}{XB}$. Note that maximization of the fitness function will ensure minimization of the XB index. Subsequently, the centers encoded in a chromosome are updated using Eq. (8.13).

8.6.2 Experimental Results

Fig. 8.16. IRS image of Calcutta in the near infrared band with histogram equalization

The effectiveness of *FVGA-clustering* in partitioning the pixels into different landcover types is demonstrated on Indian remote sensing (IRS) satellite images of parts of the cities of Calcutta and Mumbai [299] (Figs. 8.16 and 8.17 show images of the two cities, respectively, in the near infrared band). Detailed descriptions of these images are available in Appendix B.

For Calcutta image, the *FVGA-clustering* technique automatically provided four clusters for this data (shown in Fig. 8.18). From the ground knowledge, it can be inferred that these four clusters correspond to the classes turbid water (TW), pond water (PW), concrete (Concr.) and open space (OS). It may be noted that the water class has been differentiated into turbid water (the river *Hooghly*) and pond water (fisheries, etc.) because of a difference in their spectral properties. Here, the class turbid water contains sea water, river water, etc., where the soil content is more than that of pond water. *SaltLake* township has come out partially as classes concrete and open space, which appears to be correct, since this particular region is known to have several open spaces. The canal bounding *SaltLake* from the upper portion has also been correctly classified as PW. The airstrips of *Dumdum* airport have again been classified correctly as belonging to the class concrete. Presence of some small areas of PW beside the airstrips is also correct since these correspond to the several ponds that dot the region. The predominance of concrete on

Fig. 8.17. IRS image of Mumbai in the near infrared band with histogram equalization

both sides of the river, particularly towards the bottom of the image is also correct. This region corresponds to the central part of the city of Calcutta.

In order to demonstrate the performance of the genetic clustering scheme quantitatively for the Calcutta image, the variation of the best and average values of the XB-index with the number of generations is provided in Fig. 8.19. As can be seen from the figure, the best value of the XB index is attained at around generation 25, after which the value does not change anymore. (It may be noted that because of elitism, the best value of XB index can only decrease or remain the same with increase in the number of generations.) The average value of the XB index, on the other hand, shows frequent variation, although the general trend is towards reducing this as well. Note that with the introduction of variable string lengths, the diversity of the population remains high even after a significant number of generations, which is reflected correctly in the continuous variation of the average XB value. Moreover, a gradual improvement of the average value of the index over the generations indicates that this diversity is beneficial to the performance of the *VGA-classifier*.

Figure 8.20 shows the variation of the XB-index with the number of clusters when the well-known FCM algorithm is used as the underlying clustering technique for the Calcutta image. If K^* is the number of clusters provided by the proposed clustering scheme, then the number of clusters in FCM is varied from $K^* - 3$ to $K^* + 3$. In this case, since $K^* = 4$, the corresponding range is taken to be 2 to 7 (since one cluster is not practically meaningful). As can be seen from the figure, the minimum value of the XB index is again obtained

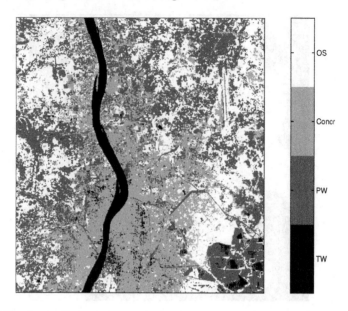

Fig. 8.18. Clustered image of Calcutta using *FVGA-Clustering*

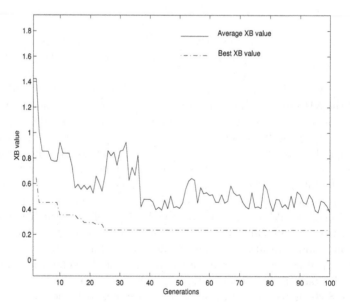

Fig. 8.19. Variation of the XB-index with the number of generations for IRS image of Calcutta

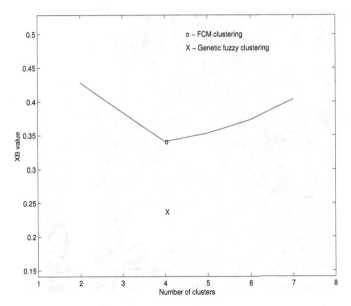

Fig. 8.20. Variation of the XB-index with the number of clusters for IRS image of Calcutta when FCM clustering is used

for four clusters with the FCM algorithm. However, the corresponding value as obtained with the genetic clustering scheme (denoted by "x" in the figure) is found to be still smaller. This not only shows that the genetic scheme found the correct number of clusters, but also demonstrates its superiority over the FCM technique, which often gets stuck at suboptimal solutions.

Figure 8.21 shows the Calcutta image partitioned into four clusters using the FCM algorithm. As can be seen, the river Hooghly as well as the city region has been incorrectly classified as belonging to the same class. Therefore, this region has been labelled as TW+Concr. Again, the entire SaltLake region, which is known to have both concrete and open space, has gone to only one class. The corresponding label is thus put as OS1 + Concr. In addition, the class corresponding to several other open spaces in the image is labelled as OS2. Therefore we can conclude that although some regions, viz., fisheries, canal bounding SaltLake, parts of the airstrip, etc., have been correctly identified, a significant amount of confusion is evident in the FCM clustering result.

The result of application of *FVGA-clustering*, using XB index as the optimizing criterion, on the Mumbai image (Fig 8.17) is shown in Fig. 8.22 [299]. The method automatically yielded seven clusters, which are labelled concrete (Concr.), open spaces (OS1 and OS2), vegetation (Veg), habitation (Hab) and turbid water (TW1 and TW2), based on the ground information available from earlier studies. Here, the class habitation refers to the regions which have concrete structures and buildings, but with relatively lower density than

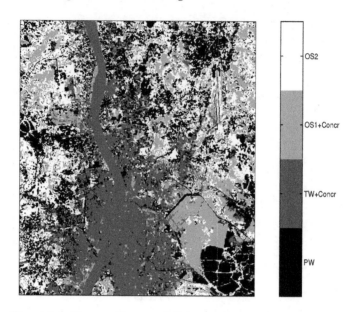

Fig. 8.21. Clustered image of Calcutta using FCM clustering

the class Concr. Thus these two classes share common properties. From the result it can be seen that the large water body of the Arabian sea has been distinguished into two classes which are named TW1 and TW2. It has been observed earlier [344] that the sea water has two distinct regions with different spectral properties. Hence the clustering result providing two partitions for this region is expected. The islands, dockyard, several road structures have mostly been correctly identified in the image. Within the islands, as expected, there is a predominance of open space and vegetation. The southern part of the city, which is heavily industrialized, has been classified as primarily belonging to habitation and concrete. Some confusion within these two classes, viz., Hab and Concr, is observed (as reflected in the corresponding label); this is expected since these two classes are somewhat similar.

Figure 8.23 demonstrates the Mumbai image partitioned into seven clusters using the FCM technique. As can be seen, the water of the Arabian sea has been partitioned into three regions, rather than two as obtained earlier. The other regions appear to be classified more or less correctly for this data. It was observed that the FCM algorithm gets trapped at local optima often enough, and the best value of the XB index was worse than that obtained using the genetic fuzzy clustering scheme.

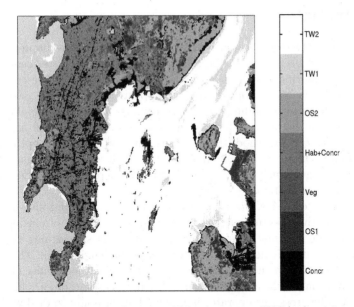

Fig. 8.22. Clustered image of Mumbai using *FVGA-clustering*

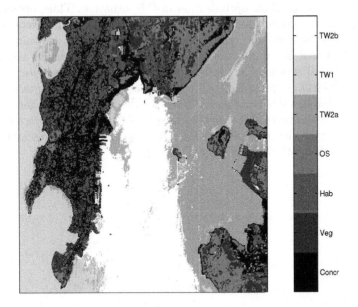

Fig. 8.23. Clustered image of Mumbai using FCM clustering

8.7 Summary

Several genetic clustering schemes have been described in this chapter. Genetic algorithms have been used to search for the appropriate cluster centers such that some clustering metric is optimized. Floating point representation of chromosomes has been adopted, since it appears to be a more natural and appropriate form for encoding the cluster centers. This entails application of a new mutation operator. When the number of clusters is fixed a priori, as in the *GA-clustering* method, the K-means criterion is used for optimization. Results demonstrate its superiority over the traditional K-means, which is often found to get stuck at suboptimal solutions depending on the choice of the initial cluster centers.

A weakness of *GA-clustering*, described in Sect. 8.4, and also of the K-means is that they assume prior knowledge of the number of clusters. Since this information may not be readily available in all cases, designing algorithms that can automatically determine this value is a challenging task. This is tackled in *VGA-clustering* and its fuzzy counterpart. Since the number of clusters is now kept variable, the concept of variable length chromosomes is introduced in the GA. The crossover operator is modified accordingly to deal with chromosomes of varying length. Moreover the K-means criterion (J in Eq. (8.2)) can no longer be used for optimization in this situation. Thus, optimization of several cluster validity indices has been studied. Their effectiveness in determining the appropriate number of clusters in a data set is evaluated and compared. A real-life application of the fuzzy genetic clustering scheme for landcover classification in satellite images yields encouraging results, given that the algorithm is provided with no information about the data set.

All the chapters described so far deal with the application of genetic algorithms for pattern recognition and learning problems. The next two chapters deal with the applications genetic learning to two emerging research areas, namely, bioinformatics and Web intelligence.

9

Genetic Learning in Bioinformatics

9.1 Introduction

Over the past few decades, major advances in the field of molecular biology, coupled with advances in genomic technologies, have led to an explosive growth in the biological information generated by the scientific community. This deluge of genomic information has, in turn, led to a requirement for computerized databases to store, organize and index the data, and for specialized tools to view and analyze the data.

The field of bioinformatics can be considered, in general, to deal with the use of computational methods to handle biological information [24]. It is an interdisciplinary field involving biology, computer science, mathematics and statistics to analyze biological sequence data, genome content and arrangement, and to predict the function and structure of macromolecules. The ultimate goal of the field is to enable the discovery of new biological insights as well as to create a global perspective from which unifying principles in biology can be derived [6]. There are three important subdisciplines within bioinformatics:

- Development of new algorithms and models to assess different relationships among the members of a large biological data set in a way that allows researchers to access existing information and to submit new information as they are produced;
- Analysis and interpretation of various types of data including nucleotide and amino acid sequences, protein domains, and protein structures;
- Development and implementation of tools that enable efficient access and management of different types of information.

Recently genetic algorithms have been gaining the attention of researchers for solving bioinformatics problems. Data analysis tools used earlier in bioinformatics were mainly based on statistical techniques like regression and estimation. GAs in bioinformatics gained significance with the need to handle large data sets in biology in a robust and computationally efficient manner. A recent review of the applications of GAs in solving different tasks in bioinformatics can be found in [345].

In this chapter, we first discuss the basic concepts of bioinformatics along with their biological basis in Sect. 9.2. The relevance of the application of GAs for solving bioinformatics problems is explained in Sect. 9.3. In Sect. 9.4 various bioinformatics tasks and different evolutionary algorithm-based methods available to address the bioinformatics tasks are explained. The next section describes some experimental results of the application of genetic learning to a specific bioinformatic task of ligand design with applications in rational drug design. Finally, Sect. 9.6 summarizes this chapter.

9.2 Bioinformatics: Concepts and Features

In this section first we introduce the basic biological concepts required to understand the various problems in bioinformatics. The relevance of GAs in bioinformatics is described in the following section.

9.2.1 Basic Concepts of Cell Biology

Deoxyribonucleic acid (DNA) and proteins are biological macromolecules built as long linear chains of chemical components. DNA strands consist of a large sequence of nucleotides, or bases. For example, there are more than 3 billion bases in human DNA sequences. DNA plays a fundamental role in different biochemical processes of living organisms in two respects. First, it contains the templates for the synthesis of proteins, which are essential molecules for any organism [426]. The second role in which DNA is essential to life is as a medium to transmit hereditary information (namely, the building plans for proteins) from generation to generation. Proteins are responsible for structural behavior.

The units of DNA are called nucleotides. One nucleotide consists of one nitrogen base, one sugar molecule (deoxyribose) and one phosphate. Four nitrogen bases are denoted by one of the letters A (adenine), C (cytosine), G (guanine) and T (thymine). A linear chain of DNA is paired to a complementary strand. The complementary property stems from the ability of the nucleotides to establish specific pairs (A–T and G–C). The pair of complementary strands then forms the double helix that was first suggested by Watson and Crick in 1953. Each strand therefore carries the entire information, and the biochemical machinery guarantees that the information can be copied over and over again, even when the "original" molecule has long since vanished.

A gene is primarily made up of sequences of triplets of the nucleotides (exons). Introns (noncoding sequences) may also be present within genes. Not all portions of the DNA sequences are coding. A coding zone indicates that it is a template for a protein. As an example, for the human genome only 3%–5% of the sequences are coding, i.e., they constitute the gene. The promoter is a region before each gene in the DNA that serves as an indication to the cellular mechanism that a gene is ahead. For example, the codon AUG codes for methionine and signals the start of a gene. Promoters are key regulatory sequences that are necessary for the initiation of transcription. Transcription is a process in which ribonucleic acid (RNA) is formed from a gene, and through translation, amino acids are formed from RNA. There are sequences of nucleotides within the DNA that are spliced out progressively in the process of transcription and translation. A comprehensive survey of the research done in this field is available in [499]. In brief, the DNA consists of three types of noncoding sequences (Fig. 9.1).

(a) Intergenic regions: Regions between genes that are ignored during the process of transcription.
(b) Intragenic regions (or introns): Regions within the genes that are spliced out from the transcribed RNA to yield the building blocks of the genes, referred to as exons.
(c) Pseudogenes: Genes that are transcribed into the RNA and stay there, without being translated, due to the action of a nucleotide sequence.

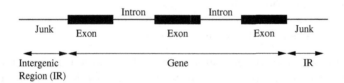

Fig. 9.1. Various parts of DNA

Proteins are polypeptides, formed within cells as a linear chain of amino acids [426]. Amino acid molecules bond with each other by eliminating water molecules and form peptides. Twenty different amino acids (or "residues") are available, which are denoted by 20 different letters of the alphabet. Each of the 20 amino acids is coded by one or more triplets (or codons) of the nucleotides making up the DNA. Based on the genetic code, the linear string of DNA is translated into a linear string of amino acids, i.e., a protein via mRNA (messenger RNA)[426]. For example, the DNA sequence GAACTA-CACACGTGTAAC codes for the amino acid sequence ELHTCN (shown in Fig. 9.2).

$$\textbf{\textit{GAA CTA CAC ACG TGT AAC}}$$
$$\textbf{\textit{E \quad L \quad H \quad T \quad C \quad N}}$$

Fig. 9.2. Coding of amino acid sequence from DNA sequence

Three-dimensional molecular structure is one of the foundations of structure-based drug design. Often, data are available for the shape of a protein and a drug separately, but not for the two together. Docking is the process by which two molecules fit together in three-dimensional space. Ligands are small molecules like a candidate drug that are used for docking to their macromolecular targets (usually proteins, and sometimes DNA).

9.2.2 Different Bioinformatics Tasks

Different biological problems considered within the scope of bioinformatics involve the study of genes, proteins, nucleic acid structure prediction and molecular design with docking. A broad classification of the various bioinformatics tasks is given below.

(a) Alignment and comparison of DNA, RNA and protein sequences
(b) Gene mapping on chromosomes
(c) Gene finding and promoter identification from DNA sequences
(d) Interpretation of gene expression and microarray data
(e) Gene regulatory network identification
(f) Construction of phylogenetic trees for studying evolutionary relationship
(g) DNA structure prediction
(h) RNA structure prediction
(i) Protein structure prediction and classification
(j) Molecular design and molecular docking

Descriptions of these tasks and their implementation in the evolutionary computing (or genetic algorithmic) framework are provided in Sect. 9.4. Before that, the relevance of GAs in bioinformatics is explained.

9.3 Relevance of Genetic Algorithms in Bioinformatics

The main advantages of using GAs for solving problems in bioinformatics are:

(a) Several tasks in bioinformatics involve optimization of different criteria (like energy, alignment score, overlap strength), thereby making the application of GAs more natural and appropriate.
(b) Problems of bioinformatics seldom need the exact optimum solution; rather they require robust, fast and close approximate solutions, which GAs are known to provide efficiently.
(c) GAs can process, in parallel, populations billions of times larger than is usual for conventional computation. The natural expectation is that larger populations can sustain larger ranges of genetic variation and thus can generate high-fitness individuals in fewer generations.

(d) Laboratory operations on DNA inherently involve errors. These are more tolerable in executing evolutionary algorithms than in executing deterministic algorithms. (To some extent, errors may be regarded as contributing to genetic diversity, a desirable property.)

Let us now discuss the relevance of GAs in bioinformatics with an example. Most of the ordering problems in bioinformatics, such as the sequence alignment problem, fragment assembly problem (FAP) and gene mapping (GM), are quite similar to the Travelling Salesman Problem (TSP, the best-known NP-hard ordering problem). The TSP can be formally defined as follows: Let $1, 2, ...n$ be the labels of n cities, and $C = [c_{i,j}]$ be an $n \times n$ cost matrix where $c_{i,j}$ denotes the cost of traveling from city i to city j. The TSP is the problem of finding the shortest closed route among n cities, having as input the complete distance matrix among all cities. A symmetric TSP (STSP) instance is any instance of the TSP such that $c_{i,j} = c_{j,i}$ for all cities i, j. An asymmetric TSP (ATSP) instance is any instance of TSP that has at least one pair of cities such that $c_{i,j} \neq c_{j,i}$. The ATSP is a special case of the problem on which we restrict the input to asymmetric instances. The total cost $A(n)$ of a TSP tour of n cities is given by

$$A(n) = \sum_{i=1}^{n-1} c_{i,i+1} + c_{n,1}. \tag{9.1}$$

The objective is to find a permutation of the n cities that has the minimum cost.

The fragment assembly problem (FAP) deals with the sequencing of DNA. Currently strands of DNA longer than approximately 500 base pairs cannot routinely be sequenced accurately. Consequently, for sequencing larger strands of DNA, they are first broken into smaller pieces. In the shotgun sequencing method, DNA is first replicated many times, and then individual strands of the double helix are broken randomly into smaller fragments. The assembly of DNA fragments into a consensus sequence corresponding to the parent sequence constitutes the "fragment assembly problem" [426]. It is a permutation problem, similar to the TSP, but with some important differences (circular tours, noise and special relationships between entities) [426]. It is NP-complete in nature.

Note that the fragmenting process does not retain either the ordering of the fragments on the parent strand of DNA or the strand of the double helix from which a particular fragment came. The only information available in the assembly stage is the base pair sequence for each fragment. Thus the ordering of the fragments must rely primarily on the similarity of fragments and how they overlap. An important aspect of the general sequencing problem is the precise determination of the relationship and orientation of the fragment. Once the fragments have been ordered, the final consensus sequence is generated from the ordering. Basic steps with four fragments are shown as an example in Fig. 9.3. Here the fragments are aligned in a fashion so that in each column all the bases are the same. As an example, the base G is selected in the sixth column after voting, to make the consensus sequence TCACTGGCTTACTAAG.

ACTGGC
     TGGCTTACT
TCACTC
        TTACTAAG
$\overline{}$
TCACTGGCTTACTAAG $\longrightarrow$ *CONSENSUS SEQUENCE*

Fig. 9.3. Alignment of DNA fragments

Formulation of the FAP as a TSP using GA: Although the endpoints of the tour of TSP are irrelevant since its solution is a circular tour of the cities, in the case of FAP the endpoints are relevant as they represent fragments on opposite ends of the parent sequence. Moreover, the cities in the TSP are not assumed to have any relationship other than the distances, and the ordering is the final solution to the problem. In FAP, the ordering referred to as "beads on a string" is only an intermediate step; the layout process uses the overlap data to position the bases within the fragments relative to each other. Here GAs can be applied. A way of using it in FAP is explained below.

Step 1:

Let $1, 2, \ldots, j, \ldots, n$ represent the indices of n fragments in the spectrum of fragments. Pairwise relationship (similarity) of a fragment with all other fragments (oligonucleotides) is calculated and kept in an $n \times n$ matrix. Dynamic programming gives the best alignment between two sequences (fragments). In this method each possible orientation is tried for the two fragments, and the overlap, orientation and alignment that maximize the similarity between fragments are chosen.

Step 2:

The indices of all the fragments are then ordered randomly with no repetition. Let $f_1, f_2, ..., f_i, ..., f_n$ be such an ordering of a sequence of n fragments, where $f_i = j$ means that fragment j (in the fragment set) appears in position i of the ordering. The fitness function of this ordered sequence can be computed using

$$F = \sum_{i=1}^{n-1} W_{f_i, f_{i+1}}, \tag{9.2}$$

where $W_{i,j}$ is the pairwise overlap strength (similarity) of fragments i and j in the ordered sequence, as obtained in the $n \times n$ matrix.

Such an ordered sequence provides a genetic representation of an individual chromosome in GA.

Step 3:

In this way P ordered sequences are generated, where P is the size of the population of GA.

Step 4:

GA is applied with this population and the following operations:

Selection: Fitness of each sequence is evaluated as in Eq. (9.2) and sequences with higher fitness are selected with roulette wheel.

Crossover: Crossover is performed between two randomly selected sequences for a given crossover rate.

Mutation: For a given mutation rate, only that mutation operator can be applied for which there will be no repetition of fragment indexes in the sequence.

Elitist Model: A new population is created at each generation of GA. The sequence with the highest fitness from the previous generation replaces randomly a sequence from the new generation, provided fitness of the fittest sequence in previous generation is higher than the best fitness in this current generation.

Step 5:

The best sequence of indices with maximum F value is obtained from the GA. From this sequence of indices, the corresponding sequence of fragments is obtained using the overlap information in the $n \times n$ matrix of Step 1.

Step 6:

This alignment of fragments is examined to determine the places where insertion or deletion error likely occurred, and gaps or bases are then inserted or deleted into the fragments to obtain their best possible alignment. The resulting sequence is called the consensus sequence.

Note: The neighboring fragments in the consensus sequence are assumed to be maximally overlapped, thereby ensuring the inclusion of as many fragments as possible. The fitness function (GA) evaluating an individual selects the best substring of oligonucleotides, or the chromosome, i.e., the one composed of the most fragments, provided its length is equal to the given length of the reference DNA sequence.

Different GA operators for the assembly of DNA sequence fragments associated with the Human Genome project were studied in [360]. The sorted order representation and the permutation representation are compared on problems ranging from 2K to 34K base pairs. It is found that edge-recombination crossover used in conjunction with several specialized operators performs the best. Other relevant investigations for solving FAP using GAs are available in [359, 361, 362].

9.4 Bioinformatics Tasks and Application of GAs

We describe below the different problems and associated tasks involved in bioinformatics, their requirements, and the ways in which computational models can be formulated for solving them. The tasks (as mentioned in Sect. 9.2.1) are first explained in this section, followed by a description of how GAs and other evolutionary techniques are applied for solving them.

9.4.1 Alignment and Comparison of DNA, RNA and Protein Sequences

An alignment is a mutual placement of two or more sequences that exhibit where the sequences are similar, and where they differ. These include alignment of DNA, RNA, and protein sequences, and fragment assembly of DNA. An optimal alignment is the one that exhibits the most correspondences and the least differences. It is the alignment with the highest score but may or may not be biologically meaningful. Basically there are two types of alignment methods, global alignment and local alignment. Global alignment [326] maximizes the number of matches between the sequences along the entire length of the sequence. Local alignment [440] gives a highest score to local match between two sequences. Global alignment includes all the characters in both sequences from one end to the other and is excellent for sequences that are known to be very similar. If the sequences being compared are not similar over their entire lengths, but have short stretches within them that have high levels of similarity, a global alignment may miss the alignment of these important regions; local alignment is then used to find these internal regions of high similarity. Pairwise comparison and alignment of protein or nucleic acid sequences is the foundation upon which most other bioinformatics tools are built. Dynamic programming (DP) is an algorithm that allows for efficient and complete comparison of two (or more) biological sequences, and the technique is known as the Smith–Waterman algorithm [440]. It refers to a programmatic technique or algorithm which, when implemented correctly, effectively makes all possible pairwise comparisons between the characters (nucleotide or amino acid residues) in two biological sequences. Spaces may need to be inserted within the sequences for alignment. Consecutive spaces are defined as gaps. The final result is a mathematically, but not necessarily biologically, optimal alignment of the two sequences. A similarity score is also generated to describe how similar the two sequences are, given the specific parameters used.

A multiple alignment arranges a set of sequences in a manner that positions believed to be homologous are placed in a common column. There are different conventions regarding the scoring of a multiple alignment. In one approach, the scores of all the induced pairwise alignments contained in a multiple alignment are simply added. For a linear gap penalty this amounts to scoring each column of the alignment by the sum of pair scores in this column [426]. Although it would be biologically meaningful, the distinctions between global, local and other forms of alignment are rarely made in a multiple alignment. A full set of optimal pairwise alignments among a given set of sequences will generally overdetermine the multiple alignment. If one wishes to assemble a multiple alignment from pairwise alignments one has to avoid "closing loops", i.e., one can put together pairwise alignments as long as no new pairwise alignment is included to a set of sequences which is already part of the multiple alignment.

Methods

GAs are used to solve the problem of multiple sequence alignment. Before we describe them, it may be mentioned that other optimization methods, such as simulated annealing [471] and Gibbs sampling [268] are also used in this regard. Simulated annealing can sometimes be very slow, although it works well as an alignment improver. Gibbs sampling is good in finding local multiple alignment blocks with no gaps, but is not suitable in gapped situations.

The use of GA to deal with sequence alignments in a general manner (without dynamic programming), was first described in Sequence Alignment by Genetic Algorithm (SAGA) [330], shortly before a similar work by Zhang et al. [512]. The population is made of alignments, and the mutations are processing programs that shuffle the gaps using complex methods. In SAGA, each individual (chromosome) is a multiple alignment of sequences. The population size is 100 with no repetitions of chromosomes in it. To create one of these alignments, a random offset is chosen for all the sequences (the typical range being from 0 to 50 for sequences 200 residues long), and each sequence is moved to the right, according to its offset. The sequences are then padded with null signs in order to have the same length. The fitness of each individual (alignment) is computed as the score of the corresponding alignment. All the individuals are ranked according to their fitness, and the weakest are replaced by new children. Only a portion (e.g., 50%) of the population is replaced during each generation. Two types of crossover, 2 types of gap insertion mutation, 16 types of block shuffling mutation, 1 block searching mutation and 2 local optimal rearrangement mutation operators are used in SAGA. During initialization of the program, all the operators have the same probability of being used, equal to 1/22. An automatic procedure (dynamic schedules proposed by Davis [112]) for selecting the operators has been implemented in SAGA. In this model, an operator has a probability of being used that is a function of the efficiency it has recently (e.g., the ten last generations) displayed at improving alignments. The credit an operator gets when performing an improvement is also shared with the operators that came before and may have played a role in this improvement. Thus, each time a new individual is generated, if it yields some improvement on its parents, the operator that is directly responsible for its creation gets the largest part of the credit (e.g., 50%). Then the operator(s) responsible for the creation of the parents also get their share of the remaining credit (50% of the remaining credit, i.e., 25% of the original credit), and so on. This report of the credit goes on for some specified number of generations (e.g., four). After a given number of generations (e.g., ten) these results are summarized for each of the operators. The credit of an operator is equal to its total credit divided by the number of children it generated. This value is taken as the usage probability and remains unchanged until the next assessment, ten generations later. To avoid the early loss of some operators that may become useful later on, all the operators are assigned a minimum probability of being used (the same for all them, typically equal to half their original probability, i.e., 1/44). The automatically assigned probabilities of usage at different stages in the alignment give a direct measure of usefulness or redundancy for a new operator. SAGA is stopped when the search has been unable to improve for some specified number of generations (typically 100). This condition is the most widely used one when working on a population with no duplicates.

Other approaches [331, 507] are similar to SAGA, where a population of multiple alignments evolves by selection, combination and mutation. The main difference between SAGA and recent algorithms has been the design of better mutation operators. A simple GA, applied in a straightforward fashion to the alignment problem was not very successful [330]. The main devices which allow GAs to efficiently reach very high quality solutions are the use of (a) a large number of mutation and crossover operators, and (b) their automatic scheduling. The GA-based methods are not very efficient at handling all types of situations. So it is necessary to invent some new operators designed specifically for the problem and to slot them into the existing scheme. Most of the investigations using GAs for sequence alignment are on different data sets and results are compared with that of CLUSTALW [459], so a clear comparison between the GA-based methods is not possible. A hybrid approach proposed in [513, 514] uses the searching ability of GAs for finding match blocks, and dynamic programming for producing close to optimum alignment of the match blocks. This method is faster and produces better results than pure GA- and DP-based approaches. Here, the population size is determined as $Q = mn/100$, where m is the average sequence length and n is the number of sequences.

In [323] it was pointed out that the combination of high-performance crossover and mutation operators does not always lead to a high performance GA for sequencing because of the negative combination effect of those two operators. A high-performance GA can be constructed by utilizing the positive combination effect of crossover and mutation. Other relevant investigations for solving multiple sequence alignment using GAs are available in [7, 184, 328, 454, 511].

9.4.2 Gene Mapping on Chromosomes

Gene mapping is defined as the determination of relative positions of genes on a chromosome and the distance between them. A gene map helps molecular biologists to explore a genome. A primary goal of the Human Genome project is to make a series of descriptive diagram maps of each human chromosome at increasingly finer resolutions. Two types of gene maps, viz., cytogenetic map and linkage map, are generally used. A cytogenetic map, also known as a physical map, offers a physical picture of the chromosome. In cytogenetic map, the chromosomes are divided into smaller fragments that can be propagated and characterized and then the fragments are ordered (mapped) to correspond to their respective locations on the chromosomes. A genetic linkage map shows the relative locations (order) of specific DNA markers along the chromosome.

Since EAs have been used for determining the genetic linkage map, it is described here briefly. The genetic markers in a linkage map are generally small but precisely defined sequences and can be expressed DNA regions (genes) or DNA segments that have no known coding function but whose inheritance pattern can be followed. DNA sequence differences are especially useful markers because they are plentiful and easy to characterize precisely [426]. A linkage map is constructed by

- producing successive generations (chromosomes) of certain organisms through crossover (recombination) and
- analyzing the observed segregation percentages of certain characteristics in each chromosomal data to find the actual gene order.

Linkage map shows the order and relative distance between genes but has two drawbacks [426]. First, it does not tell the actual distance of genes, and second, if genes are very close, one can not resolve their order, because the probability of separation is so small that observed recombinant frequency are all zero. The closer two genes are, the lower is the probability that they will be separated during DNA repair or replication process, and hence the greater the probability that they will be inherited together. For example, suppose a certain stretch of DNA has been completely sequenced, giving us a sequence S. If we know which chromosome S came from, and if we have a physical map of this chromosome, one could try to find one of the map's markers in S. If the process succeeds one can locate the position of S in the chromosome. The best criterion to quantify how well a map explains the data set is the multipoint maximum likelihood (exploiting the data on all markers simultaneously) of the map. Given a probabilistic model of recombination for a given family structure, a genetic map of a linkage group and the set of available observations on markers of the linkage group, one can define the probability that the observations may have occurred given the map. This is termed the likelihood of the map. The likelihood is only meaningful when compared to the likelihood of other maps.

The problem of finding a maximum likelihood genetic map can be described as a double optimization problem: For a given gene order, there is the problem of finding recombination probabilities (crossover probabilities) that yields a maximum multipoint likelihood, then one must find an order that maximizes this maximum likelihood. The first problem is solved by using the Expectation Maximization (EM) algorithm. The second problem is more difficult because the number of possible orders to consider for N markers is $N!/2$. This type of combinatorial problem can be handled by evolutionary algorithms efficiently. The problem of finding an order of genes that maximizes the maximum multipoint likelihood is equivalent to the problem of symmetric TSP. One can simply associate one imaginary city to each marker, and define as the distance between two cities the inverse of the elementary contribution to the log-likelihood defined by the corresponding pair of markers.

Methods

The method of genetic mapping, described in [159], is embodied in a hybrid framework that relies on the statistical optimization algorithms (e.g., expectation maximization) to handle the continuous variables (recombination probabilities) while GAs handle the ordering problem of genes. The efficiency of the approach lies critically in the introduction of greedy local search in the fitness evaluation of the GA, using a neighborhood structure inspired by the TSP. A population size ranging from 25 to 250 has been used for number of markers between 10 to 29.

In gene mapping problem, Gunnels et al. [179] compared GAs with simulated annealing (SA) and found that the GA-based method always converges to a good solution faster since its population based nature allows it to take advantage of the extra information to construct good local maps that can then be used to construct good global maps.

It is difficult to design a map in canonical GAs without a priori knowledge of the solution space. This is overcome in [322], where GAs using a coevolutionary approach are utilized for exploring not only within a part of the solution space defined by the genotype–phenotype map but also the map itself. Here the genotype–phenotype map is improved adaptively during the searching process for solution candidates. The algorithm is applied to three-bit deceptive problems as a kind of typical combinatorial optimization problem. The difficulty with canonical GAs can be controlled by the genotype–phenotype map, and the output shows fairly good performance. Another relevant investigation for gene mapping using GAs is available in [145].

9.4.3 Gene Finding and Promoter Identification from DNA Sequences

Automatic identification of the genes from the large DNA sequences is an important problem in bioinformatics [146]. A cell mechanism recognizes the beginning of a gene or gene cluster with the help of a promoter and are necessary for the initiation of transcription. The promoter is a region before each gene in the DNA that serves as an indication to the cellular mechanism that a gene is ahead. For example, the codon AUG (which codes for methionine) also signals the start of a gene. Recognition of regulatory sites in DNA fragments has become particularly popular because of the increasing number of completely sequenced genomes and mass application of DNA chips. Recently a genome-wide map of active promoters in human fibroblast cells, determined by experimentally locating the sites of preinitiation complex binding throughout the human genome has been described [242]. This map defines 10,567 active promoters corresponding to 6,763 known genes and at least 1,196 unannotated transcriptional units. Features of the map suggest extensive use of multiple promoters by the human genes and widespread clustering of active promoters in the genome.

Methods

Using GA, Kel et al. [235] designed sets of appropriate oligonucleotide probes capable of identifying new genes belonging to a defined gene family within a cDNA or genomic library. One of the major advantages of this approach is the low homology requirement to identify functional families of sequences with little homology.

Levitsky et al. [272] described a method for recognizing promoter regions of eukaryotic genes with an application on Drosophila melanogaster. Its novelty lies in realizing the GA to search for an optimal partition of a promoter region into local nonoverlapping fragments and selection of the most significant dinucleotide frequencies for the aforesaid fragments.

The method of prediction of eukaryotic PolII promoters from DNA sequence [248], takes advantage of a combination of elements similar to neural networks and GAs to recognize a set of discrete subpatterns with variable separation as one pattern: a promoter. The neural networks use, as input, a small window of DNA sequence, as well as the output of other neural networks. Through the use of GAs, the weights in the neural networks are optimized to discriminate maximally between promoters and nonpromoters. Genetic algorithms are also used in conjunction with neural networks in [53] for the detection of promoters. Two levels of GAs are used. In the first level, the outer GA, the chromosome encodes a predetermined number of input features sampled from a larger pool, along with the architecture and connectivity of the neural network. The inner GA is used for training the network so encoded in a chromosome of the outer GA. The fitness of the chromosomes in the inner GA is equal to the predictive accuracy, A, on the training set, defined as

$$A = \frac{(TP/|Pos| + TN/|Neg|}{2}, \qquad (9.3)$$

where TP and TN are the number of correctly classified positive and negative examples, and $|Pos|$ and $|Neg|$ are the total number of positive and negative examples. The fitness of the chromosome of the outer GA is equal to the accuracy defined in Eq. (9.3), but computed on the test set. ClusterScan [236] is another approach where GAs are used to detect optimal combinations of binding sites from the TRANSFAC database [139].

9.4.4 Interpretation of Gene Expression and Microarray Data

Gene expression is the process by which a gene's coded information is converted into the structures present and operating in the cell. Expressed genes include those that are transcribed into mRNA and then translated into protein and those that are transcribed into RNA but not translated into protein (e.g., transfer and ribosomal RNAs). Not all genes are expressed, and gene expression involves the study of the expression level of genes in the cells under different conditions. Conventional wisdom is that gene products that interact with each other are more likely to have similar expression profiles than if they do not [286].

Microarray technology [382] allows expression levels of thousands of genes to be measured at the same time. A microarray is typically a glass (or some other material) slide, on to which DNA molecules are attached at fixed locations (spots). There may be tens of thousands of spots on an array, each containing a huge number of identical DNA molecules (or fragments of identical molecules), of lengths from twenty to hundreds of nucleotides. Each of these molecules ideally should identify one gene or one exon in the genome. The spots are either printed on the microarrays by a robot, or synthesized by photolithography (similarly as in computer chip productions) or by ink-jet printing.

Many unanswered and important questions could potentially be answered by correctly selecting, assembling, analyzing and interpreting microarray data. Clustering is commonly used in microarray experiments to identify groups of genes that share similar expressions. Genes that are similarly expressed are often coregulated and involved in the same cellular processes. Therefore, clustering suggests functional relationships between groups of genes. It may also help in identifying promoter sequence elements that are shared among genes. In addition, clustering can be used to analyze the effects of specific changes in experimental conditions and may reveal the full cellular responses triggered by those conditions.

A good solution of a gene ordering problem (i.e., finding optimal order of DNA microarray data) will have similar genes grouped together, in clusters. A notion of distance must thus be defined in order to measure similarity among genes. A simple measure is the Euclidean distance (other options are possible using Pearson correlation, absolute correlation, Spearman rank correlation, etc.). One can thus construct a matrix of intergene distances. Using this matrix one can calculate the total distance between adjacent genes and find that permutation of genes for which the total distance is minimized (similar to what is done in the TSP using GA (Sect. 9.3)).

Methods

Finding the optimal order of microarray data is known to be NP complete. Tsai et al. [462] formulated this as the TSP and applied family competition GA (FCGA) to solve it. The edge assembly crossover (EAX) is combined with the family competition concept and neighbor join mutation (NJ). In [463], a modified EAX and NJ are used in EA for efficiently optimizing the clustering and ordering of genes, ranging in size from 147 to 6221. Chromosomes in EAs are represented as a permutation of genes. The size of the population is taken equal to the number of genes in problems that involved fewer than 1000 genes, and half of the number of genes in larger problems. Fitness of chromosomes are evaluated from Eq. (9.1), and the distance matrix is formed using Pearson correlation. Crossover and mutation rates are set to one. Microarray data analysis is a competitive field, and no decisive measure of the performance of methods is available, so methods using EAs for microarray are compared in the TSP framework [463].

Garibay et al. [497] introduced a proportional GA (PGA) that relies on the existence or nonexistence of genes to determine the information that is expressed. The information represented by a PGA individual depends only on what is present in the individual and not on the order in which it is present. As a result, the order of the encoded information is free to evolve in response to factors other than the value of the solution.

9.4.5 Gene Regulatory Network Identification

Inferring a gene regulatory network from gene expression data obtained by DNA microarray is considered as one of the most challenging problems in the field of bioinfomatics [4]. An important and interesting question in biology, regarding the variation of gene expression levels is how genes are regulated. Since almost all cells in a particular organism have an identical genome, differences in gene expression and not the genome content are responsible for cell differentiation during the life of the organism.

For gene regulation, an important role is played by a particular type of protein, called the transcription factors [426]. The transcription factors bind to specific parts of the DNA, called transcription factor binding sites (i.e., specific, relatively short combinations of A, T, C or G), which are located in promoter regions. Specific promoters are associated with particular genes and are generally not too far from the respective genes, though some regulatory effects can be located as far as 30,000 bases away, which makes the definition of the promoter difficult.

Transcription factors control gene expression by binding to the gene's promoter and either activating (switching on) the gene or repressing it (switching it off). Transcription factors are gene products themselves, and therefore in turn can be controlled by other transcription factors. Transcription factors can control many genes, and some (probably most) genes are controlled by combinations of transcription factors. Feedback loops are possible. Therefore we can talk about gene regulation networks. Microarrays and computational methods are playing a major role in attempts to reverse engineer gene networks from various observations.

Methods

In gene network inference problems the objective is to predict a regulating network structure of the interacting genes from the observed data, i.e., expression pattern. The gene expressions are regulated in discrete state transitions, such that the expression levels of all genes are updated simultaneously. In [11] each real-valued chromosomes (in GAs) represents the expression level of all the genes. Each gene has a specific expression level for another gene; so, for N genes there are N^2 expression levels. Fitness of the chromosomes are evaluated by absolute error with generated expression pattern (sum of all expressions) from the target expression pattern. Population sizes of 2500, 5000 and 7000 are taken for 5, 7 and 10 genes, respectively. The GA run for 150 generations with a crossover and mutation rate of 0.99 and 0.01, respectively. Relevant investigations in this regard using GAs are also available in [10, 12, 52, 460].

9.4.6 Construction of Phylogenetic Trees for Studying Evolutionary Relationship

All species on earth undergo a slow transformation process called evolution. To explain the evolutionary history of today's species and how species relate to one another in terms of common ancestors, trees are constructed, whose leaves represent the present-day species and interior nodes represent the hypothesized ancestors. These kinds of labelled binary trees are called phylogenetic trees [426]. Phylogenetic analysis is used to study evolutionary relationships.

Phylogenies are reconstructed based on comparisons between present-day objects. The term object is used to denote the units for which one wants to reconstruct the phylogeny. Input data required for constructing phylogeny are classified into two main categories [426]: (a) Discrete character, such as beak shape, number of fingers, and presence or absence of a molecular restriction site. Each character can have a finite number of states. The data relative to these characters are placed in an objects character matrix called character state matrix. (b) Comparative numerical data, called distances between objects. The resulting matrix is called the distance matrix.

Given the data (character state matrix or distance matrix) for n taxa (objects), the phylogenetic tree reconstruction problem is to find the particular permutation of taxa that optimize the criteria (distance). The problem is equivalent to the problem of TSP. One can simply associate one imaginary city to each taxon, and define as the distance between two cities the data obtained from the data matrix for the corresponding pair of taxa.

Methods

Exhaustive search of the space of phylogenetic trees is generally not possible for more than 11 taxa, and so algorithms for efficiently searching the space of trees must be developed. Phylogeny reconstruction is a difficult computational problem, because the number of possible solutions (permutations) increases with the number of included taxa (objects) [273]. Branch-and-bound methods can reasonably be applied for up to about 20 taxa, so scientists generally rely on heuristic algorithms, such as stepwise addition and star decomposition methods. However, such algorithms generally involve a prohibitive amount of computation time for large problems and often find trees that are only locally optimal. Heuristic search strategies using GAs [233, 269, 273, 297] can overcome the aforesaid problems by faster reconstruction of the optimal trees with less computing power.

In [297] each chromosome in GA is encoded as a permutation of 15 taxa (same as TSP), and selection, crossover and mutation operations are performed to minimize the distance among the taxa. Here each taxon is an amino acid sequence taken from the GenBank, and distance between them is computed as an alignment score using CLUSTAL W [459]. The GA population consisted of 20 trial trees. Crossover probability of 0.5 and mutation probability of 0.2 has been used. Optimal trees are obtained after 138 generations. The only difference with TSP is that the endpoints of the chromosome (GA) are relevant in phylogenetic trees as they represent the starting and the endpoints of evolutionary relationship. GAs has also been used [436] for automatic self-adjustment of the parameters of the optimization algorithm of phylogenetic trees.

9.4.7 DNA Structure Prediction

DNA structure plays an important role in a variety of biological processes. Different dinucleotide and trinucleotide scales have been described to capture various aspects of DNA structure, including base stacking energy, propeller twist angle, protein deformability, bendability and position preference [23]. Three-dimensional DNA structure and its organization into chromatin fibres is essential for its functions and is applied in protein binding sites, gene regulation, triplet repeat expansion diseases, etc. DNA structure depends on the exact sequence of nucleotides and largely on interactions between neighboring base pairs. Different sequences can have different intrinsic structures. Periodic repetitions of bent DNA in phase with the helical pitch will cause DNA to assume a macroscopically curved structure. Flexible or intrinsically curved DNA is energetically more favorable to wrap around histones than rigid and unbent DNA.

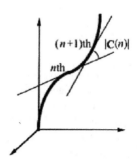

Fig. 9.4. Representation of the DNA curvature in terms of angular deviation between the local helical axes of the turn centered on the nth and $(n+1)$th base pairs [15]

The curvature of a space line is defined as the derivative, dt/dl, of the tangent vector t, along the line l. Its modulus is the inverse of the curvature radius, and its direction is that of the main normal to the curve [266]. In the case of DNA, the line corresponds to the helical axis, and the curvature is a vectorial function of the sequence. The curvature represents the angular deviation ($|C(n)|$) between the local helical axes of the nth and $(n+1)$th base pairs (Fig. 9.4). Under similar external conditions, the intrinsic curvature function represents the differential behavior of different DNA tracts and corresponds to the most stable superstructure. The physical origin of curvature is still a matter of debate [15]; it is, however, a result of the chemical and consequently stereochemical inhomogeneity of the sequence, which gives rise to different macroscopic manifestations. These manifestations change with the thermodynamic conditions, such as pH, the ionic force, the kind of counterions, and obviously the temperature as a result of perturbations on the intrinsic curvature depending on the sequence-dependent bendability. Therefore, it is generally found useful to characterize a DNA superstructure with the so-called intrinsic curvature function [15].

Methods

The three-dimensional spatial structure of a methylene-acetal-linked thymine dimer present in a 10 base pair (bp) sense–antisense DNA duplex was studied in [51] with a GA designed to interpret Nuclear Overhauser Effect

(NOE) interproton distance restraints. Trial solutions (chromosomes in GAs) are encoded on bit strings which represent torsion angles between atoms. From these torsion angles, atomic coordinates, which are needed for fitness function, are calculated using the DNA Evolutionary Nuclear Overhauser Enhancement Interpretation System for Structure Evaluation, or, DENISE, program. The problem is to find a permutation of torsion angles (eight torsion angles for each nucleotide in DNA) that minimizes the atomic distance between protons of nucleotides. The GA minimizes the difference between distances in the trial structures and distance restraints for a set of 63 proton–proton distance restraints defining the methylene-acetal-linked thymine dimer. The torsion angles were encoded by Gray coding, and the GA population consisted of 100 trial structures. Uniform crossover with a probability of 0.9 and mutation rate of 0.04 was used. It was demonstrated that the bond angle geometry around the methylene-acetal linkage plays an important role in the optimization.

A hybrid technique involving artificial neural network (ANN) and GA is described in [358] for optimization of DNA curvature characterized in terms of the reliability (RL) value. In this approach, first an ANN approximates (models) the nonlinear relationship(s) existing between its input and output example data sets. Next, the GA searches the input space of the ANN with a view to optimize the ANN output. Using this methodology, a number of sequences possessing high RL values have been obtained and analyzed to verify the existence of features known to be responsible for the occurrence of curvature.

9.4.8 RNA Structure Prediction

An RNA molecule is considered as a string of n characters $R = r_1 r_2 \ldots r_n$ such that $r_i \in \{A, C, G, U\}$. Typically n is in the hundreds, but it could also be in the thousands. The secondary structure of the molecule is a collection S of a set of stems, each stem consisting of a set of consecutive base pairs $(r_i r_j)$ (e.g., GU, GC, AU). Here $1 \leq i \leq j \leq n$, and r_i and r_j are connected through hydrogen bonds. If $(r_i, r_j) \in S$, in principle we should require that r_i be a complement to r_j and that $j - i > t$, for a certain threshold t (because it is known that an RNA molecule does not fold too sharply on itself). With such an assumption [426], the total free energy E of a structure S is given by

$$E(s) = \sum_{(r_i, r_j) \in S} \alpha(r_i, r_j), \qquad (9.4)$$

where $\alpha(r_i, r_j)$ gives the free energy of base pair (r_i, r_j). Generally the adopted convention is $\alpha(r_i, r_j) < 0$, if $i \neq j$, and $\alpha(r_i, r_j) = 0$, if $i = j$.

Attempts to predict automatically the RNA secondary structure can be divided into essentially two general approaches. The first one involves the overall free energy minimization by adding contributions from each base pair, bulged base, loop and other elements [2]. EAs are found to be suitable for this purpose. Chromosomes in EAs are encoded to represent the RNA structure, and fitness of each chromosome is evaluated in terms of the free energy (Eq. (9.4)). The second type of approach [486] is more empirical and it involves searching for a combination of nonexclusive helices with a maximum number of base pairings, satisfying the condition of a tree-like structure for the biomolecule. Within the latter, methods using dynamic programming (DP) are the most common [486, 523]. While DP can accurately compute the minimum energy within a given thermodynamic model, the natural fold of RNA is often in a suboptimal energy state and requires soft computing (EAs) rather than hard computing.

RNA may enter intermediate conformational states that are key to its functionality. These states may have a significant impact on gene expression. The biologically functional states of RNA molecules may not correspond to their minimum energy state, and kinetic barriers may exist that trap the molecule in a local minimum. In addition, folding often occurs during transcription, and cases exist in which a molecule will undergo transitions between one or more functional conformations before reaching its native state. Thus, methods for simulating the folding pathway of an RNA molecule and locating significant intermediate states are important for the prediction of RNA structure and its associated function.

Methods

The possibilities of using GAs for the prediction of RNA secondary structure were investigated in [46, 177]. The implementations used a binary representation for the solutions (chromosomes in GAs). The algorithm, using the procedure of stepwise selection of the most fit structures (similarly to natural evolution), allows different models of fitness for determining RNA structures. The analysis of free energies for intermediate foldings suggests that in some RNAs the selective evolutionary pressure suppresses the possibilities for alternative structures that could form in the course of transcription. The algorithm had inherent incompatibilities of stems due to the binary representation of the solutions.

Wiese et al. [494] used GAs to predict the secondary structure of RNA molecules, where the secondary structure is encoded as a permutation similar to path representation in TSP (each helix is associated to one imaginary city) to overcome the inherent incompatibilities of binary representation for RNA molecule structure prediction. They showed that the problem can be decomposed into a combinatorial problem of finding the subset of helices from a set of feasible helices leading to a minimum energy (using Eq. (9.4)) in the molecule. More specifically, the algorithm predicts the specific canonical base pairs that will form hydrogen bonds and build helices. Different combinations of crossover and mutation probabilities ranging from 0.0 to 1.0 in increments of 0.01 and 0.1 were tested for 400 generations with a population size of 700 (maximum). Results on RNA sequences of lengths 76, 210, 681 and 785 nucleotides were provided. It was shown that the Keep-Best Reproduction operator has similar benefits as in the TSP domain. A comparison of several crossover operators was also provided.

A massively parallel GA for the RNA folding problem has been used in [428, 429, 430]. The authors demonstrated that the GA with improved mutation operator predicts more correct (true-positive) stems and more correct base pairs than what could be predicted with a DP algorithm.

9.4.9 Protein Structure Prediction and Classification

Identical protein sequences result in identical three-dimensional structures. So it follows that similar sequences may result in similar structures, and this is usually the case. The converse, however, is not true: Identical three-dimensional structures do not necessarily indicate identical sequences. It is because of this that there is a distinction between "homology" and "similarity". There are examples of proteins in the databases that have nearly identical three-dimensional structures, and are therefore homologous, but do not exhibit significant (or detectable) sequence similarity. Pairwise comparisons do not readily show positions that are conserved among a whole set of sequences and tend to miss subtle similarities that become visible when observed simultaneously among many sequences. Thus one wants to simultaneously compare several sequences.

Structural genomics is the prediction of the three-dimensional structure of a protein from the primary amino acid sequence [89]. This is one of the most challenging tasks in bioinformatics. The five levels of protein structure are described below. Four of them are illustrated in Fig. 9.5.

- Primary structure is the sequence of amino acids that compose the protein (Fig. 9.5(a)).
- The secondary structure of a protein is the spatial arrangement of the atoms constituting the main protein backbone (Fig. 9.5(b)). Linus Pauling was the first to develop a hypothesis for different potential protein secondary structures. He developed the α-helix structure and later the β-sheet structure for different proteins. An α-helix is a spiral arrangement of the protein backbone in the form of a helix with hydrogen bonding between side-chains. The β-sheets consist of parallel or antiparallel strands of amino acids linked to adjacent strands by hydrogen bonding. Collagen is an example of a protein with β-sheets serving as its secondary structure.
- The super-secondary structure (or motif) is the local folding patterns built up from particular secondary structures. For example the EF-hand motif consists of an α-helix, followed by a turn, followed by another α-helix.
- Tertiary structure is formed by packing secondary structural elements linked by loops and turns into one or several compact globular units called domains, i.e., the folding of the entire protein chain (Fig. 9.5(c)).

- Final protein may contain several protein subunits arranged in a quaternary structure (Fig. 9.5(d)).

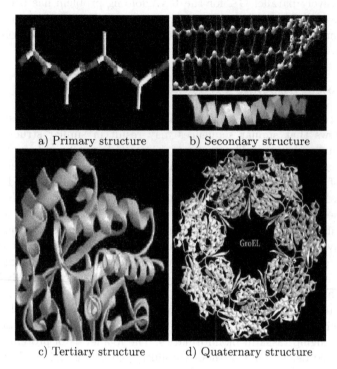

a) Primary structure   b) Secondary structure

c) Tertiary structure   d) Quaternary structure

Fig.   9.5.   Different   levels   of   protein   structures   (Source-http://swissmodel.expasy.org//course/course-index.htm, Chaps. 1-4)

Protein sequences almost always fold into the same structure in the same environment. Hydrophobic interaction, hydrogen bonding, electrostatic and other Van der Waals-type interactions also contribute to determine the structure of the protein. Many efforts are underway to predict the structure of a protein, given its primary sequence. A typical computation of protein folding would require computing all the spatial coordinates of atoms in a protein molecule starting with an initial configuration and working up to a final minimum-energy folding configuration [426]. Sequence similarity methods can predict the secondary and tertiary structures based on homology to known proteins. Secondary structure predictions methods include Chou-Fasman [89], neural network [381, 395], nearest neighbor methods [408, 409] and Garnier–Osguthorpe–Robson [158]. Tertiary structure prediction methods are based on energy minimization, molecular dynamics, and stochastic searches of conformational space.

Proteins clustered together into families are clearly evolutionarily related. Generally, this means that pairwise residue identities between the proteins are 30% or greater. Proteins that have low sequence identities, but whose structural and functional features suggest that a common evolutionary origin is probable, are placed together in superfamilies.

Methods

The work of Unger and Moult [467, 468, 469] is one of the earlier investigations that discussed the reduced three-dimensional lattice protein folding problem for determining tertiary structure of protein in a GA framework. In this model, the energy function of protein chains is optimized. The encoding proposed by Unger et al. is a direct encoding of the direction of each peptide from the preceding peptide (5 degrees of freedom, disallowing back move). Peptides are represented as single-point units without side chains. Each peptide is represented by 3 bits to encode 5 degrees of freedom. The evaluation function solely evaluates nonsequential hydrophobe-to-hydrophobe contacts and is stated as a negative value (-1 per contact) with larger negative values indicating better energy conformations (thus stating the problem in terms of minimization). The algorithm begins with a population of identical unfolded configurations. Each generation begins with a series of K mutations being applied to each individual in the population, where K is equal to the length of the encoding. These mutations are filtered using a Monte Carlo acceptance algorithm which disallows lethal configurations (those with back moves), always accepts mutations resulting in better energy, and accepts increased energy mutations based upon a threshold on the energy gain which becomes stricter over time. One-point crossover with an additional random mutation at the crossover point follows, producing a single offspring for each selected pair of parents; however, lethal configurations are rejected. In this situation, the crossover operation is retried for a given pair of parents until a nonlethal offspring can be located. Offspring are accepted using a second Monte Carlo filter which accepts all reduced energy conformations and randomly accepts increased energy offspring, again using a cooling threshold on the energy gain. The algorithm uses 100% replacement of all individuals in a generation through crossover except the single best, elitist individual. Test data consisted of a series of 10 randomly produced 27-length sequences and 10 randomly produced 64-length sequences. The algorithm operated on each of the 27- and 64-length sequences for roughly 1.2 million and 2.2 million function evaluations, respectively, using a population size of 200. Performance comparisons were given between the above algorithm and a pure Monte Carlo approach, which greatly favored the former. While the encoding and evaluation function proposed by Unger and Moult are fairly straightforward, the algorithm differs from a standard GA approach in several aspects. Most notable are the nonrandom initialization, the high level of mutation, and the

Monte Carlo filtering of both the mutation and crossover results, which resembles typical simulated annealing approaches.

Patton et al. [363] determined tertiary structures of proteins based on the concept of Unger and Moult. They enlarged the representation from 3 to 7 bits per peptide in order to encode one of the 120 permutations of the 5 allowable directions for each. It was shown that the GA indeed appears to be effective for determining the tertiary structure with much less computational steps than that reported by Unger et al.

Krasnogor et al. [257, 258] investigated the impact of several algorithmic factors for simple protein structure prediction (PSP) problem: the conformational representation, the energy formulation and the way in which infeasible conformations are penalized. Their analysis leads to specific recommendations for both GAs and other heuristic methods for solving PSP on the HP model. A detailed comparison between the work of Unger et al. and Patton et al. and an algorithm using GAs to overcome their limitations has also been presented [258].

A hill-climbing GA for simulation of protein folding has been described in [100]. The program builds a set of Cartesian points to represent an unfolded polypeptide's backbone. The dihedral angles determining the chain's configuration are stored in an array of chromosome structures that is copied and then mutated. The fitness of the mutated chain's configuration is determined by its radius of gyration. A four-helix bundle was used to optimize the simulation conditions. The program ran 50% faster than the other GA programs, and tests on 100 nonredundant structures produced results comparable to that of other GAs.

In [26], features are extracted from protein sequences using a position-specific weight matrix. Thereafter, a GA-based fuzzy clustering scheme [299] is used for generating prototypes of the different superfamilies. Finally, superfamily classification of new sequences is performed by using the nearest neighbor rule.

Other investigations on protein structure prediction are available in [50, 81, 99, 178, 205, 239, 251, 301, 334, 366, 386, 414, 451]. An overview and state-of-the-art of the applications of EAs for protein folding problem is described in [422].

9.4.10 Molecular Design and Docking

When two molecules are in close proximity, it can be energetically favorable for them to bind together tightly. The molecular docking problem is the prediction of energy and physical configuration of binding between two molecules. A typical application is in drug design, in which one might dock a small molecule that is a described drug to an enzyme one wishes to target. For example, HIV protease is an enzyme in the AIDS virus that is essential for its replication. The chemical action of the protease takes place at a localized active site on its surface. HIV protease-inhibitor drugs are small molecules that bind to the active site in HIV protease and stay there, so that the normal functioning of the enzyme is prevented. Docking software allows us to evaluate a drug design by predicting whether it will be successful in binding tightly to the active site in the enzyme. Based on the success of docking and the resulting docked configuration, designers can refine the drug molecule [271].

Molecular design and docking is a difficult optimization problem, requiring efficient sampling across the entire range of positional, orientational and conformational possibilities [501]. The major problem in molecular binding is that the search space is very large and the computational cost increases tremendously with the growth of the degrees of freedom. A docking algorithm must deal with two distinct issues: a sampling of the conformational degrees of freedom of molecules involved in the complex, and an objective function (OF) to assess its quality.

For molecular design the structure of a flexible molecule is encoded by an integer-valued or real-valued chromosome in GA, the ith element of which contains the torsion angle for the ith rotatable bond. The energy for the specified structure (conformation) can be calculated using standard molecular modelling package, and this energy is used as the fitness function for the GA. GAs try to identify a set of torsion angle values that minimize the calculated energy. So GA is becoming a popular choice for the heuristic search method in molecular design and docking applications [489]. Both canonical GAs and evolutionary programming methods are found to be successful in drug design and docking. Some of them are described below.

Methods

A novel and robust automated docking method that predicts the bound conformations (structures) of flexible ligands to macromolecular targets has been developed [317]. The method combines GAs with a scoring function that estimates the free energy change upon binding. This method applies a Lamarckian model of genetics, in which environmental adaptations of an individual's phenotype are reverse transcribed into its genotype and become inheritable traits. Three search methods, viz., Monte Carlo simulated annealing, a traditional GA and the Lamarckian GA (LGA) were considered, and their performance was compared in docking of seven protein–ligand test systems having known three-dimensional structure. The chromosome is composed of a string of real-valued genes: three Cartesian coordinates for the ligand translation; four variables defining a quaternion specifying the ligand orientation; and one real-value for each ligand torsion, in that order. The order of the genes that encode the torsion angles is defined by the torsion tree created by AUTO-TORS (a preparatory program used to select rotatable bonds in the ligand). Thus, there is a one-to-one mapping from the ligand's state variables to the genes of the individual's chromosome. An individual's fitness is the sum of the intermolecular interaction energy between the ligand and the protein, and the intramolecular interaction energy of the ligand. In the GA and LGA docking, an initial population of 50 random individuals, a maximum number of 1.5×10^6 energy evaluations, a maximum number of 27,000 generations, a mutation rate of 0.02 and a crossover rate of 0.80 were used. Proportional selection was used, where the average of the worst energy was calculated over a window of the previous ten generations.

An evolutionary approach for designing a ligand molecule that can bind to the active site of a target protein is presented in [20, 29]. Here, a two-dimensional model of the ligand was considered. Variable string length genetic algorithm (VGA) was used for evolving an appropriate arrangement of the basic functional units of the molecule to be designed. The method is found to be superior to fixed string length GA for designing a ligand molecule to target several receptor proteins. More details regarding this method and some experimental results are provided in Sect. 9.5.

Chen et al. [86] derived a population based annealing genetic algorithm (PAG) using GAs and simulated annealing (SA). They applied it to find binding structures for three drug protein molecular pairs, including an anti-cancer drug methotrexate (MTX). All of the binding results keep the energy at low levels, and have a promising binding geometrical structure in terms of number of hydrogen bonds formed. One of the design methods of PAG, which incorporates an annealing scheme with the normal probability density function as the neighbor generation method, was described in [481]. The algorithm was used for computer-aided drug design. Using a dihydrofolate reductase enzyme with the anti-cancer drug methotrexate and two analogs of antibacterial drug trimethoprim, PAGs can find a drug structure within a couple of hours. A similar work is available in [482].

Christopher et al. [399] evaluated the use of GAs with local search in molecular docking. They investigated several GA-local search hybrids and compared results with those obtained from simulated annealing in terms of optimization success, and absolute success in finding the true physical docked configuration.

Other relevant investigations are available in [90, 120, 201, 219, 220, 303, 335, 474, 489, 506]. A survey on the application of GAs for molecular modelling, docking of flexible ligands into protein active sites and for de novo ligand design is described in [495]. Advantages and limitations of GAs are mentioned for the aforesaid tasks.

9.5 Experimental Results

This section describes some experimental results of a GA-based technique for designing small ligand molecules that can bind to the active sites of target proteins [20, 29]. Here, a two-dimensional model of the system is used, and the ligand molecule is assumed to adopt a tree-like structure on both sides of the pharmacophore, whose location is fixed. Each node of the tree is filled up by a group selected from among a set of seven groups, namely, *Alkyl 1C, Alkyl 1C Polar, Alkyl 3C, Alkyl 3C Polar, Polar, Aromatic* and *Aromatic Polar*. The scheme employs variable string length genetic algorithm (VGA) for encoding the tree, which is in contrast to the method in [163], where the size of the tree was kept fixed, and hence a fixed string length genetic algorithm was used. Figure 9.6 shows a schematic diagram of a tree structure of the ligand molecule. The van der Waals interaction energy among the functional groups of the ligand and the closest residue of the active site of the target protein is taken as the minimizing criterion.

Several protein target receptors, namely HIV-I Nef protein, HIV-I Integrase, Retroviral protease, HIV-I capsid protein and Rhinoviral protein obtained from human rhinovirus strain 14, were considered in [29]. Here, the result for HIV-I Nef protein is provided for illustration. The coordinates of the proteins were obtained from Protein Data Bank [54]. For the genetic algorithm, the probabilities of mutation and single-point crossover were fixed at 0.01 and 0.8, respectively. Population size was taken to be equal to 200 and the algorithm was executed for 500 generations. Results were taken for different possible positions of the pharmacophore within the active site, and the evolved ligand having the lowest energy value is reported as the solution.

It was found that the energy value of the ligands and also the interaction energy with the target protein of the molecules evolved using VGA was lower than the corresponding value for the ligand evolved using fixed length GA. Moreover, the configuration of the former was expected to be more stable. Figure 9.7 shows the molecule evolved by the proposed method and also in a three-dimensional view (which is obtained using Insight II [MSI/Accelrys, San Diego, CA, USA]) appropriately docked into the active site of the corresponding proteins (where the docking is obtained using GOLD [221]), respectively. The target protein is shown in ribbon form, while the ligand is represented in Corey–Pauling–Koltun (CPK) form, where the balls denote the atoms. A stability study of the evolved molecules was also made by considering similar molecules from the Cambridge Crystallographic Datacentre (CSD) [5]. The molecules with code ACOYUM and ALITOL with energy values of -24.29 KCal/mol and -41.87 Kcal/mol were found to be similar to the molecules evolved using fixed GA and VGA, respectively. Again, ALITOL (which is more similar to the VGA-evolved ligand) is associated with a lower energy value, and hence is expected to be more stable.

9.6 Summary

The increasing availability of annotated genomic sequences has resulted in the introduction of computational genomics and proteomics — large-scale analysis of complete genomes and the proteins that they encode for relating specific genes to diseases. The rationale for applying computational approaches to facilitate the understanding of various biological processes mainly includes:

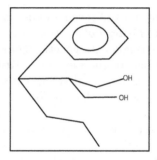

Fig. 9.6. Schematic diagram of the tree structure of the ligand molecule [29]

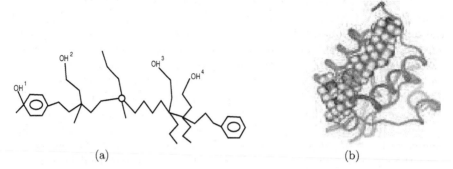

(a)   (b)

Fig. 9.7. The VGA-based method for HIV-I Nef protein [29]: (a) Evolved molecule; (b) Interaction of the ligand (represented in CPK) with the target protein (represented as a ribbon)

- to provide a more global perspective in experimental design
- to capitalize on the emerging technology of database mining: the process by which testable hypotheses are generated regarding the function or structure of a gene or protein of interest by identifying similar sequences in better characterized organisms.

GAs appear to be a very powerful artificial intelligence paradigm to handle these issues. This chapter provided an overview of different bioinformatics tasks and the relevance of GAs to handle them efficiently.

Even though the current approaches in biocomputing using EAs are very helpful in identifying patterns and functions of proteins and genes, the output results are still far from perfect. There are three general characteristics that might appear to limit the effectiveness of GAs. First, the basic selection, crossover and mutation operators are common to all applications, so research is now focussed to design problem-specific operators to get better results. Second, a GA requires extensive experimentation for the specification of several parameters so that appropriate values can be identified. Third, GAs involve a large degree of randomness and different runs may produce different results, so it is necessary to incorporate problem-specific domain knowledge into GA to reduce randomness and computational time and current research is going on in this direction also. The methods are not only time-consuming, requiring Unix workstations to run on, but might also lead to false interpretations and assumptions due to necessary simplifications. It is therefore still mandatory to use biological reasoning and common sense in evaluating the results delivered by a biocomputing program. Also, for evaluation of the trustworthiness of the output of a program it is necessary to understand its mathematical/theoretical background to finally come up with a useful analysis.

10

Genetic Algorithms and Web Intelligence

10.1 Introduction

With the continued increase in the usage of the World Wide Web (WWW), Web mining has been established as an important area of research. The WWW is a vast repository of unstructured information, in the form of inter-related files, distributed on numerous Web servers over wide geographical regions. Web mining deals with the discovery and analysis of useful information from the WWW. Given that an extremely large amount of information is available on the Web (the well-known search engine Google alone indexes about 20–30 terrabytes of data), it has become imperative that new and highly efficient Web mining techniques are developed, and specialized data mining techniques are tailored for utilizing this huge repository of information in a potentially useful manner.

Web pages are typically maintained on Web servers, which are accessed by different users in client–server transactions. The Web is queried, in general, hundreds of million times each day. The access patterns, user profiles and other data are maintained at the servers and/or the clients. The primary challenges in Web mining are designing search and retrieval algorithms that can find the relevant information quickly, creating new knowledge out of the information already available, customizing or personalizing this information based on user preferences, and also learning about the behavior of the consumers or the users.

To proceed towards Web intelligence, obviating the need for human intervention, we need to incorporate and embed artificial intelligence into Web tools. The necessity of creating server-side and client-side intelligent systems, that can effectively mine knowledge both across the Internet and, in particular, Web localities, is drawing the attention of researchers from the domains of information retrieval, knowledge discovery, machine learning and AI, among others. However, the problem of developing automated tools in order to find, extract, filter and evaluate the users' desired information from unlabelled, distributed and heterogeneous Web data is far from being solved.

Since researchers are actively developing Web mining techniques, which involves mining Web data, this technology is expected to play an important role for developing the intelligent Web. To handle the typical characteristics of the Web, and to overcome some of the limitations of existing methodologies, soft computing seems to be a good candidate. In particular, genetic algorithms (GAs) are seen to be useful for prediction and description, efficient search, and adaptive and evolutionary optimization of complex objective functions in the dynamic and complex environments of the Web. Searching the Web is very similar to an optimization problem. The Web can be considered to be a set of Web pages S structured as a graph with a neighborhood relation [74]. A fitness function

$$f : S \to R^+ \tag{10.1}$$

can evaluate Web pages (as in Google [73]). The task of a search engine is to output those pages which maximize f. Thus many researchers have been active in the application of GAs for making Web-related tasks more efficient.

Solving the different research problems in Web mining is a crucial step towards making the Web intelligent. The objective of this chapter is to provide an outline of Web mining, its various classifications, subtasks and challenges, and to provide a perspective about the potential of applying GAs to its different components. Mention of some commercially available systems is made. Broad guidelines for future research on Web mining, in general, are outlined. The chapter is organized as follows: Section 10.2 discusses the different issues in Web mining, its components and methodologies, Web mining categories, and its challenges and limitations. The use of GAs in different Web mining tasks is described in Sect. 10.3. Finally, Sect. 10.4 concludes the chapter.

10.2 Web Mining

Web mining deals with the application of data mining techniques to the Web for extracting interesting patterns and discovering knowledge [84]. Web mining, though essentially an integral part of data mining, has emerged as an important and independent research direction due to the typical nature of the Web, e.g., its diversity, size, and dynamic and link-based nature. A few important characteristics of Web data are that it is:

- unlabelled
- huge and distributed
- heterogeneous (mixed media)
- semistructured
- time varying
- high-dimensional

Some recent reviews on Web mining are available in [252, 287, 355].

Typically, the Web may be viewed as a directed graph (Fig. 10.1) that is highly dynamic and immensely large. Here the nodes represent the Web pages, and the edges represent the hyperlinks. The hyperlinks contain important information which can be utilized for efficient information retrieval. For example, in Fig. 10.1 the information that several hyperlinks (edges) point to page A may indicate that A is an *authority* [244] on some topic. Again, based on the structure of the Web graph, it may be possible to identify Web communities [149]. A Web community is described as a collection of Web pages such that each member of the collection contains many more links to other members in the community than outside it.

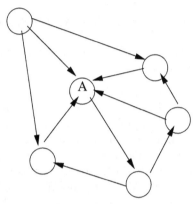

Fig. 10.1. Example of a part of the Web viewed as a directed graph

Mining of the Web data can be viewed in different levels — mining the Web content, its structure and its usage. The first task in Web content mining is the automatic creation of page summaries. The contents of a Web page may be varied, e.g., text, images, HTML, tables or forms. Accordingly, the pages may be classified based on content, and retrieval algorithms can be designed. Mining the search result is also an integral part of Web content mining. This may involve summarization of the results, clustering them into a hierarchy, categorizing the documents using phrases in title and snippets, and also combining and reorganizing the results of several search engines. The last task is also known as metasearching.

Mining the structure of the Web involves extracting knowledge from the interconnections of the hypertext documents in the WWW. This results in discovery of Web communities, and also pages that are authoritative. Moreover, the nature of relationship between neighboring pages can be discovered by structure mining. Important applications of structure mining include finding important pages, related pages and homepages. It is known that the contents of a page alone do not constitute the sole information about it. Information in a page is characterized by the content in that page, as well as that in some of the neighboring pages. Thus content and structure mining of the Web are essentially complementary in nature.

Web usage mining essentially consists of analyzing the data pertaining to the use of the Web. This can be viewed from two perspectives, viz., that of the user and that of the Web site. On the client (or the user) side, usage data are collected through the browsing history, while on the server side, data are collected through the request logs. Web usage mining involves discovering the category of users accessing pages of a certain type, the time and duration of access, as well as the order of page references. This information can be extremely useful for organizing the pages in a manner that will make future accesses more efficient and faster.

10.2.1 Web Mining Components and Methodologies

As mentioned in [140], Web mining can be viewed as consisting of four tasks, shown in Fig. 10.2. These are

- *Information Retrieval (IR)* deals with the automatic retrieval of all relevant documents, while at the same time ensuring that the irrelevant ones are fetched as rarely as possible. The IR process mainly includes document representation, indexing and searching for documents. Some popularly used search engines that query and retrieve information stored in databases (fully structured), HTML pages (semistructured) and free text (unstructured) are Google, AltaVista and WebCrawler.
- *Information Extraction (IE)* deals with automatically (without human interaction) extracting knowledge once the documents have been retrieved. The IE process mainly deals with the task of identifying specific fragments of a single document that constitute its core semantic content. Harvest [75], FAQFinder [181], OCCAM [265] and Parasite [446] are some IE systems.
- *Generalization* deals with the task of learning some general concepts from a given set of documents. In this phase, pattern recognition and machine learning techniques, like classification, clustering and association rule mining, are usually used on the extracted information. For example, learning to distinguish a homepage from other Web pages is a generalization task.
- *Analysis* deals with the task of understanding, visualizing and interpreting the patterns once they are discovered in the *Generalization* phase. Analysis is a data-driven problem which presumes that there is sufficient data available so that potentially useful information can be extracted and analyzed. Some typical analysis tools are Webviz system [377] and WEBMINER [314].

10.2.2 Web Mining Categories

The information contained in the Web can be broadly categorized into:

- *Web content*: the component that consists of the stored facts, e.g., text, image, sound clips, structured records like lists and tables

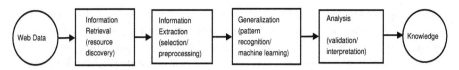

Fig. 10.2. Web mining subtasks [355]

- *Web structure*: the component that defines the connections within the content of the Web, e.g., hyperlinks and tags
- *Web usage*: that describes the users interaction with the Web, e.g., HTTP logs and application server logs

Depending on which category of Web data is being mined, Web mining has been classified as:

- *Web content mining*
- *Web structure mining*
- *Web usage mining*

Figure 10.3 shows a hierarchical description of the different tasks.

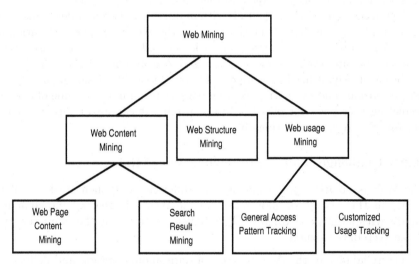

Fig. 10.3. Hierarchical description of the tasks in Web mining

Web content mining (WCM) is the process of analyzing and extracting information from the contents of Web documents. Research in this direction involves using techniques of other related fields, e.g., information retrieval, text mining, image mining and natural language processing.

In WCM, the data is preprocessed for extracting text from HTML documents, eliminating the stop words, identifying the relevant terms and computing some measures like the term frequency (TF) and document frequency (DF). The next issue in WCM involves adopting a strategy for representing the documents in such a way that the retrieval process is facilitated. Here the common information retrieval techniques are used. The documents are generally represented as a sparse vector of term weights; additional weights are given to terms appearing in title or keywords. The common data mining techniques applied on the resulting representation of the Web content are

- classification, where the documents are assigned to one or more existing categories
- clustering, where the documents are grouped based on some similarity measure (the dot product between two document vectors being the more commonly used measure of similarity)
- association, where association between the documents is identified

Other issues in WCM include topic identification, tracking and drift analysis, concept hierarchy creation and computing the relevance of the Web content.

In Web structure mining (WSM) the structure of the Web is analyzed in order to identify important patterns and inter-relations. For example, Web structure may reveal information about the quality of a page ranking, page classification according to topic and related/similar pages [354].

Web usage mining (WUM) deals with mining such data in order to discover meaningful patterns like associations among the Web pages, categorization of users. An example of discovered associations could be that 60% of users who accessed the site /webmining also accessed /webmining/software/. WUM can be effectively utilized in commercial applications, like designing new product promotions and evaluating existing ones, determining the value of clients, predicting user behavior based on users' profile. It can also be used in reorganizing the Web more rationally.

10.2.3 Challenges and Limitations in Web Mining

The Web is creating new challenges to different component tasks of Web mining as the amount of information on the Web is increasing and changing rapidly without any control [355]. The challenges stem from the characteristics of the Web data, namely

(a) low reliability of the data because of data anonymization and deliberate data corruption
(b) presence of noise
(c) extremely dynamic and transient nature
(d) heterogeneity and lack of structure
(e) semantic ambiguity
(f) requirement of personalization (i.e., though the query may be the same, the response may be different depending the user posing the query, e.g., response to the query "cricket" will be different to a sports enthusiast and a biologist)

(g) high redundancy in content

(h) nonnumeric nature of the data where definition of similarity itself is problematic

among others. As a consequence, the existing systems find difficulty in handling the newly emerged problems during information retrieval and extraction, generalization (clustering and association) and analysis. Some of these are described below.

In information retrieval (IR) or resource discovery, it is necessary for the IR system to be able to detect what information is really relevant. However, this is an extremely difficult and subjective task. Most of the IR systems simply follow a blind keyword-matching strategy that improves efficiency in terms of retrieval time at the expense of accuracy. Moreover, since relevance itself is difficult to define and is not a hard concept, ranking the retrieved pages according to their relevance becomes difficult. At the same time, taking a soft approach for rejecting pages (rather than the traditionally used hard decision) is advisable, though seldom implemented. Lack of sufficient data is a major problem faced during personalization of search. To achieve effective personalization, we need to combine demographic data, psychographic data, as well as user access patterns and interests. This leads to a feature space of an extremely high dimensionality. However, this type of data is hard to collect for a large number of customers [259].

Current Web mining methods are often limited due to dynamism, massive scale and heterogeneity of Web data. Because of the time-varying nature of Web data, many of the documents returned by the search engines are outdated, irrelevant and unavailable in the future, and hence users have to query different indices several times before getting a satisfactory response. Regarding the scaling problem, Etzioni [141] has studied the effect of data size on precision of the results obtained by the search engine. Current IR systems are not able to index all the documents present on the Web, which leads to the problem of low recall. The heterogenous nature of Web documents demands a separate mining method for each type of data.

Most of the information extraction algorithms used by different tools are based on the "wrapper" technique. Wrappers are procedures for extracting a particular information from Web resources. Their biggest limitation is that each wrapper is an IE system customized for a particular site and is not universally applicable. Also, source documents are designed for people, and few sites provide machine-readable specifications of their formatting conventions. Here ad hoc formatting conventions used in one site are rarely relevant elsewhere.

In the generalization phase, the challenges to Web mining arise from the fact that Web data is distributed, inherently high dimensional and noisy. As an example, consider the task of clustering in the context of Web mining. Although clustering is a well-developed topic in pattern recognition and data mining, the above-mentioned characteristics of the Web data demand much more robust and noise-resistant methods. Moreover, the clusters would be overlapping in high-dimensional space, and even the number of clusters cannot be stated a priori. Some attempts for developing robust clustering algorithms can be found in [222, 259]. However, presently most of the search engines use some basic form of clustering, e.g., Google currently supports simple host name-based clustering. The presence of a large number of outliers, or observations that do not conform to the general definition, presents an added level of complexity to all the machine learning algorithms in the generalization phase. For example, several users often traverse paths that do not correspond to their stated interests. These can generate misleading data and outliers. The data being inherently distributed, detection of outliers is by itself a difficult problem. Understanding, interpreting and visualizing the patterns/trends discovered during the generalization phase also presents new challenges in Web mining because of its context sensitive nature, and inherent uncertainty and imprecision.

Given the above challenges that are present in Web mining, a growing breed of researchers has become interested in the application of soft computing to Web mining problems. A recent review [355] is a testimony in this regard. In the following section, we discuss some such applications of GAs, an important component of soft computing, for solving different search and optimization tasks in Web mining.

10.3 Genetic Algorithms in Web Mining

Search and retrieval, query optimization, document representation and distributed mining are the primary areas where GAs and related techniques have been mainly applied for Web mining [355]. These are discussed below.

10.3.1 Search and Retrieval

A GA-based search to find other relevant homepages, given some user-supplied homepages, has been implemented in GSearch [104]. Web document retrieval by genetic learning of importance factors of HTML tags has been described in [241]. Here, the method learns the importance of tags from a training text set. GAs have also been applied for the purpose of feature selection in text mining [293]. A search engine, complementary to standard ones, based on GAs is proposed in [375]. It is called GeniMiner. The search problem is formulated as follows: Given a fixed number of Web page evaluations (downloading, content analysis, etc.), what strategy can maximize the users' gain (i.e., which pages/links should be explored in order to find interesting documents that will minimize the time devoted to manual analysis)?

Genetic algorithms are used in [291] to design a form of metasearch, where the results of standard search engines are combined in an optimal way so as to generate the most interesting pages for the users. The fitness function F of the Web pages is defined in the following way:

(a) Link quality $F(L)$:

$$F(L) = \sum_{i=1}^{n} \#K_i, \tag{10.2}$$

where n is the total number of input keywords given by the user, and $\#K_i$ denotes the number of occurrence of keyword K_i in the link L.

(b) Page quality $F(P)$:

$$F(P) = \sum_{i=1}^{m} F(L_j), \tag{10.3}$$

where m is the total number of links per page.

(c) Mean quality function M_q:

$$M_q = \frac{F_{max}(P) + F_{min}(P)}{2}, \tag{10.4}$$

where $F_{max}(P)$ and $F_{min}(P)$ are the maximum and minimum values of the pages qualities, respectively, after applying the GA. It should be noted that the upper value of F_{max} is $m * n$, and the least value of $F_{min}(P)$ is zero.

For each problem the authors used 300 downloaded pages from 4 search engines, namely Yahoo, Google, AltaVista and MSN. Crossover was applied on selected pages by interchanging the links on the parents after the crossover point.

In [143], genetic programming (GP) is used for automatically generating ranking functions that are context specific. In information retrieval, a ranking function is used to order the documents in terms of their predicted relevance to the given query. However, in general, these functions are fixed, regardless of the context in which the query is posed. Fan et al. used GP in [143] to automatically discover the best ranking function using a set of training documents, and a set of validation documents. A ranking function is represented as a tree, and this is taken as the chromosome of the GP which evolves the best tree. The average of precisions every time a new relevant document is found, normalized by the total number of relevant documents in an entire collection, is taken as a measure of the fitness function. The method is found to outperform two other retrieval systems. A comparative study of seven different fitness functions that can be used with GP has been carried out in [142]. This enhances a previous attempt by Lopez-Pujalte et al. [282], where a number of commonly used fitness functions (both with and without the relevance order information) were compared in GA studies and it was found that the *order-based* ones provide more favorable results. The *order-based fitness functions* take into account both the number of relevant documents and the order in which they appear in the ranked list. For example, consider two ranking functions both retrieving six relevant documents in the top ten results. Their relevance information (1 being relevant and 0 being irrelevant) in the ranking list are shown as follows:

Rank list 1: 1 0 1 1 0 0 1 0 1 1
Rank list 2: 1 1 0 1 1 1 0 1 0 0

If ranking order is ignored and only the number of relevant documents returned is considered, the performance of these two rank lists is the same. However, the second rank list should be preferred over the first because the second rank list presents relevant documents sooner (i.e., has higher precision). The study in [142] complements that in [282] by comparing more order-based fitness functions in the Web search context using a much larger text corpus. Further enhancements to the task of discovering ranking functions are proposed in [144].

An important component of search engines is Web crawlers. Optimization of these crawlers can significantly improve the search performance. In [433] an intelligent crawler called Gcrawler is described that uses a genetic algorithm for improving its crawling performance. Gcrawler estimates the best path for crawling as well as expands its initial keywords by using a genetic algorithm during the crawling. It acts intelligently without any relevance feedback or training and with no direct interaction with the users.

10.3.2 Query Optimization and Reformulation

An important technique for query reformulation is based on the relevance feedback approach. The purpose of relevance feedback is to use relevant and irrelevant document sets to modify a user query, making it more similar to the set of relevant documents. One way in which this can be achieved is to add the terms found in documents considered to be relevant in earlier searches (corresponding to the links that were followed) to the current query. At the same time, terms found in documents that were not considered to be relevant earlier (corresponding to the links that were not followed) are subtracted. This is expected now to retrieve more relevant documents.

GAs have also been extensively used for query reformulation and optimization. It is used in [329] to build user profiles by monitoring people's browsing behavior over time. They have been used to directly modify a user query, where the document and query are represented as vectors. Each individual represents a query vector and the individual genes can represent the weights of the keywords, or the presence or absence of keywords. In [68], Boughanem et al. have developed a query reformulation technique in which a GA generates several queries that explore different areas of the document space and determines the optimal one. Yang et al. [504] have presented an evolutionary algorithm for query optimization by reweighting the document indexing without query expansion. Another similar attempt is reported in [505]. A comparison of GAs with relevance feedback, ID3 and simulated annealing using a database with 8000 records was made in [88], where the authors showed that the GA outperformed the other three methods. In a subsequent study where the authors compared a GA with the best first search for spidering relevant Web pages [87], they found that recall (defined as the number of relevant documents retrieved as fraction of all relevant documents) was higher with the GA; precision (defined as the number of relevant documents retrieved as a fraction all the documents in retrieved by the system) was equal to best first search. An additional interesting finding was that the Web pages from the two algorithms were not overlapping but were largely complementary.

Kraft et al. [256, 374] apply genetic programming in order to improve weighted Boolean query formulation. Here the individuals represent Boolean queries. Each gene represents a subset of a query. Experiments were conducted with a small document collection of 483 abstracts taken from the *Communications of the ACM*, and different seeding strategies and fitness functions were tested. Although unable to draw conclusions about different fitness functions, it was found that seeding the initial population with terms taken from a predefined set of relevant documents gave better results than seeding with terms based on the distribution of the terms in the complete collection.

In [270], Leroy et al. show that query optimization based on a relevance feedback or genetic algorithm using dynamic query contexts can help casual users search the Internet more efficiently. It is reasoned that for casual users low success in Internet searching is due to the usage of a small number of keywords, around two [447], too few to retrieve a subset of relevant pages from a collection of millions. However since most of the users (around 77%) engage in multiple search sessions on the same topic, using the same search engine, valuable information regarding the implicit user interests from their behavior can be extracted during consecutive searches. Based on this information, the user queries can be more efficiently handled. In [270], both positive and negative expansion was used. In positive expansion, terms are added to the users' keywords with a Boolean "and", while in negative expansion terms are added to the users' keywords with a Boolean "not". The authors added five keywords using either "and" (positive expansion) or "not" (negative expansion), depending on whether the keyword was taken from a relevant context or an irrelevant one. In the genetic implementation, each chromosome was designed to encode a query and had five free slots to be filled with appropriate keywords. For computing the fitness of a chromosome i, the corresponding encoded query was sent to the search engine, and the top ten retrieved documents were considered. Let the current search be denoted by x. Two Jaccard scores, J_{Rix} and J_{Nix}, based on the words in the results and their similarity to the relevant and irrelevant contexts, were computed. The final score of the chromosome was computed as:

$$J_{ix} = \frac{(J_{Rix} - J_{Nix}) + 1}{2}. \tag{10.5}$$

The traditional selection, crossover and mutation operators were used. The relevance feedback algorithm and the GA was examined for three user groups, high, middle and low achievers, who were classified according to their overall performance. The interactions of users with different levels of expertise with different expansion types or algorithms were evaluated. The genetic algorithm with negative expansion tripled recall and doubled precision for low achievers, but high achievers displayed an opposite trend and seemed to be hindered in this condition. The effect of other conditions was less substantial. The advantage of the method in [270] is that it does not require users to submit a profile or to exert any additional effort, and does not invade privacy by gathering and retaining user information. Furthermore, by using only information of a particular search to build a dynamic query context, the user interest can change for every search without affecting the algorithms.

10.3.3 Document Representation and Personalization

Gordon [173] adopted a GA to derive better descriptions of documents. Here each document is assigned N descriptions where each description is a set of indexing terms. Then genetic operators and relevance judgements are applied to these descriptions in order to determine the best one in terms of classification performance in response to a specific query. Automatic Web page categorization and updating can also be performed using GAs [281].

GAs have been applied for classifying the students performance in a Web-based educational system in [311]. Methods for extracting useful and interesting patterns from large databases of students using online educational resources and their recorded paths within the system are investigated. The purpose is to find if there exists groups of students who use these online resources in a similar way, and if so, then can the class for any individual student be predicted. This information can then be used to help a student use the resources in a better manner. GA is used here to weight the features in the best possible way, and then an ensemble of classifiers with voting is adopted as the final classification tool. Results show a significant improvement over the version where GA is not used for feature weighting.

10.3.4 Distributed Mining

Gene Expression Messy Genetic Algorithm (GEMGA) [227], which is a subquadratic, highly parallel evolutionary search algorithm, is specially found suitable for distributed data mining applications, including the Web. The foundation of GEMGA is laid on the principles of both decomposing black box search into iterative construction of partial ordering and performing selection operation in the relation, class and sample spaces. The GEMGA is designed based on an alternate perspective of evolutionary computation proposed by the SEARCH framework that emphasizes the role of gene expression or intracellular flow of genetic information. The research on the use of GEMGA in distributed data mining is rapidly growing, and it deals with the problem of finding patterns in data in an environment where both the data and the computational resource are distributed. In [231], Kargupta et al. suggest Collective Data Mining (CDM) as a new approach towards distributed data mining (DDM) from heterogeneous sites. It is pointed out that naive approaches to distributed data analysis in a heterogeneous environment may lead to an ambiguous situation. CDM processes information locally and builds a global model in situations where it is not possible, as in the case of Web, to collect all the data in a common data warehouse and then process.

10.4 Summary

In order to utilize the full potential of the Web, and to make it intelligent, its services need to be improved, and it should be more comprehensible and usable for the users [182]. The continuing research in Web mining and its allied fields is expected to play a key role towards this goal. Web mining is a fast growing field, and new methodologies are being developed using both classical and soft computing approaches concurrently. In this chapter, the basic components of Web mining and its different tasks have been summarized. The challenges in this domain have been stated. A detailed discussion on the use of genetic algorithms for tackling different problems in Web mining, particularly search and retrieval, query optimization and reformulation, and document representation and personalization, is provided.

It may be noted in this regard that although application of GAs in Web mining has a lot of potential, the literature in this regard is poor, leaving scope for more research in this domain. A future potential application area of genetic algorithms in Web mining lies in the construction of adaptive Web sites. These are sites which automatically improve their organization and presentation by learning from visitor access patterns [141]. It focuses on the problem of index page synthesis, where an index page is a page consisting of a set of links that cover a particular topic. Its basic approach is to analyze Web access logs and find groups of pages that often occur together in user visits (and hence represent coherent topics in users' minds) and to convert them into "index" pages. Here genetic algorithms may be used for prediction of user preferences, dynamic optimization and evolution of Web pages. Comparison to other agent-based search methods, the use of parallel GAs to speed up the search, self-adaptation of parameters which will improve the GA performances, clustering of the results and interactive personalization of the search are other areas of further work in this regard.

Resource Description Framework (RDF) is becoming a popular encoding language for describing and interchanging metadata of Web resources. An Apriori-based algorithm for mining association rules from RDF documents is provided in [33]. In addition to RDF, user behavior analysis, distributed Web mining, Web visualization and Web services [295, 315, 445, 510] are some of the recent research directions in Web mining. Semantic Webs, where the stored documents have attached semantics, are also a recent development, and hence semantic Web mining is also a promising area. Application of genetic algorithms in these domains provides a prospective and potential direction of future research.

A

ϵ-Optimal Stopping Time for GAs

A.1 Introduction

The proof for the limiting nature of elitist GA (EGA) has been provided in Chap. 2. According to Theorem 2.2, an EGA will result in the best string as the number of iterations n goes to ∞. But the process needs to be stopped for some finite value of n. The value of n at which the process is stopped is called the stopping time of the process. The objective here is to determine the value of n which is in some sense "optimal". More specifically, the problem is to find a value for n which will provide the best solution. The following discussion is available in detail in [324].

A.2 Foundation

Here the same notation as used in Chap. 2 is followed. Note that, for a starting population Q, GAs can result in many populations with different probabilities after n iterations. Thus, if the process is to be stopped after a specific number of iterations, it may not always be guaranteed that the optimal string is obtained. In fact, the lemma stated below [324] indicates that there is a positive (> 0) probability of not obtaining the optimal solution after a finite number of iterations.

Lemma A.1. *Let the fitness function be such that the string S_0 is not an optimal string (there always exists such an S_0 since the function is not constant). Let us consider the population Q such that Q contains P copies of S_0. Then, the probability $p_{Q.Q}^{(n)}$ that the process will remain in Q in n steps is positive (> 0).*

Proof. Note that selection and crossover operations have no impact on the generation process in the first iteration since they do not change the configuration of Q. Then, the probability $p_{Q.Q}^{(n)}$, $n = 1, 2, 3, \cdots$ is given by the following expression:

$$p_{Q.Q}^{(1)} = p_{Q.Q} = (1 - \mu_m)^{Pl} > 0,$$

$$p_{Q.Q}^{(2)} = \sum_{Q_1 \in \mathcal{Q}} p_{Q.Q_1} p_{Q_1.Q},$$

$$\geq (p_{Q.Q})^2 = ((1 - \mu_m)^{Pl})^2 > 0.$$

Here μ_m is the mutation probability, P is the population size and l is the length of a string. Similarly,

$$p_{Q.Q}^{(n)} \geq (p_{Q.Q})^n = ((1 - \mu_m)^{Pl})^n > 0. \tag{A.1}$$

From the above lemma, it follows that no finite stopping time can guarantee the optimal solution. On the other hand, the process is to be terminated after finitely many iterations with the expectation that the process has achieved the optimal solution. Thus any decision regarding the stopping time should necessarily be probabilistic, since GA is a stochastic process.

The literature on stopping time for stochastic processes is vast [432]. Definition for ϵ-optimal stopping time is also available in literature [432]. The existing definition for ϵ-optimal stopping time, in the context of genetic algorithms, is stated below [324].

Let $Q_{ij} \in E_i$. Let $g_{ij}^{(n)}$ denote the fitness function value of the population that is obtained at the end of the nth iteration of the genetic algorithm with the starting population as Q_{ij}. Let $\mathcal{E}(g_{ij}^{(n)})$ denote the expected value of $g_{ij}^{(n)}$, then

$$\mathcal{E}(g_{ij}^{(n)}) = \sum_{k=1}^{i} F_k p_{ij.k}^{(n)}. \tag{A.2}$$

Now the ϵ-optimal stopping time of GA can be defined as follows [324].

Definition A.2. *Let $\epsilon \geq 0$ and N_0 be a positive integer. Then N_0 is said to be an ϵ-optimal stopping time for a GA if*

$$n \geq N_0 \implies \mathcal{E}(g_{ij}^{(n)}) \geq F_1 - \epsilon \ \forall i \text{ and } j$$

In particular, if $\epsilon = 0$, N_0 is called 0-optimal stopping time or simply optimal stopping time [432]. Note that the above definition cannot be used directly for GAs, since the value of F_1 is not known. Thus the above definition of ϵ-optimal stopping time for a GA has been modified slightly as follows:

Definition A.3. ϵ-**Optimal Stopping Time.** *Let $\epsilon \geq 0$. Let N be a positive integer. Then N is said to be an ϵ-optimal stopping time for a GA if*

$$n \geq N \implies \mathcal{E}(g_{ij}^{(n)}) \geq F_1(1 - \epsilon) \ \forall i \text{ and } j. \tag{A.3}$$

Note that F_1 is used in the above definition too. But, the following manipulations remove the dependency of F_1 on the ϵ-optimal stopping time.

$$\mathcal{E}(g_{ij}^{(n)}) = \sum_{k=1}^{i} F_k p_{ij.k}^{(n)} \tag{A.4}$$
$$\geq F_1 p_{ij.1}^{(n)} \quad \forall i \text{ and } j.$$

Thus, from (A.3) and (A.4), we would like to find N such that

$$n \geq N \implies p_{ij.1}^{(n)} \geq 1 - \epsilon \ \forall i \text{ and } j. \tag{A.5}$$

Note that, if N is an ϵ-optimal stopping time for a GA, then any $N_1 > N$ is also an ϵ-optimal stopping time for the GA. Thus for a given $\epsilon \geq 0$, the task is to find N_0 such that N_0 is an ϵ-optimal stopping time for GA and $N < N_0 \implies N$ is not an ϵ-optimal stopping time for the GA. In other words, the minimal ϵ-optimal stopping time for GA needs to be found to reduce the computation cost.

Definition A.4. Minimal ϵ-optimal time. *For an $\epsilon \geq 0$, N_0 is said to be minimal ϵ-optimal time of a GA if N_0 is an ϵ-optimal stopping time and $N < N_0 \implies N$ is not an ϵ-optimal stopping time for the GA.*

Definition A.5. ϵ-optimal string. *The best string obtained at the end of N iterations is called ϵ-optimal string where N is an ϵ-optimal stopping time for the GA*

Note also that ϵ can not take the value zero for a finite ϵ-optimal stopping time (Lemma A.1). Thus, from now onwards, ϵ is taken to be strictly greater than zero.

We have $p_{ij.1}^{(n)} > 0$ and $\lim_{n \to \infty} p_{ij.1}^{(n)} = 1$ (from Theorem 2.2). This implies that for a given ϵ (> 0) there exists an N such that $p_{ij.1}^{(n)} \geq 1 - \epsilon$ for $n \geq N$. Hence, for a given $\epsilon > 0$, finite ϵ-optimal stopping time exists for a GA.

Now, the problem is to find the values for $p_{ij.1}^{(n)}$ for any initial population Q_{ij}. In other words, in solving a problem using GAs, once the values for $p_{ij.1}^{(n)}$ are known, the ϵ-optimal stopping time N can be determined. Since, the transition probabilities depend on the characteristics of the fitness function [60], we shall briefly discuss the fitness function below.

A.3 Fitness Function

Extensive literature is available on the characteristics of the fitness function to expedite the search using GAs [164, 308]. Often, in implementing a GA, some additional measures are taken into consideration. Among various such mechanisms, linear scaling, sigma truncation and power scaling are the most popular [308]. However, the most noticeable problem associated with the characteristics of the function under consideration involves differences in the relative fitness values.

Moreover, the convergence depends also on the number of points or strings at which the maximum value occurs. It can be intuitively seen that if the number of strings having the maximum fitness function value is more, the chance of fast convergence is high. This is formally explained in the following theorem.

Theorem A.6. *Let f_1 be a fitness function, which assumes s distinct values $F_1, F_2, \cdots, F_s$, defined on Σ. Let S_1 and S_2 be two different strings such that $f_1(S_1) = f_1(S_2)$ and $f_1(S) \leq f_1(S_1) \; \forall S \in \Sigma$. Let us now define f_2 on Σ as follows:*

$$f_2(S) = f_1(S), \quad \forall S \neq S_2 \text{ and } f_2(S_2) = F_i, \quad \text{for some } i > 1.$$

Let p^1 and p^2 be the transition probabilities for the functions f_1 and f_2, respectively. Let $p_{Q.1}^{1\,(n)}$ be the probability that at the end of n iterations, GA results in a population containing the best string from a population $Q \in \mathcal{Q}$ for the fitness function f_1. Similarly, $p_{Q.1}^{2\,(n)}$ is that for f_2. Then for any population $Q \in \mathcal{Q}, \quad p_{Q.1}^1 \geq p_{Q.1}^2$, and in general $p_{Q.1}^{1\,(n)} \geq p_{Q.1}^{2\,(n)}$.

Proof. Let Q be any population and Q' be a population containing one of the best strings. Then

$$
\begin{aligned}
p_{Q.1}^1 &= \sum_{Q' \in E_1} p_{Q.Q'}^1, \\
&= \sum_{Q' \in E_1, S_2 \notin Q'} p_{Q.Q'}^1 + \sum_{Q' \in E_1, S_2 \in Q'} p_{Q.Q'}^1, \\
&\geq \sum_{Q' \in E_1, S_2 \notin Q'} p_{Q.Q'}^1, \\
&\geq p_{Q.1}^2.
\end{aligned}
$$

Note that the equality holds when $S_2 \in Q$.
Now,

$$
\begin{aligned}
p_{Q.1}^{1\,(2)} &= \sum_{Q_1} p_{Q.Q_1}^1 p_{Q_1.1}^1, \\
&= \sum_{Q_1} p_{Q.Q_1}^1 \Big(\sum_{Q' \in E_1, S_2 \notin Q'} p_{Q_1.Q'}^1 + \sum_{Q' \in E_1, S_2 \in Q'} p_{Q_1.Q'}^1 \Big), \\
&\geq \sum_{Q_1} p_{Q.Q_1}^1 \sum_{Q' \in E_1, S_2 \notin Q'} p_{Q_1.Q'}, \\
&\geq p_{Q.1}^{2\,(2)}.
\end{aligned}
$$

Similarly,

$$p_{Q.1}^{1\,(n)} \geq p_{Q.1}^{2\,(n)}.$$

It is now clear from the theorem that the probabilities in reaching to a population containing one of the best strings will be higher if the number of strings having the highest fitness function value is more. That is, the ϵ-optimal stopping time for the function f_2 is also an ϵ-optimal stopping time for f_1. Here we shall deal with the functions which has exactly one optimum string. These fitness functions are termed as single optimal functions. Let $W = \{fit : fit$ is a fitness function and it has exactly one optimal string $\}$.

Definition A.7. *A fitness function, fit, is said to be a single optimal function if $fit \in W$.*

Another important problem is related to the distribution of the fitness values over the search space (set of 2^l strings). The inherent assumption in the implementation of GAs is that the variations in the fitness function values are less if the Hamming distance between the strings is less. But this assumption is not always true for certain fitness functions (e.g., the minimum deceptive problem). For convenience, let us demonstrate a function in which the above assumption is not true.

Example A.8. Let $S^* = 111111 \cdots 111$, and

$$f(S^*) = l + 1, \text{ and}$$
$$fit(S) = D(S, S^*); S \neq S^*,$$

where the Hamming distance $D(S, S^*)$ between S and S^* is the number of bits at which the alleles are different. Note that fit takes only $l + 1$ values. Once the operators of GAs generate a population whose fitness value is $fit(000000 \cdots 000)$, high mutation probability will probably help the process to obtain the global optimal solution.

A.4 Upper Bound for Optimal Stopping Time

In this subsection an upper bound for ϵ-optimal stopping time is obtained for an EGA. The fitness function under consideration is a single optimal function. Let the fitness function be fit, which has only one optimal string $S^* \in \Sigma$, i.e.,

$$fit(S^*) > fit(S), \quad \forall S \in \Sigma; \quad S \neq S^*.$$

Consider the population Q' consisting of P copies of a string $\overline{S^*}$, the complement of S^* where S^* is the optimal string, i.e., $D(S^*, \overline{S^*}) = l$.

Theorem A.9. *Let $Q \neq Q'$ be a population of size P then $p_{Q.1} \geq p_{Q'.1}$.*

Proof. Since $Q \neq Q'$, there exists at least one string $S \in Q$ such that $D(S, S^*) = d \leq l - 1$.

Let the event $R =$ after selection and crossover on Q, there exists no string whose Hamming distance from the optimal string is $\leq l - 1$. Let the probability of R be τ.

If R occurs, then the probability of reaching the optimal string in one step is $p_{Q'.1}$, since all the strings after selection and crossover will possess Hamming distance $= l$, i.e., the collection of strings in hand is nothing but Q'.

If R^c occurs, then after selection and crossover there exists at least one string whose Hamming distance with respect to the optimal is $\le l - 1$. Then the probability of obtaining the optimal string with that collection of strings is $\ge p_{Q'.1}$, since $\mu_m \le 1 - \mu_m$.

Thus,

$$p_{Q.1} \ge \tau p_{Q'.1} + (1 - \tau)p_{Q'.1} = p_{Q'.1},$$

and hence

$$\min_{Q \in \mathcal{Q}} p_{Q.1} = p_{Q'.1} = 1 - \delta,$$

where δ is as defined in Eq. (2.5) in Chap. 2.

Now for Q', $p_{Q'.1}$ can be obtained by computing the probability of mutating each bit position of at least one string of Q'.

$$p_{Q'.1} = \binom{P}{1}(\mu_m^l) - \binom{P}{2}(\mu_m^l)^2 + \cdots + (-1)^P \binom{P}{P}(\mu_m^l)^P$$

$$= \sum_{r=1}^{P} \binom{P}{r}(\mu_m^l)^r(-1)^{r+1} \qquad (A.6)$$

$$= 1 - (1 - \mu_m^l)^P.$$

Note: The minimum probability of reaching a population containing the best string is independent of the characteristics of the fitness function values since one can always assume without loss of generality that there exists the complement of $S^\star$. It is also to be mentioned here that the expression for $p_{Q'.1}$ is independent of the probability of crossover p and the selection procedure.

Theorem A.10. *Let fit be a fitness function with exactly one optimal string $S^\star$. Let $\overline{S^\star}$ be the complement string of $S^\star$ and Q' be a population consisting of P copies of $\overline{S^\star}$, where P is the population size. Let Q_0 be any population of size P. Then*

$$p_{Q_0.1}^{(n)} \ge 1 - (1 - \mu_m^l)^{Pn}.$$

Proof. We have $p_{Q'.1} = 1 - (1 - \mu_m^l)^P$. Now,

$$p_{Q'.1} = 1 - \sum_{Q \notin E_1} p_{Q'.Q}^{(n)}$$

$$\ge 1 - (1 - p_{Q'.1})^n.$$

Now let $Q_0 \notin E_1$ and $Q_0 \ne Q'$. Then,

$$p_{Q_0.1}^{(n)} = 1 - \sum_{Q \notin E_1} p_{Q_0.Q}^{(n)},$$

Now

$$p_{Q_0.1}^{(n)} \geq 1 - \delta^n \quad \text{where} \quad \delta = \max_Q(1 - p_{Q.1}),$$
$$= 1 - (1 - \min_Q p_{Q.1})^n,$$
$$= 1 - (1 - p_{Q'.1})^n.$$

If $Q_0 \in E_1$ then $p_{Q_0.1}^{(n)} = 1$. Hence the theorem.

Upper bound for ϵ-optimal stopping time: It follows from the above theorem that

$$\begin{aligned} p_{Q_0.1}^{(n)} &\geq 1 - (1 - \mu_m^l)^{Pn}, \quad \text{for } Q_0 \notin E_1, \\ &= 1, \quad \text{for } Q_0 \in E_1. \end{aligned} \tag{A.7}$$

Thus, $p_{Q.1}^{(n)} \geq 1 - (1 - \mu_m^l)^{Pn} \; \forall Q \in \mathcal{Q}$. Let $0 < \epsilon < 1$. Now if $1 - (1 - \mu_m^l)^{Pn} \geq 1 - \epsilon$ for $n \geq N$, then N is an ϵ-optimal stopping time for the GA.

Note that,

$$1 - \epsilon \leq 1 - (1 - \mu_m^l)^{Pn},$$
$$\iff \quad \epsilon \geq (1 - \mu_m^l)^{Pn},$$
$$\iff \quad n \geq \frac{\log \frac{1}{\epsilon}}{P \log \frac{1}{1-\mu_m^l}}. \tag{A.8}$$

Let $N(\epsilon, P, \mu_m, l) = \dfrac{\log \frac{1}{\epsilon}}{P \log \frac{1}{1-\mu_m^l}}$.

Note that given ϵ, P, μ_m and l, $N(\epsilon, P, \mu_m, l)$ is an upper bound for ϵ-optimal stopping time because of the following reasons.

(a) Let N_0 be the minimal ϵ-optimal stopping time for GA for given P, μ_m and l. Then $N_0 \leq N(\epsilon, P, \mu_m, l)$, since $N(\epsilon, P, \mu_m, l)$ is an ϵ-optimal stopping time.

(b) From Theorem A.10, note that $p_{Q'.1}^{(n)} \geq 1 - (1 - \mu_m^l)^{Pn}$ and $1 - (1 - \mu_m^l)^{Pn}$ is taken to be greater than or equal to $1 - \epsilon$ to find $N(\epsilon, P, \mu_m, l)$.

Thus, within $N(\epsilon, P, \mu_m, l)$ iterations an ϵ-optimal string will be obtained for the GA.

Note that the knowledge of minimal ϵ-optimal stopping time for GAs is not available. Hence, in the subsequent analysis it is assumed that $N(\epsilon, P, \mu_m, l)$ iterations will be performed during the process.

Remarks:

(a) For a given ϵ, P, μ_m and l, $N(\epsilon, P, \mu_m, l)$ is independent of the characteristics of the fitness function, crossover probability and the selection procedure.

(b) Note that, if the starting population is not Q' then also ϵ-optimal string will be obtained in $N(\epsilon, P, \mu_m, l)$ iterations.

(c) Note that $N(\epsilon, P, \mu_m, l)$ is inversely proportional to P, i.e., if P increases $N(\epsilon, P, \mu_m, l)$ decreases, which coincides with the intuition that as the population size increases, the number of required iterations decreases.

(d) Given ϵ, μ_m and l, $N(\epsilon, P, \mu_m, l)P$ (the product of $N(\epsilon, P, \mu_m, l)$ and P) is a constant. Note that, $N(\epsilon, P, \mu_m, l)P$ provides the number of strings searched upto $N(\epsilon, P, \mu_m, l)$ iterations. This means that the number of strings to be searched to obtain ϵ-optimal string is independent of the population size P.

(e) It can be seen that for a given P, μ_m and l,

$$\epsilon_1 < \epsilon_2 \implies N(\epsilon_1, P, \mu_m, l) > N(\epsilon_2, P, \mu_m, l).$$

It implies that the number of iterations required is more to obtain a more accurate solution.

(f) It is also clear that, for a given ϵ, μ_m and l,

$$l_1 > l_2 \implies N(\epsilon, P, \mu_m, l_1) > N(\epsilon, P, \mu_m, l_2).$$

This also coincides with the intuition that for a fixed ϵ if the length of the string increases the required number of iterations also increases.

(g) Note that the expression for $N(\epsilon, P, \mu_m, l)$ has been derived from the following inequalities.

$$p_{Q.1}^{(n)} \geq p_{Q'.1}^{(n)} \geq 1 - (1 - p_{Q'.1})^n.$$

$1 - (1 - p_{Q'.1})^n$ is taken to be $\geq 1 - \epsilon$. Note that Q' is a pessimistic choice of an initial population and the stopping time has been derived from the above inequality. Thus the stopping time derived is also a pessimistic one, and we shall denote it by $N_{pes}(\epsilon, P, \mu_m, l)$.

So far we have discussed the stopping time keeping the mutation probability μ_m as fixed. In the next section, we shall study the behavior of ϵ-optimal stopping time while varying the mutation probability.

A.5 Mutation Probability and ε-Optimal Stopping Time

In the previous section, the behavior of $N(\epsilon, P, \mu_m, l)$ has been studied keeping the mutation probability μ_m as constant. Here, first of all, we shall study the behavior of $N(\epsilon, P, \mu_m, l)$ while varying μ_m. Then, we shall try to find the modified expression for ϵ-optimal stopping time with the assumption that the characteristics of the fitness function are well related to the Hamming distance of the strings with the best string $S^\star$. That is, the differences in the fitness function are such that they decrease with the reduction in $D(S, S^\star)$, the Hamming distance between S and the optimal string $S^\star$.

$N_{pes}(\epsilon, P, \mu_m, l)$ represents an upper bound for ϵ-optimal time. Here we shall try to find the optimal value of μ_m which will minimize $N_{pes}(\epsilon, P, \mu_m, l)$.

Theorem A.11. For $\mu_{m1} < \mu_{m2} \leq 0.5$, $N_{pes}(\epsilon, P, \mu_{m1}, l) > N_{pes}(\epsilon, P, \mu_{m2}, l)$.

Proof. We have, $N_{pes}(\epsilon, P, \mu_m, l) = \dfrac{\log \frac{1}{\epsilon}}{P \log \frac{1}{1-\mu_m^l}}$. It can be easily shown that

for $0 \le \mu_m \le 0.5$, $N_{pes}(\epsilon, P, \mu_m, l)$ is a decreasing function of μ_m and hence the theorem.

It follows from Theorem A.11 that the optimal value of μ_m for $N_{pes}(\epsilon, P, \mu_m, l)$ is 0.5. Note that 0.5 is a very high value for μ_m in the implementation of a GA.

It has been found above that $N_{pes}(\epsilon, P, \mu_m, l)$ is minimum when $\mu_m = 0.5$. Note that the mutation probability μ_m is indeed very high ($\mu_m = 0.5$) for making $N_{pes}(\epsilon, P, \mu_m, l)$ minimum. In practice, the researchers take μ_m to be very low in order to ape the usual genetic systems. Thus μ_m being very high would go against the practice of the usual genetic system. But one can still find $\mu_m = 0.5$ to be useful, provided the number of strings searched up to $N_{pes}(\epsilon, P, \mu_m, l)$ iterations is less than 2^l. We shall show below that the number of strings searched up to $N_{pes}(\epsilon, P, \mu_m, l)$ iterations is greater than 2^l.

Note that,

$$P.N_{pes}(\epsilon, P, \mu_m, l) = \frac{\log \frac{1}{\epsilon}}{\log \frac{1}{1-0.5^l}}. \tag{A.9}$$

Now substituting $P.N$ with 2^l, we get

$$\frac{\log \frac{1}{\epsilon}}{\log \frac{1}{1-0.5^l}} \le 2^l \tag{A.10}$$

$$\Longleftrightarrow \quad \log \frac{1}{\epsilon} \le 2^l \log \frac{1}{1-0.5^l},$$
$$\Longleftrightarrow \quad \epsilon \ge \left(1 - \frac{1}{2^l}\right)^{2^l}. \tag{A.11}$$

Note that, as $l \to \infty$, $(1 - 0.5^l)^{2^l} \to e^{-1} \ge 0.37$. Which implies, that for $\mu_m = 0.5$ and for sufficiently large values of l, if the number of strings searched is 2^l then it can be stated that the fitness value obtained will be at least equal to $F_1(1 - 0.37) = 0.63F_1$.

It is now clear that the number of strings to be searched for $\mu_m = 0.5$ may be greater than 2^l in order to obtain a reasonable accuracy (i.e., ϵ is small). Note that pessimistic stopping time always assumes that the starting population for any iteration is Q'. This is not the case in practice. Note also that for many fitness function, even if the initial population is Q', it will result in some other population after a few iterations excepting for minimum deceptive problem. Observe that, usually, for minimum deceptive problem, $D(S, S^\star)$ is less $\Longleftrightarrow$ difference between $fit(S)$ and $fit(S^\star)$ is high. On the other hand, for the general optimization problems, $D(S, S^\star)$ is less $\Longleftrightarrow$ difference between $fit(S)$ and $fit(S^\star)$ is also less.

Thus from the above observation, we shall study the stopping time for EGA under the following assumption:

Assumption: Let $S^\star$ be the optimal string. Let there exist an integer d, $1 \le d \le l-1$ such that for every two strings S_1 and S_2, $S_1 \ne S_2$, $D(S_1, S^\star) \le d$, $D(S_2, S^\star) > d$ implies $|fit(S_1) - fit(S^\star)| < |fit(S_2) - fit(S^\star)|$

Remarks:

(a) The above assumption provides a cut-off point d for the Hamming distance between the optimum string and the other strings. Note that, if the fitness function satisfies the above assumption for a specific d and if a string S is present in a starting population for an EGA such that $D(S, S^\star) \le d$, then the best string of any subsequent iteration will possess the Hamming distance $\le d$ with the optimal.

(b) The above assumption holds good for many practical problems. It is not valid for minimum deceptive problems.

(c) Note that d is taken to be a positive integer, and it is strictly less than l. If there does not exist any d which satisfies the above assumption (or, $d = 0$ or $d = l$) then the fitness function is deceptive.

(d) Suppose a fitness function fit satisfies the above assumption for any $d < d_1$. In that case, the fitness function will be regarded as well related with respect to the Hamming distance.

In order to study the behavior of an EGA, we shall define below the Hamming distance between a string S and a population Q.

Definition A.12. *The Hamming distance $D(S, Q)$ between the string S and the population Q is defined as*

$$D(S, Q) = \min_{S' \in Q} D(S, S'),$$

where $D(S, S')$ is the Hamming distance between S and S'.

Let $Q \notin E_1$ be the population at the start of an iteration and $S^\star$ be the best string, and $D(S^\star, Q) = d$ Then,

$$p_{Q.1} \ge 1 - [1 - \mu_m^d (1 - \mu_m)^{l-d}]^P,$$

and,

$$p_{Q.1}^{(n)} \ge 1 - [1 - \mu_m^d (1 - \mu_m)^{l-d}]^{Pn}.$$

Now, ϵ-optimal string will be obtained if

$$n \ge \frac{\log \frac{1}{\epsilon}}{P \log \frac{1}{1 - \mu_m^d (1 - \mu_m)^{l-d}}}, \quad \text{for } d \le \frac{l}{2}. \tag{A.12}$$

Assuming l, P and ϵ as constants the minimum value of n is obtained when $\mu_m = \frac{d}{l}$.

It is also to be noted that the value for μ_m does not exceed 0.5 [60] and hence the optimal value for μ_m is $\min(0.5, \frac{d}{l})$. Thus the number of iterations required to obtain an ϵ-optimal string is

$$n \geq \frac{\log \frac{1}{\epsilon}}{P \log \frac{1}{1-(\frac{d}{l})^d (\frac{l-d}{l})^{l-d}}}, \quad \text{for } d \leq \frac{l}{2}. \tag{A.13}$$

This expression is obtained with an optimistic assumption and hence, we shall term it as an optimistic ϵ-optimal stopping time and denote it by N_{op}, i.e.,

$$N_{op}(\epsilon, P, d, l) = \frac{\log \frac{1}{\epsilon}}{P \log \frac{1}{1-(\frac{d}{l})^d (\frac{l-d}{l})^{l-d}}}. \tag{A.14}$$

Remarks:

(a) Unlike N_{pes}, $N_{op}(\epsilon, P, d, l)$ is not independent of the characteristics of the fitness function for given ϵ, P, d and l. But it is independent of crossover probability and the selection procedure.

(b) Note that, $N_{op}(\epsilon, P, d, l)$ is inversely proportional to P, i.e., if P increases then N_{op} decreases, which coincides with the intuition that as the population size increases the number of iterations required decreases.

(c) Given ϵ, d and l, $N_{op} P$ (the product of $N_{op}(\epsilon, P, d, l)$ and P) is a constant. Note that, $N_{op}(\epsilon, P, \mu_m, l)P$ provides the number of strings searched upto $N_{op}(\epsilon, P, \mu_m, l)$ iterations. This means that the number of strings to be searched to obtain the ϵ-optimal string in N_{op} many iterations is independent of the population size P.

(d) It can be seen that for given P, d and l,

$$\epsilon_1 < \epsilon_2 \implies N_{op}(\epsilon_1, P, d, l) > N_{op}(\epsilon_2, P, d, l).$$

It implies that the number of iterations required is more to obtain a more accurate solution.

(e) It is also clear that, for given ϵ, d and P,

$$l_1 > l_2 \implies N_{op}(\epsilon, P, d, l_1) > N_{op}(\epsilon, P, d, l_2).$$

This also coincides with the intuition that for a fixed ϵ if the length of the string increases the required number of iterations also increases.

(f) Now, we are in a position to explain the affinity of researchers to assume μ_m to be very small. Let the fitness function be well related to the Hamming distance between the best string and the best obtained so far. Since the characteristics of the fitness function are completely unknown, the users will appreciate any improvement in the fitness function value due to even a single change in any bit position of the string. More specifically, the intention of GAs users is to make a small change to improve the result in hand. Then to obtain a better string compared to the string obtained so far, one needs to change $d < l$ (say) bit positions (i.e., reduce the Hamming distance by d) and assume mutation probability μ_m to be $\frac{d}{l}$. For example, if a user expects an improvement by changing only 1 bit then the suitable value for μ_m will be 0.01 for $l = 100$ and $d = 1$. μ_m will be 0.001 for $l = 1000$ and $d = 1$ and so on.

(g) It can also be seen that the number of strings searched to get an ϵ-optimal string in N_{op} iterations is $\leq 2^l$ for sufficiently large l and for any $\epsilon > 0$. To estimate the value of ϵ for $P.N \leq 2^l$, let us substitute $P.N$ by 2^l in Eq. (A.13), and we get

$$\log \tfrac{1}{\epsilon} \leq 2^l \log \frac{1}{1-(\frac{d}{l})^d(\frac{l-d}{l})^{l-d}},$$
$$\text{or} \quad \epsilon \geq [1 - (\tfrac{d}{l})^d(\tfrac{l-d}{l})^{l-d}]^{2^l}. \tag{A.15}$$

Let us consider $\theta = \frac{d}{l}$, then from Eq. (A.15) we have

$$\epsilon \geq (1 - \theta^{\theta l} \cdot (1 - \theta)^{(1-\theta)l})^{2^l},$$
$$\text{or} \quad (1 - \epsilon^{\frac{1}{2^l}})^{\frac{1}{l}} \leq \theta^\theta \cdot (1 - \theta)^{1-\theta}.$$

Note that $\theta^\theta \cdot (1-\theta)^{1-\theta}$ is minimum when $\theta = 0.5$ and the minimum value of $\theta^\theta \cdot (1 - \theta)^{1-\theta}$ is 0.5. Thus if $\theta < 0.5$ then $\theta^\theta \cdot (1 - \theta)^{1-\theta} > 0.5$. Let us assume

$$0.5 < \theta^\theta \cdot (1 - \theta)^{1-\theta} < 0.5 + \rho, \text{ for } 0 < \rho < 0.5. \tag{A.16}$$

Then, note that $(1-(0.5+\rho)^l)^{2^l} \longrightarrow 0$ as $l \longrightarrow \infty$ and $\epsilon \geq (1-(0.5+\rho)^l)^{2^l}$. For an appropriate value of μ_m one can find an ϵ-optimal string (for any given ϵ) by searching $\leq 2^l$ strings for sufficiently large value of l. It is to be noted that if μ_m decreases then ρ increases and consequently $(1 - (0.5 + \rho)^l)^{2^l}$ decreases and hence ϵ can be taken to be a small value. But it should be remembered that μ_m being small implies that the fitness function is well behaved in the sense of convergence of GAs. That is, it has been shown theoretically that GAs are indeed useful in searching optimal solutions and the number of strings to be searched is less than 2^l.

B

Data Sets Used for the Experiments

A description of the data sets used for the experiments is provided here for convenience. The sources for the data sets, and the chapters in which these data sets are used, are mentioned.

ADS 1

This artificial data set, shown in Fig. 3.7, has 557 points in 2 classes, of which 460 points are in class 1 and 97 points are in class 2. The class boundary of the data sets is seen to be highly nonlinear, although the classes are separable. This data has been used in Chaps. 3 and 5. It can be found in http://www.isical.ac.in/~miu.

Data_2_2

This is a two-dimensional data set where the number of clusters is two. It has ten points. This data set has been used in Chap. 8, and can be found in http://www.isical.ac.in/~sanghami.

Data_3_2

This is a two-dimensional data set where the number of clusters is three. It has 76 points. Figure 8.1 shows the data set. This data set has been used in Chap. 8, and can be found in http://www.isical.ac.in/~sanghami.

AD_5_2

This is a two-dimensional data with five clusters. It has 250 points, with 50 points in each cluster. Figure 8.2 shows the data set. The clusters are found to be overlapping in nature. This data set has been used in Chap. 8, and can be found in http://www.isical.ac.in/~sanghami.

AD_10_2

This is a two-dimensional data with ten clusters. It has 500 points. Figure 8.3 shows the data set. Some of the clusters are found to be overlapping. The number of clusters is also large. This data set has been used in Chap. 8, and can be found in http://www.isical.ac.in/∼sanghami.

AD_4_3

This is a three-dimensional data with four clusters having two noisy points situated exactly between two distinct clusters. It has 402 points. Figure 8.4 shows the data set. This data set has been used in Chap. 8, and can be found in http://www.isical.ac.in/∼sanghami.

Data Set 1

This is a two-dimensional data set, generated using a triangular distribution of the form shown in Fig. 4.1 for the two classes, 1 and 2. The range for class 1 is $[0,2] \times [0,2]$ and that for class 2 is $[1,3] \times [0,2]$ with the corresponding peaks at $(1,1)$ and $(2,1)$, respectively. Figure 4.1 shows the distribution along the X axis since only this axis has discriminatory capability. The distribution along the X axis may be formally quantified as

$$
\begin{aligned}
f_1(x) &= \quad 0, \quad for\ x \leq 0, \\
f_1(x) &= \quad x, \quad for\ 0 < x \leq 1, \\
f_1(x) &= 2 - x,\ for\ 1 < x \leq 2, \\
f_1(x) &= \quad 0, \quad for\ x > 2,
\end{aligned}
$$

for class 1. Similarly for class 2

$$
\begin{aligned}
f_2(x) &= \quad 0, \quad for\ x \leq 1, \\
f_2(x) &= x - 1,\ for\ 1 < x \leq 2, \\
f_2(x) &= 3 - x,\ for\ 2 < x \leq 3, \\
f_2(x) &= \quad 0, \quad for\ x > 3.
\end{aligned}
$$

The distribution along the Y axis for both the classes is

$$
\begin{aligned}
f(y) &= \quad 0, \quad for\ y \leq 0, \\
f(y) &= \quad y, \quad for\ 0 < y \leq 1, \\
f(y) &= 2 - y,\ for\ 1 < y \leq 2, \\
f(y) &= \quad 0, \quad for\ y > 2.
\end{aligned}
$$

If P_1 is the a priori probability of class 1 then using elementary mathematics, it can be shown that the Bayes' classifier will classify a point to class 1 if its X coordinate is less than $1 + P_1$. This indicates that the Bayes' decision boundary is given by

$$ x = 1 + P_1. \tag{B.1} $$

This data set has been used in Chap. 4, and is available on request.

Data Set 2

This is a normally distributed data consisting of two classes. The mean (μ_1, μ_2) and covariance values (Σ_1, Σ_2) for the two classes are:

$$\mu_1 = (0.0, 0.0), \ \mu_2 = (1.0, 0.0), \text{ and}$$

$$\Sigma_1 = \begin{pmatrix} 1.0 \ 0.0 \\ 0.0 \ 1.0 \end{pmatrix},$$

$$\Sigma_2 = \begin{pmatrix} 4.0 \ 0.5 \\ 0.5 \ 4.0 \end{pmatrix},$$

respectively.

The classes are assumed to have equal a priori probability $(= 0.5)$. Mathematical analysis shows that the Bayes' decision boundary for such a distribution of points will be of the following form:

$$a_1 x_1^2 + a_2 x_2^2 + 2a_3 x_1 x_2 + 2b_1 x_1 + 2b_2 x_2 + c = 0. \qquad (B.2)$$

The data set and the Bayes' decision boundary are also shown in Fig. 4.2. This data set has been used in Chap. 4, and is available on request.

Data Set 3

This data has nine classes, generated using a triangular distribution. All the classes are assumed to have equal a priori probabilities $(= 1/9)$. The $X - Y$ ranges for the nine classes are as follows:

Class 1: [-3.3, -0.7] × [0.7, 3.3]
Class 2: [-1.3, 1.3] × [0.7, 3.3]
Class 3: [0.7, 3.3] × [0.7, 3.3]
Class 4: [-3.3, -0.7] × [-1.3, 1.3]
Class 5: [-1.3, 1.3] × [-1.3, 1.3]
Class 6: [0.7, 3.3] × [-1.3, 1.3]
Class 7: [-3.3, -0.7] × [-3.3, -0.7]
Class 8: [-1.3, 1.3] × [-3.3, -0.7]
Class 9: [0.7, 3.3] × [-3.3, -0.7]

Thus the domain for the triangular distribution for each class and for each axis is 2.6. Consequently, the height will be $\frac{1}{1.3}$ (since $\frac{1}{2} * 2.6 * height = 1$). Note that (i) each class has some overlap with each of its adjacent classes, and (ii) the class boundaries are represented in a simple way (e.g., $x = 1$). In any direction the nonoverlapping portion is 1.4 units, the overlapping portion is 1.2 units and the length of the Bayes' decision region is 2 units. The resulting Bayes' boundary along with the data set is shown in Fig. 4.3. This data set has been used in Chap. 4, and is available on request.

Data Set 4

This two-class, two-dimensional data set is normally distributed with the following parameters:

$$\mu_1 = (0.0, 0.0), \; \mu_2 = (1.0, 0.0), \text{ and}$$

$$\Sigma_1 = \Sigma_2 = \Sigma = \begin{pmatrix} 1.0 & 0.0 \\ 0.0 & 1.0 \end{pmatrix}.$$

Since $\Sigma_1 = \Sigma_2$, the Bayes' boundary will be linear [155] of the following form:

$$(\mu_2 - \mu_1)^T \Sigma^{-1} X + \frac{1}{2}(\mu_1^T \Sigma^{-1} \mu_1 - \mu_2^T \Sigma^{-1} \mu_2). \tag{B.3}$$

This data set has been used in Chap. 4, and may be available on request.

Vowel Data

This data consists of 871 Indian Telugu vowel sounds. These were uttered in a consonant–vowel–consonant context by three male speakers in the age group 30–35 years. The data set has three features, F_1, F_2 and F_3, corresponding to the first, second and third formant frequencies, and six vowel classes $\{\delta, a, i, u, e, o\}$. The details of the method of extraction of formant frequencies through spectrum analysis are described in [350]. It is known that two features, namely F_1 and F_2, are more important in characterizing the classes than F_3 [350]. Figure 3.8 shows the distribution of the six classes in the $F_1 - F_2$ plane. The boundaries of the classes are seen to be ill-defined and overlapping. This data has been used in Chaps. 3 and 5. It can be found in http://www.isical.ac.in/~miu.

Iris Data

This data represents different categories of irises. It has four feature values that represent the sepal length, sepal width, petal length and the petal width in centimeters [148]. It has three classes, Setosa, Versicolor and Virginica (labelled 1, 2 and 3), with 50 samples per class. It is known that two classes Versicolor and Virginica have a large amount of overlap, while the class Setosa is linearly separable from the other two. This data has been used in Chaps. 3, 5 and 8. It is available in www.ics.uci.edu/~mlearn/MLRepository.html.

Cancer Data

This breast cancer database was obtained from the University of Wisconsin Hospital, Madison [289]. It has 683 samples belonging to two classes *Benign* (class 1) and *Malignant* (class 2), and nine features corresponding to *clump thickness, cell size uniformity, cell shape uniformity, marginal adhesion, single epithelial cell size, bare nuclei, bland chromatin, normal nucleoli* and *mitoses*. This data has been used in Chap. 8. It is available in www.ics.uci.edu/~mlearn/MLRepository.html.

Mango Data

The mango data set [62] consists of 18 measurements taken of the leaves of 3 different kinds of mango trees. It has 166 samples with 18 features each. It has three classes representing three kinds of mangoes. The feature set consists of measurements like Z-value, area, perimeter, maximum length, maximum breadth, petiole, K-value, S-value, shape index, upper midrib/ lower midrib and perimeter upper half/ perimeter lower half. The terms *upper* and *lower* are used with respect to the maximum breadth position.

SPOT Image of Calcutta

The French satellites SPOT (Systems Probataire d'Observation de la Terre) [394], launched in 1986 and 1990, carried two imaging devices that consisted of a linear array of charge coupled device (CCD) detectors. Two imaging modes were possible, the multispectral and panchromatic modes. The 512×512 *SPOT* image of a part of the city of Calcutta, taken from this satellite, is available in three bands in the multispectral mode. These bands are:

Band 1: green band of wavelength $0.50 - 0.59$ μm

Band 2: red band of wavelength $0.61 - 0.68$ μm

Band 3: near infrared band of wavelength $0.79 - 0.89$ μm.

The seven classes considered in this image are *turbid water, concrete, pure water, vegetation, habitation, open space* and *roads* (including bridges). Figure 6.5 shows the image in the near infrared band.

Some important landcovers of Calcutta are present in the image. Most of these can be identified, from a knowledge about the area, more easily in the near infrared band of the input image (Fig. 6.5). These are the following: The prominent black stretch across the figure is the river *Hooghly*. Portions of a bridge (referred to as the *second bridge*), which was under construction when the picture was taken, protrude into the *Hooghly* near its bend around the center of the image. There are two distinct black, elongated patches below the river, on the left side of the image. These are water bodies, the one to the left being *Garden Reach lake* and the one to the right being *Khidirpore dockyard*. Just to the right of these water bodies, there is a very thin line, starting from the right bank of the river, and going to the bottom edge of the picture. This is a canal called the *Talis nala*. Above the *Talis nala*, on the right side of the picture, there is a triangular patch, the *race course*. On the top, right hand side of the image, there is a thin line, stretching from the top edge, and ending on the middle, left edge. This is the *Beleghata canal* with a road by its side. There are several roads on the right side of the image, near the middle and top portions. These are not very obvious from the images. A bridge cuts the river near the top of the image. This is referred to as the *first bridge*. This image has been used in Chap. 6, and is available on request.

IRS Image of Calcutta

This image was acquired from Indian Remote Sensing Satellite (IRS-1A) [1] using the *LISS-II* sensor that has a resolution of 36.25 m×36.25 m. The image is contained in four spectral bands, namely blue band of wavelength 0.45 – 0.52 μm, green band of wavelength 0.52 – 0.59 μm, red band of wavelength 0.62 – 0.68 μm, and near infrared band of wavelength 0.77 – 0.86 μm. Figure 8.16 in Chap. 8 shows the Calcutta image in the near infrared band. Some characteristic regions in the image are the river *Hooghly* cutting across the middle of the image, several fisheries observed towards the lower right portion, a township, *SaltLake*, to the upper left-hand side of the fisheries. This township is bounded on the top by a canal. Two parallel lines observed towards the upper right-hand side of the image correspond to the airstrips in the *Dumdum* airport. Other than these there are several water bodies, roads, etc. in the image. This image has been used in Chap. 8, and is available on request.

IRS Image of Mumbai

As for the Calcutta image, the IRS image of Mumbai was also obtained using the LISS-II sensor. It is available in four bands, viz., blue, green, red and near infrared. Figure. 8.17 in Chap. 8 shows the *IRS* image of a part of Mumbai in the near infrared band. As can be seen, the elongated city area is surrounded on three sides by the Arabian sea. Towards the bottom right of the image, there are several islands, including the well-known *Elephanta islands*. The dockyard is situated on the south eastern part of Mumbai, which can be seen as a set of three finger-like structures. This image has been used in Chap. 8, and is available on request.

C

Variation of Error Probability with P_1

In Chap. 4, it is mentioned that the Bayes' error probability a is given by (see Eq. (4.6))

$$a = \sum_{i=1}^{K} P_i \int_{S_{0i}^c} p_i(\mathbf{x})d\mathbf{x},$$

where S_{0i} is the region for class i. For a two class problem a may be written as

$$a = P_1 \int_{S_{01}^c} p_1(\mathbf{x})d\mathbf{x} + P_2 \int_{S_{02}^c} p_2(\mathbf{x})d\mathbf{x}.$$

Since $P_2 = 1 - P_1$, we get

$$a = P_1 \int_{S_{01}^c} p_1(\mathbf{x})d\mathbf{x} + (1 - P_1) \int_{S_{02}^c} p_2(\mathbf{x})d\mathbf{x}. \tag{C.1}$$

For the triangular distribution mentioned in Appendix B for generating *Data Set 1*, using Eq. (B.1) we may write

$$a = P_1 \int_{1+P_1}^{2} (2 - \mathbf{x})d\mathbf{x} + (1 - P_1) \int_{1}^{1+P_1} (\mathbf{x} - 1)d\mathbf{x}.$$

Solving for a, we get

$$a = P_1 \frac{(1 - P_1)}{2}.$$

Obviously, this is a symmetric function with minimum values at $P_1 = 0$ or 1, and maximum value at $P_1 = 0.5$. Thus the recognition score of the Bayes' classifier should be minimum for $P_1 = 0.5$, increasing symmetrically on both sides.

For normal distribution, it is very difficult to obtain a closed form expression for a in terms of P_1. An analysis presented in [155] indicates that the risk r associated with a particular decision is maximum for some value of $P_1 = P_1^*$, decreasing on both sides of this value when the regions associated with each class change with the class a priori probabilities.

Alternatively, one can also derive bounds on the error probabilities. One such bound for normal distribution is given by [155]

$$a \leq \sqrt{P_1 P_2} \ \exp^{-\mu(1/2)}, \tag{C.2}$$

where $\mu(1/2)$ is called the *Bhattacharyya distance*. Let us define the upper bound of a by a', i.e., $a' = \sqrt{P_1 P_2} \ \exp^{-\mu(1/2)}$. Or,

$$a' = \sqrt{P_1(1 - P_1)} \ \exp^{-\mu(1/2)},$$

$$\frac{da'}{dP_1} = \frac{1}{2} \frac{1 - 2P_1}{\sqrt{P_1(1 - P_1)}} \exp^{-\mu(1/2)}.$$

This shows that the error bound is maximum when $P_1 = P_2 = 0.5$.

References

1. IRS data users handbook. Technical Report IRS/NRSA/NDC/HB-01/86, NRSA, Hyderabad, 1986.
2. J. P. Adrahams and M. Breg. Prediction of RNA secondary structure including pseudoknotting by computer simulation. *Nucleic Acids Research*, 18:3035–3044, 1990.
3. M. Affenzeller and S. Wagner. Offspring selection: A new self-adaptive selection scheme for genetic algorithms. In *Adaptive and Natural Computing Algorithms*, Computer Science, pages 218–221. Springer, Berlin Heidelberg New York, 2005.
4. T. Akutsu, S. Miyano, and S. Kuhara. Identification of genetic networks from a small number of gene expression patterns under the boolean network model. In *Proceedings of the Pacific Symposium on Biocomputing*, volume 99, pages 17–28, 1999.
5. F. H. Allen. The Cambridge structural database: A quarter of a million crystal structures and rising. *Acta Crystallography B*, 58:283–438, 2002.
6. R. B. Altman, A. Valencia, S. Miyano, and S. Ranganathan. Challenges for intelligent systems in biology. *IEEE Intelligent Systems*, 16(6):14–20, 2001.
7. L. A. Anbarasu, P. Narayanasamy, and V. Sundararajan. Multiple molecular sequence alignment by island parallel genetic algorithm. *Current Science*, 78(7):858–863, 2000.
8. M. R. Anderberg. *Cluster Analysis for Application*. Academic Press, 1973.
9. T. W. Anderson. *An Introduction to Multivariate Statistical Analysis*. Wiley, New York, 1958.
10. S. Ando and H. Iba. Quantitative modeling of gene regulatory network: Identifying the network by means of genetic algorithms. *Genome Informatics*, 11:278–280, 2000.
11. S. Ando and H. Iba. Inference of gene regulatory model by genetic algorithms. In *Proceedings of the Congress on Evolutionary Computation*, volume 1, pages 712–719, 2001.
12. S. Ando, E. Sakamoto, and H. Iba. Evolutionary modeling and inference of gene network. *Information Sciences —Informatics and Computer Science: An International Journal*, 145(3–4):237–259, 2002.
13. H. C. Andrews. *Mathematical Techniques in Pattern Recognition*. Wiley Interscience, New York, 1972.

14. C. A. Ankerbrandt, B. P. Buckles, and F. E. Petry. Scene recognition using genetic algorithms with semantic nets. *Pattern Recognition Letters*, 11:285–293, 1990.

15. C. Anselmi, G. Bocchinfuso, P. De Santis, M. Savino, and A. Scipioni. A theoretical model for the prediction of sequence-dependent nucleosome thermodynamic stability. *Journal of Biophysics*, 79(2):601–613, 2000.

16. T. M. Apostol. *Mathematical Analysis*. Narosa Publishing House, New Delhi, 1985.

17. T. Bäck. The interaction of mutation rate, selection and self adaptation within a genetic algorithm. In R. Manner and B. Manderick, editors, *Proceedings of the Parallel Problem Solving from Nature*, pages 85–94. North Holland, Amsterdam, 1992.

18. T. Bäck. Optimal mutation rates in genetic algorithms. In S. Forrest, editor, *Proceedings of the 5th International Conference on Genetic Algorithms*, pages 2–8. Morgan Kaufmann, San Mateo, 1993.

19. T. Bäck, editor. *Evolutionary Algorithms in Theory and Practice: Evolution Strategies, Evolutionary Programming and Genetic Algorithms*. Oxford University Press, New York, 1996.

20. A. Bagchi, S. Bandyopadhyay, and U. Maulik. Determination of molecular structure for drug design using variable string length genetic algorithm. *Workshop on Soft Computing, High Performance Computing (HiPC) Workshops 2003: New Frontiers in High-Performance Computing*, Hyderabad:145–154, 2003.

21. S. Bagchi, S. Uckun, Y. Miyabe, and K. Kawamura. Exploring problem-specific recombination operators for job shop scheduling. In Rick Belew and Lashon Booker, editors, *Proceedings of the Fourth International Conference on Genetic Algorithms*, pages 10–17, San Mateo, CA, 1991. Morgan Kaufmann.

22. J. E. Baker. Adaptive selection methods for genetic algorithms. In J. J. Grefenstette, editor, *Proceedings of the 1st International Conference on Genetic Algorithms*, pages 101–111. Lawrence Erlbaum Associates, Hillsdale, 1985.

23. P. Baldi and P. F. Baisnee. Sequence analysis by additive scales: DNA structure for sequences and repeats of all lengths. *Bioinformatics*, 16:865–889, 2000.

24. P. Baldi and S. Brunak. *Bioinformatics: The Machine Learning Approach*. MIT Press, Cambridge, MA, 1998.

25. S. Bandyopadhyay. *Pattern Classification Using Genetic Algorithms*. PhD thesis, Machine Intelligence Unit, Indian Statistical Institute, Calcutta, India, 1998.

26. S. Bandyopadhyay. An efficient technique for superfamily classification of amino acid sequences: Feature extraction, fuzzy clustering and prototype selection. *Fuzzy Sets and Systems*, 152:5–16, 2005.

27. S. Bandyopadhyay. Satellite image classification using genetically guided fuzzy clustering with spatial information. *International Journal of Remote Sensing*, 26(3):579–593, 2005.

28. S. Bandyopadhyay. Simulated annealing using reversible jump Markov chain Monte Carlo algorithm for fuzzy clustering. *IEEE Transactions on Knowledge and Data Engineering*, 17(4):479–490, 2005.

29. S. Bandyopadhyay, A. Bagchi, and U. Maulik. Active site driven ligand design: An evolutionary approach. *Journal of Bioinformatics and Computational Biology*, 3(5):1053–1070, 2005.

30. S. Bandyopadhyay, H. Kargupta, and G. Wang. Revisiting the GEMGA: Scalable evolutionary optimization through linkage learning. In *Proceedings of the International Conference on Evolutionary Computation*, pages 603–608. IEEE Press, Alaska, USA, 1998.

31. S. Bandyopadhyay and U. Maulik. Non-parametric genetic clustering: Comparison of validity indices. *IEEE Transactions on Systems, Man and Cybernetics Part-C*, 31(1):120–125, 2001.

32. S. Bandyopadhyay and U. Maulik. An evolutionary technique based on k-means algorithm for optimal clustering in R^N. *Info. Sc.*, 146:221–237, 2002.

33. S. Bandyopadhyay, U. Maulik, L. Holder, and D. J. Cook, editors. *Advanced Methods for Knowledge Discovery from Complex Data*. Springer, Berlin Heidelberg New York, 2005.

34. S. Bandyopadhyay, U. Maulik, and M. K. Pakhira. Clustering using simulated annealing with probabilistic redistribution. *International Journal of Pattern Recognition and Artificial Intelligence*, 15(2):269–285, 2001.

35. S. Bandyopadhyay, C. A. Murthy, and S. K. Pal. Pattern classification using genetic algorithms. *Pattern Recognition Letters*, 16:801–808, 1995.

36. S. Bandyopadhyay, C. A. Murthy, and S. K. Pal. Pattern classification using genetic algorithms: Determination of H. *Pattern Recognition Letters*, 19(13):1171–1181, 1998.

37. S. Bandyopadhyay, C. A. Murthy, and S. K. Pal. Theoretical performance of genetic pattern classifier. *Journal of the Franklin Institute*, 336:387–422, 1999.

38. S. Bandyopadhyay, C. A. Murthy, and S. K. Pal. *VGA-Classifier*: Design and application. *IEEE Transactions on Systems, Man and Cybernetics-B*, 30(6):890–895, 2000.

39. S. Bandyopadhyay and S. K. Pal. Pattern classification with genetic algorithms: Incorporation of chromosome differentiation. *Pattern Recognition Letters*, 18:119–131, 1997.

40. S. Bandyopadhyay and S. K. Pal. Relation between VGA-classifier and MLP: Determination of network architecture. *Fundamenta Informaticae*, 37:177–196, 1999.

41. S. Bandyopadhyay and S. K. Pal. Pixel classification using variable string genetic algorithms with chromosome differentiation. *IEEE Transactions on Geoscience and Remote Sensing*, 39(2):303–308, 2001.

42. S. Bandyopadhyay, S. K. Pal, and B. Aruna. Multi-objective gas, quantitative indices and pattern classification. *IEEE Transactions Systems, Man and Cybernetics - B*, 34(5):2088–2099, 2004.

43. S. Bandyopadhyay, S. K. Pal, and U. Maulik. Incorporating chromosome differentiation in genetic algorithms. *Information Sciences*, 104(3/4):293–319, 1998.

44. S. Bandyopadhyay, S. K. Pal, and C. A. Murthy. Simulated annealing based pattern classification. *Information Sciences*, 109:165–184, 1998.

45. A. Bardossy and L. Samaniego. Fuzzy rule-based classification of remotely sensed imagery. *IEEE Transactions on Geoscience and Remote Sensing*, 40(2):362–374, February 2002.

46. V. Batenburg, A. P. Gultyaev, and C. W. A. Pleij. An APL-programmed genetic algorithm for the prediction of RNA secondary structure. *Journal of Theoretical Biology*, 174(3):269–280, 1995.

47. R. Battiti. Using mutual information for selecting features in supervised neural net learning. *IEEE Transactions on Neural Networks*, 5:537–550, 1994.

48. U. Baumgartner, Ch. Magele, and W. Renhart. Pareto Optimality and Particle Swarm Optimization. *IEEE Transactions on Magnetics*, 40(2):1172–1175, March 2004.

49. A. Baykasoglu. Goal Programming using Multiple Objective Tabu Search. *Journal of the Operational Research Society*, 52(12):1359–1369, December 2001.

50. M. J. Bayley, G. Jones, P. Willett, and M. P. Williamson. Genfold: A genetic algorithm for folding protein structures using NMR restraints. *Protein Science*, 7(2):491–499, 1998.

51. M. L. Beckers, L. M. Buydens, J. A. Pikkemaat, and C. Altona. Application of a genetic algorithm in the conformational analysis of methylene-acetal-linked thymine dimers in DNA: Comparison with distance geometry calculations. *Journal of Biomol NMR*, 9(1):25–34, 1997.

52. N. Behera and V. Nanjundiah. Trans gene regulation in adaptive evolution: A genetic algorithm model. *Journal of Theoretical Biology*, 188:153–162, 1997.

53. R. G. Beiko and R. L. Charlebois. GANN: Genetic algorithm neural networks for the detection of conserved combinations of features in dna. *BMC Bioinformatics*, 6(36), 2005.

54. H. M. Berman, J. Westbrook, Z. Feng, G. Gilliland, T. N. Bhat, H. Weissig, I. N. Shindyalov, and P. E. Bourne. The protein data bank. *Nucleic Acids Research*, 28:235–242, 2000.

55. M. Bessaou, A. Ptrowski, and P. Siarry. Island model cooperating with speciation for multimodal optimization. In *Parallel Problem Solving from Nature*, pages 437–446, Berlin Heidelberg New York, 2000. Springer.

56. H.-G. Beyer and H.-P. Schwefel. Evolution strategies — a comprehensive introduction. *Natural Computing: An International Journal*, 1(1):3–52, 2002.

57. J. C. Bezdek. *Pattern Recognition with Fuzzy Objective Function Algorithms*. Plenum, New York, 1981.

58. J. C. Bezdek and N. R. Pal. Some new indexes of cluster validity. *IEEE Transactions on Systems, Man and Cybernetics*, 28:301–315, 1998.

59. J. C. Bezdek and S. K. Pal, editors. *Fuzzy Models for Pattern Recognition: Methods that Search for Structures in Data*. IEEE Press, New York, 1992.

60. D. Bhandari, C. A. Murthy, and S. K. Pal. Genetic algorithm with elitist model and its convergence. *International Journal of Pattern Recognition and Artificial Intelligence*, 10:731–747, 1996.

61. D. Bhandari, N. R. Pal, and S. K. Pal. Directed mutation in genetic algorithms. *Information Sciences*, 79:251–270, 1994.

62. A. Bhattacharjee. *Some aspects of mango (Mangifora Indica L) leaf growth features in varietal recognition*. PhD thesis, University of Calcutta, Calcutta, India, 1986.

63. N. J. Bhuyan, V. V. Raghavan, and K. E. Venkatesh. Genetic algorithms for clustering with an ordered representation. In *Proceedings of the Fourth International Conference Genetic Algorithms*, pages 408–415, 1991.

64. C.M. Bishop. *Neural Networks for Pattern Recognition*. Clarendon Press, Oxford, 1995.

65. L. B. Booker, D. E. Goldberg, and J. H. Holland. Classifier systems and genetic algorithms. *Artificial Intelligence*, 40:235–282, 1989.

66. S. Bornholdt and D. Graudenz. General asymmetric neural networks and structure design by genetic algorithms. *Neural Networks*, 5:327–334, 1992.

67. P. A. N. Bosman and D. Thierens. The balance between proximity and diversity in multiobjective evolutionary algorithms. *IEEE Transactions Evolutionary Computation*, 7(2):174–188, 2003.

68. M. Boughanem, C. Chrisment, J. Mothe, C. S. Dupuy, and L. Tamine. Connectionist and genetic approaches for information retrieval. In F. Crestani and G. Pasi, editors, *Soft Computing in Information Retrieval: Techniques and Applications*, volume 50, pages 102–121. Physica-Verlag, Heidelberg, 2000.

69. P. Bradley, U. Fayyad, and C. Reina. Scaling EM (expectation maximization) clustering to large databases. Technical Report MSR-TR-98-35, Microsoft, 1998.

70. M. F. Bramlette. Initialization, mutation and selection methods in genetic algorithms for function optimization. In R. K. Belew and L. B. Booker, editors, *Proceedings of the 4th International Conference on Genetic Algorithms*, pages 100–107. Morgan Kaufmann, San Mateo, 1991.

71. J. Branke. Evolutionary algorithms in neural network design and training – A review. In J. T. Alander, editor, *Proc. of the First Nordic Workshop on Genetic Algorithms and their Applications (1NWGA)*, pages 145–163, Vaasa, Finland, 1995.

72. J. Branke. Evolutionary approaches to dynamic optimization problems – updated survey. In J. Branke and T. Bäck, editors, *GECCO Workshop on Evolutionary Algorithms for Dynamic Optimization Problems*, pages 27–30, San Francisco, California, USA, 2001.

73. S. Brin and L. Page. The anatomy of a large-scale hypertextual Web search engine. *Computer Networks and ISDN Systems*, 30:107–117, 1998.

74. A. Broder, R. Kumar, F. Maghoul, P. Raghavan, S. Rajagopalan, R. Stata, A. Tomkins, and J. Wiener. Graph structure in the Web. In *Proceedings of the Ninth International World Wide Web Conference*. Elsevier, 2000.

75. C. M. Brown, D. Hardy, B. B. Danzig, U. Manber, and M. F. Schwartz. The harvest information discovery and access system. In *Proceedings of Second International WWW conference on Distributed Environments*, 1994.

76. B. G. Buchanan. Can machine learning offer anything to expert systems? *Machine Learning*, 4:251–254, 1989.

77. B. P. Buckles, F. E. Petry, D. Prabhu, and M. Lybanon. Mesoscale feature labeling from satellite images. In S. K. Pal and P. P. Wang, editors, *Genetic Algorithms for Pattern Recognition*, pages 167–177. CRC Press, Boca Raton, 1996.

78. E. Cantú-Paz. *Efficient and Accurate Parallel Genetic Algorithms*. Kluwer Academic Publishers, USA, 2000.

79. E. Cantú-Paz. Migration policies, selection pressure, and parallel evolutionary algorithms. *Journal of Heuristics*, 7(4):311–334, 2001.

80. E. Cantú-Paz. Order statistics and selection methods of evolutionary algorithms. *Information Processing Letters*, 8(1):15–22, 2002.

81. C. A. Del Carpio. A parallel genetic algorithm for polypeptide three dimensional structure prediction: A transputer implementation. *Journal of Chemical Information and Computer Sciences*, 36(2):258–269, 1996.

82. D. R. Carvalho and A. A. Freitas. A hybrid decision tree/ genetic algorithm for coping with the problem of small disjuncts in data mining. In D. Whitley, D. E. Goldberg, E. Cantú-Paz, L. Spector, I. Parmee, and H. G. Beyer, editors, *Proceedings of the Genetic and Evolutionary Computation Conference*

GECCO-2000, volume I, pages 1061–1068, Las Vegas, Nevada, US, 2000. Morgan Kaufmann Publishers.

83. D. J. Cavicchio. Reproductive adaptive plans. In *Proceedings of the ACM 1972 Annual Conference*, pages 1–11, 1972.

84. S. Chakrabarti. *Mining the Web: Discovering Knowledge from Hypertext Data*. Morgan Kaufmann, 2002.

85. R. Chandrasekharam, S. Subhramanian, and S. Chaudhury. Genetic algorithm for node partitioning problem and applications in VLSI design. *IEE Proceedings, Part E: Computers and Digital Techniques*, 140(5):255–260, 1993.

86. C. Chen, L. H. Wang, C. Kao, M. Ouhyoung, and W. Chen. Molecular binding in structure-based drug design: A case study of the population-based annealing genetic algorithms. *IEEE International Conference on Tools with Artificial Intelligence*, pages 328–335, 1998.

87. H. Chen, Y.-M. Chung, and M. Ramsey. A smart itsy bitsy spider for the Web. *Journal of the American Society for Information Science and Technology*, 49(7):604–618, 1998.

88. H. Chen, G. Shankaranarayanan, and L. She. A machine learning approach to inductive query by examples: An experiment using relevance feedback, ID3, genetic algorithms, and simulated annealing. *Journal of the American Society for Information Science and Technology*, 49(8):693–705, 1998.

89. P. Chou and G. Fasmann. Prediction of the secondary structure of proteins from their amino acid sequence. *Advances in Enzymology*, 47:145–148, 1978.

90. D. E. Clark and D. R. Westhead. Evolutionary algorithms in computer-aided molecular design. *Journal of Computer-Aided Molecular Design*, 10(4):337–358, 1996.

91. T. Cleghorn, P. Baffes, and L. Wang. Robot path planning using a genetic algorithm. In *Proceedings of the SOAR*, pages 81–87, Houston, 1988.

92. F. A. Cleveland and S. F. Smith. Using genetic algorithms to schedule flow shop releases. In J. D. Schaffer, editor, *Proceedings of the 3rd International Conference on Genetic Algorithms*, pages 160–169. Morgan Kaufmann, San Mateo, 1989.

93. C. A. Coello Coello. An updated survey of GA-based multiobjective optimization techniques. Technical Report Lania-RD-98-08, Laboratorio Nacional de Informatica Avanzada, Xalapa, Veracruz, Mexico, 1998.

94. C. A. Coello Coello. A comprehensive survey of evolutionary-based multiobjective optimization techniques. *Knowledge and Information Systems*, 1(3):129–156, 1999.

95. C. A. Coello Coello. Constraint-handling using an evolutionary multiobjective optimization technique. *Civil Engineering and Environmental Systems*, 17:319–346, 2000.

96. C. A. Coello Coello and A. D. Christiansen. Multiobjective optimization of trusses using genetic algorithms. *Computers and Structures*, 75(6):647–660, 2000.

97. R. G. Congalton, R. G. Oderwald, and R. A. Mead. Assessing Landsat classification accuracy using discrete multivariate analysis statistical techniques. *Photogrammetric Engineering and Remote Sensing*, 49:1671–1678, 1983.

98. C. B. Congdon. Classification of epidemiological data: A comparison of genetic algorithm and decision tree approaches. In *Proceedings of the 2000 Congress on Evolutionary Computation*, pages 442–449, Piscataway, NJ, 2000. IEEE Service Center.

99. B. Contreras-Moreira, P. W. Fitzjohn, M. Offman, G. R. Smith, and P. A. Bates. Novel use of a genetic algorithm for protein structure prediction: Searching template and sequence alignment space. *Proteins*, 53(6):424–429, 2003.

100. L. Cooper, D. Corne, and M. Crabbe. Use of a novel hill-climbing genetic algorithm in protein folding simulations. *Computational Biology and Chemistry*, 27(6):575–580, 2003.

101. M. G. Cooper and J. J. Vidal. Genetic design of fuzzy controllers. In S. K. Pal and P. P. Wang, editors, *Genetic Algorithms for Pattern Recognition*, pages 283–298. CRC Press, Boca Raton, 1996.

102. O. Cord, F. Herrera, F. Hoffmann, and L. Magdalena. *Genetic Fuzzy Systems: Evolutionary Tuning and Learning of Fuzzy Knowledge Bases*. World Scientific, Singapore, 2001.

103. T. M. Cover and P. E. Hart. Nearest neighbor pattern classification. *IEEE Transactions on Information Theory*, IT-13:21–27, 1967.

104. F. Crestani and G. Pasi, editors. *Soft Computing in Information Retrieval: Techniques and Application*, volume 50. Physica-Verlag, Heidelberg, 2000.

105. N. Cristianini and J. Shawe-Taylor. *An Introduction to Support Vector Machines (and Other Kernel-based Learning Methods)*. Cambridge University Press, UK, 2000.

106. R. Das and D. Whitley. The only challenging problems are deceptive: Global search by solving order-1 hyperplane. In R. K. Belew and L. B. Booker, editors, *Proceedings of the 4th International Conference on Genetic Algorithms*, pages 166–173. Morgan Kaufmann, San Mateo, 1991.

107. D. Dasgupta and D. R. McGregor. Nonstationary function optimization using the structured genetic algorithm. In R. Männer and B. Manderick, editors, *Parallel Problem Solving from Nature 2 (Proc. 2nd Int. Conf. on Parallel Problem Solving from Nature, Brussels 1992)*, pages 145–154, Amsterdam, 1992. Elsevier.

108. V. N. Davidenko, V. M. Kureichik, and V. V. Miagkikh. Genetic algorithm for restrictive channel routing problem. In Thomas Bäck, editor, *Proceedings of the Seventh International Conference on Genetic Algorithms (ICGA97)*, pages 636–642, San Francisco, CA, 1997. Morgan Kaufmann.

109. Y. Davidor. Robot programming with a genetic algorithm. In *Proceedings of the 1990 IEEE International Conference on Computer Systems and Software Engineering*, pages 186–191, Tel Aviv, Israel, 1990. IEEE Computer Society Press, Los Alamitos, CA.

110. D. L. Davies and D. W. Bouldin. A cluster separation measure. *IEEE Transactions on Pattern Analysis and Machine Intelligence*, 1:224–227, 1979.

111. L. Davis, editor. *Genetic Algorithms and Simulated Annealing*. Pitman, London, 1987.

112. L. Davis. Adapting operator probabilities in genetic algorithms. In J. D. Schaffer, editor, *Proceedings of the Third International Conference on Genetic Algorithms*, pages 61–69. Morgan Kaufmann, San Mateo, 1989.

113. L. Davis, editor. *Handbook of Genetic Algorithms*. Van Nostrand Reinhold, New York, 1991.

114. T. E. Davis and J. C. Principe. A simulated annealing-like convergence theory for the simple genetic algorithm. In R. K. Belew and L. B. Booker, editors, *Proceedings of the 4th International Conference on Genetic Algorithms*, pages 174–181. Morgan Kaufmann, San Mateo, 1991.

115. J. E. Dayhoff. *Neural Network Architectures An Introduction*. Van Nostrand Reinhold, New York, 1990.
116. S. De, A. Ghosh, and S. K. Pal. Incorporating ancestors' influence in genetic algorithms. *Applied Intelligence*, 18:7–25, 2003.
117. K. De Jong. *An Analysis of the Behaviour of a Class of Genetic Adaptive Systems*. PhD thesis, Dept. of Computer and Communication Science, Univ. of Michigan, Ann Arbor, 1975.
118. K. De Jong. Learning with genetic algorithms: An overview. *Machine Learning*, 3:121–138, 1988.
119. K. De Jong and W. M. Spears. An analysis of the interacting roles of population size and crossover in genetic algorithms. In *Proceedings of the Parallel Problem Solving from Nature*, pages 38–47. Springer, Berlin Heidelberg New York, 1990.
120. D. M. Deaven and K. O. Ho. Molecular-geometry optimization with a genetic algorithm. *Physical Review Letters*, 75(2):288–291, 1995.
121. K. Deb. *Multi-objective Optimization Using Evolutionary Algorithms*. John Wiley and Sons, Ltd, England, 2001.
122. K. Deb, S. Agrawal, A. Pratap, and T. Meyarivan. A Fast Elitist Non-Dominated Sorting Genetic Algorithm for Multi-Objective Optimization: NSGA-II. In *Proceedings of the Parallel Problem Solving from Nature VI Conference*, pages 849–858, Paris, France, 2000. Springer. Lecture Notes in Computer Science No. 1917.
123. K. Deb and D. E. Goldberg. An investigation of niche and species formation in genetic function optimization. In J. D. Schaffer, editor, *Proceedings of the 3rd International Conference on Genetic Algorithms*, pages 42–50. Morgan Kaufmann, USA, 1989.
124. K. Deb, L. Thiele, M. Laumanns, and E. Zitzler. Scalable multi-objective optimization test problems. In *Proc. Congress on Evolutionary Computation*, pages 825–830, Honolulu, USA, 2002. IEEE Press.
125. T. Denoeux. A k-nearest neighbor classification rule based on Dempster-Shafer theory. *IEEE Transactions on Systems, Man and Cybernetics*, 25:804–813, 1995.
126. K. Dev and C. R. Murthy. A genetic algorithm for the knowledge base partitioning problem. *Pattern Recognition Letters*, 16:873–879, 1995.
127. P. A. Devijver and J. Kittler. *Pattern Recognition: A Statistical Approach*. Prentice-Hall, London, 1982.
128. K. Dmitry and V. Dmitry. An algorithm for rule generation in fuzzy expert systems. In *Proceedings of the International Conference on Pattern Recognition (ICPR04)*, pages I: 212–215, 2004.
129. R. C. Dubes and A. K. Jain. Clustering techniques: The user's dilemma. *Pattern Recognition*, 8:247–260, 1976.
130. R. O. Duda and P. E. Hart. *Pattern Classification and Scene Analysis*. Wiley, New York, 1973.
131. R. O. Duda, P. E. Hart, and D. G. Stork. *Pattern Classification (2nd ed.)*. John Wiley and Sons, 2001.
132. J. C. Dunn. A fuzzy relative of the ISODATA process and its use in detecting compact well-separated clusters. *Journal of Cybernetics*, 3:32–57, 1973.
133. C. Emmanouilidis, A. Hunter, and J. MacIntyre. A multiobjective evolutionary setting for feature selection and a commonality-based crossover operator. In *Proceedings of the 2000 Congress on Evolutionary Computation CEC00*, pages 309–316, La Jolla, California, USA, 2000. IEEE Press.

134. L. J. Eshelman. The CHC adaptive search algorithm: How to have safe search when engaging in nontraditional genetic recombination. In G. J. E. Rawlins, editor, *Foundations of Genetic Algorithms*, pages 265–283. Morgan Kaufmann, San Mateo, 1991.

135. L. J. Eshelman and J. D. Schaffer. Preventing premature convergence by preventing incest. In *Proceedings of the 4th International Conference on Genetic Algorithms*, pages 115–122. Morgan Kaufmann, San Mateo, 1991.

136. L. J. Eshelman and J. D. Schaffer. Real-coded genetic algorithms and interval schemata. In L.D. Whitley, editor, *Foundations of Genetic Algorithms 2*, pages 187–202. Morgan Kaufmann, San Mateo, CA, 1993.

137. M. Ester, H.-P. Kriegel, J. Sander, and X. Xu. Density-based algorithm for discovering clusters in large spatial databases. In *Proceedings of the Second International Conference on Data Mining KDD-96*, pages 226–231, Portland, Oregon, 1996.

138. M. Ester, H.-P. Kriegel, and X. Xu. Knowledge discovery in large spatial databases: Focusing techniques for efficient class identification. In *Proceedings of the 4th International Symposium on Large Spatial Databases (SSD'95)*, pages 67–82, Portland, Maine, 1995.

139. V. Matys et al. Transfac: transcriptional regulation, from patterns to profiles. *Nucleic Acids Research*, 31(1):374–378, 2003.

140. O. Etzioni. The World Wide Web: Quagmire or gold mine? *Communications of the ACM*, 39(11):65–68, 1996.

141. O. Etzioni and M. Perkowitz. Adaptive Web sites: An AI challenge. In *Proceedings of Fifteenth National Conference on Artificial Intelligence*, 1998.

142. W. Fan, E. A. Fox, P. Pathak, and H. Wu. The effects of fitness functions on genetic programming-based ranking discovery for web search. *Journal of the American Society for Information Science and Technology*, 55(7):628–636, 2004.

143. W. Fan, M. D. Gordon, and P. Pathak. Discovery of context-specific ranking functions for effective information retrieval using genetic programming. *IEEE Transactions on Knowledge and Data Engineering*, 16(4):523–527, 2004.

144. W. Fan, M. D. Gordon, and P. Pathak. Genetic programming based discovery of ranking functions for effective Web search. *Journal of Management Information Systems*, 21(4):37–56, 2005.

145. J. Fickett and M. Cinkosky. A genetic algorithm for assembling chromosome physical maps. In *Proceedings of the Second International Conference on Bioinformatics, Supercomputing, and Complex Genome Analysis, World Scientific*, pages 272–285, 1993.

146. J. W. Fickett. Finding genes by computer: The state of the art. *Trends in Genetics*, 12(8):316–320, 1996.

147. J. L. R. Filho, P. C. Treleaven, and C. Alippi. Genetic algorithm programming environments. *IEEE Computer*, pages 28–43, June 1994.

148. R. A. Fisher. The use of multiple measurements in taxonomic problems. *Annals of Eugenics*, 3:179–188, 1936.

149. G. W. Flake, S. Lawrence, and C. L. Giles. Efficient identification of the web communities. In *Proceedings on the 6th ACM SIGKDD Conference on Knowledge Discovery and Data Mining*, pages 150–160. ACM, 2000.

150. C. M. Fonseca and P. J. Fleming. Genetic algorithms for multi-objective optimization: Formulation, discussion and generalization. In Forrest S, editor,

Proceedings of the Fifth International Conference on Genetic Algorithms, pages 416–423. Morgan Kaufmann, San Mateo, CA, 1993.

151. C. M. Fonseca and P. J. Fleming. An overview of evolutionary algorithms in multiobjective optimization. *Evolutionary Computation*, 3:1–16, 1995.

152. M. Friedman and A. Kandel. *Introduction to Pattern Recognition, statistical, structural, neural and fuzzy logic approaches.* World Scientific, Singapore, 1999.

153. K. S. Fu. *Syntactic Pattern Recognition and Applications.* Academic Press, London, 1982.

154. T. Fukuda, Y. Komata, and T. Arakawa. Recurrent neural network with self-adaptive GAs for biped locomotion robot. In *Proceedings of the IEEE International Conference on Neural Networks*, pages 1710–1715, Houston, 1997.

155. K. Fukunaga. *Introduction to Statistical Pattern Recognition (2nd ed.).* Academic Press, New York, 1990.

156. C. Garcia-Martinez, O. Cordon, and F. Herrera. An empirical analysis of multiple objective ant colony optimization algorithms for the bi-criteria TSP. In *Proceedings of Ant Colony Optimization and Swarm Intelligence, Lecture Notes in Computer Science*, volume 3172, pages 61–72, Berlin Heidelberg New York, 2004. Springer.

157. I. Garibay, O. Garibay, and A. S. Wu. Effects of module encapsulation in repetitively modular genotypes on the search space. In *Proceedings of Genetic and Evolutionary Computation Conference - GECCO 2004*, volume 1, pages 1125–1137, Seattle, USA, 2004.

158. J. Garnier, J. F. Gibrat, and B. Robson. GOR method for predicting protein secondary structure from amino acid sequence. *Methods Enzymol.*, 266:540–553, 1996.

159. C. Gaspin and T. Schiex. Genetic algorithms for genetic mapping. In *Proceedings of the Third European Conference on Artificial Evolution*, pages 145–156, 1997.

160. E. S. Gelsema, editor. *Special Issue on Genetic Algorithms, Pattern Recognition Letters*, volume 16, no. 8. Elsevier Science, Inc., 1995.

161. E. S. Gelsema and L. Kanal, editors. *Pattern Recognition in Practice II.* North Holland, Amsterdam, 1986.

162. F. Ghannadian, C. Alford, and R. Shonkwiler. Application of random restart to genetic algorithms. *Information Sciences*, 95:81–102, 1996.

163. G. Goh and J. A. Foster. Evolving molecules for drug design using genetic algorithm. In *Proceedings of the International Conference on Genetic and Evolvable Computing*, pages 27–33. Morgan Kaufmann, 2000.

164. D. E. Goldberg. *Genetic Algorithms in Search, Optimization and Machine Learning.* Addison-Wesley, New York, 1989.

165. D. E. Goldberg. Sizing populations for serial and parallel genetic algorithms. In J. D. Schaffer, editor, *Proceedings of the 3rd International Conference on Genetic Algorithms*, pages 70–79. Morgan Kaufmann, San Mateo, 1989.

166. D. E. Goldberg. *The Design of Innovation: Lessons from and for Competent Genetic Algorithms.* Kluwer Academic Publishers, Boston, MA, USA, 2002.

167. D. E. Goldberg, K. Deb, H. Kargupta, and G. Harik. Rapid, accurate optimization of difficult problems using fast messy genetic algorithms. In S. Forrest, editor, *Proceedings of the 5th International Conference on Genetic Algorithms*, pages 56–64. Morgan Kaufmann, San Mateo, 1993.

168. D. E. Goldberg, K. Deb, and B. Korb. Messy genetic algorithms: Motivation, analysis, and first results. *Complex Systems*, 3:493–530, 1989.

169. D. E. Goldberg, K. Deb, and B. Korb. Do not worry, be messy. In R. K. Belew and L. B. Booker, editors, *Proceedings of the 4th International Conference on Genetic Algorithms*, pages 24–30. Morgan Kaufmann, San Mateo, 1991.

170. D. E. Goldberg and J. J. Richardson. Genetic algorithms with sharing for multimodal function optimization. In *Proc. 2nd International Conference on Genetic Algorithms*, pages 41–49. Lawrence Erlbaum Associates, Hillsdale, 1987.

171. D. E. Goldberg and P. Segrest. Finite Markov chain analysis of genetic algorithms. In *Proc. 2nd International Conference on Genetic Algorithms*, pages 1–8. Lawrence Erlbaum Associates, Hillsdale, 1987.

172. R. C. Gonzalez and M. G. Thomason. *Syntactic Pattern Recognition: An Introduction.* Addison-Wesley, Reading, 1978.

173. M. D. Gordon. Probabilistic and genetic algorithms for document retrieval. *Communications of the ACM*, 31(10):208–218, 1988.

174. J. J. Grefenstette. Optimization of control parameters for genetic algorithms. *IEEE Transactions on Systems, Man and Cybernetics*, 16:122–128, 1986.

175. J. J. Grefenstette. Incorporating problem specific information in genetic algorithms. In L. Davis, editor, *Genetic Algorithms and Simulated Annealing*, pages 42–60. Pitman, London, 1987.

176. J. J. Grefenstette, R. Gopal, B. Rosmaita, and D. Van Gucht. Genetic algorithms for the traveling salesman problem. In J. J. Grefenstette, editor, *Proceedings of the 1st International Conference on Genetic Algorithms*, pages 160–168. Lawrence Erlbaum Associates, Hillsdale, 1985.

177. A. P. Gultyaev, V. Batenburg, and C. W. A. Pleij. The computer simulation of RNA folding pathways using a genetic algorithm. *Journal of Molecular Biology*, 250:37–51, 1995.

178. J. R. Gunn. Sampling protein conformations using segment libraries and a genetic algorithm. *Journal of Chemical Physics*, 106:4270–4281, 1997.

179. J. Gunnels, P. Cull, and J. L. Holloway. Genetic algorithms and simulated annealing for gene mapping. In *Proceedings of the First IEEE Conference on Evolutionary Computation*, pages 385–390, USA, 1994. Lawrence Erlbaum Associates, Inc.

180. M. Halkidi, Y. Batistakis, and M. Vazirgiannis. Cluster validity methods: Part I. *SIGMOD Rec.*, 31(2):40–45, 2002.

181. K. Hammond, R. Burk, C. Martin, and S. Lytinen. Faqfinder: a case based approach to knowledge navigation. In *Working Notes of AAAI Spring Symposium on Information Gathering from Heterogeneous Distributed Environments*, 1995.

182. J. Han and K. C. Chang. Data mining for Web intelligence. *IEEE Computer*, pages 54–60, November 2002.

183. J. Han and M. Kamber. *Data Mining: Concepts and Techniques.* Morgan Kaufmann Publishers, San Francisco, USA, 2000.

184. K. Hanada, T. Yokoyama, and T. Shimizu. Multiple sequence alignment by genetic algorithm. *Genome Informatics*, 11:317–318, 2000.

185. S. A. Harp and T. Samad. Genetic synthesis of neural network architecture. In L. Davis, editor, *Handbook of Genetic Algorithms*, pages 202 – 221. Van Nostrand Reinhold, New York, 1991.

186. W. E. Hart, N. Krasnogor, and J. E. Smith, editors. *Recent Advances in Memetic Algorithms*, volume 166 of *Studies in Fuzziness and Soft Computing*. Springer, Berlin Heidelberg New York, 2005.

288   References

187. J. A. Hartigan. *Clustering Algorithms*. Wiley, 1975.
188. S. Haykin. *Neural Networks, A Comprehensive Foundation*. Macmillan College Publishing Company, New York, 1994.
189. D. O. Hebb. *The Organization of Behaviour*. Wiley, New York, 1949.
190. R. Hecht-Nielsen. Kolmogorov's mapping neural network existence theorem. In *Proceedings of the 1st IEEE International Conference on Neural Networks*, volume 3, pages 11–14, San Diego, 1987.
191. A. Hernández Aguirre, S. Botello Rionda, C. A. Coello Coello, G. Lizárraga Lizárraga, and E. Mezura Montes. Handling Constraints using Multiobjective Optimization Concepts. *International Journal for Numerical Methods in Engineering*, 59(15):1989–2017, April 2004.
192. F. Herrera, M. Lozano, and J. L. Verdegay. Generating fuzzy rules from examples using genetic algorithms. In *Proceedings of the IPMU'94 (5th International Conference on Information Processing and Management of Uncertainty in Knowledge-Based Systems)*, pages 675–680, 1994.
193. A. Hill and C. J. Taylor. Model-based image interpretation using genetic algorithms. *Image and Vision Computing*, 10:295–300, 1992.
194. S. Hill, J. Newell, and C. O'Riordan. Analysing the effects of combining fitness scaling and inversion in genetic algorithms. In *Proceedings of the 16th IEEE International Conference on Tools with Artificial Intelligence (ICTAI'04)*, pages 380–387, 2004.
195. J. H. Holland. *Adaptation in Natural and Artificial Systems*. Ann Arbor: The University of Michigan Press, 1975.
196. A. Homaifar. A new approach on the traveling salesman problem by genetic algorithms. In Stephanie Forrest, editor, *Proceedings of the Fifth International Conference on Genetic Algorithms*, San Mateo, CA, 1993. Morgan Kaufmann.
197. A. Homaifar and E. McCormick. Simultaneous design of membership functions and rule sets for fuzzy controllers using genetic algorithms. *IEEE Transactions on Fuzzy Systems*, 3:129–139, 1995.
198. J. Hong, Y. Lin, and W. Lai. Type-2 fuzzy neural network systems and learning. *International Journal of Computational Cognition*, 1(4):79–90, 2003.
199. J. Horn. Finite Markov chain analysis of genetic algorithms with niching. In S. Forrest, editor, *Proceedings of the 5th International Conference on Genetic Algorithms*, pages 110–117. Morgan Kaufmann, San Mateo, 1993.
200. J. Horn and N. Nafpliotis. Multiobjective optimization using the niched Pareto genetic algorithm. Technical Report IlliGAL Report 93005, University of Illinois at Urbana-Champaign, Urbana, Illinois, USA, 1993.
201. T. Hou, J. Wang, L. Chen, and X. Xu. Automated docking of peptides and proteins by using a genetic algorithm combined with a tabu search. *Protein Engineering*, 12:639–647, 1999.
202. H. V. Hove and A. Verschoren. Genetic algorithms and recognition problems. In S. K. Pal and P. P. Wang, editors, *Genetic Algorithms for Pattern Recognition*, pages 145–166. CRC Press, Boca Raton, 1996.
203. J. Hu, E. Goodman, K. Seo, Z. Fan, and R. Rosenberg. The hierarchical fair competition (HFC) framework for continuing evolutionary algorithms. *Evolutionary Computation*, 13(2):241–277, 2005.
204. M. Hulin. Analysis of schema distribution. In R. K. Belew and L. B. Booker, editors, *Proceedings of the 4th International Conference on Genetic Algorithms*, pages 190–196. Morgan Kaufmann, San Mateo, 1991.

205. H. Iijima and Y. Naito. Incremental prediction of the side-chain conformation of proteins by a genetic algorithm. In *Proceedings of the IEEE Conference on Evolutionary Computation*, volume 1, pages 362–367, 1994.

206. H. Ishibuchi and T. Murata. A Multi-Objective Genetic Local Search Algorithm and its Application to Flowshop Scheduling. *IEEE Transactions on Systems, Man, and Cybernetics-Part C: Applications and Reviews*, 28:392–403, 1998.

207. H. Ishibuchi, T. Murata, and H. Tanaka. Construction of fuzzy classification systems with linguistic if-then rules using genetic algorithms. In S. K. Pal and P. P. Wang, editors, *Genetic Algorithms for Pattern Recognition*, pages 227–251. CRC Press, Boca Raton, 1996.

208. H. Ishibuchi, M. Nii, and T. Murata. Linguistic rule extraction from neural networks and genetic algorithm based rule selection. In *Proceedings of the IEEE International Conference on Neural Networks*, pages 2390–2395, Houston, 1997.

209. H. Ishibuchi, K. Nozaki, N. Yamamoto, and H. Tanaka. Acquisition of fuzzy classification knowledge using genetic algorithms. In *Proceedings of the 3rd IEEE International Conference on Fuzzy Systems*, pages 1963–1968, Orlando, 1994.

210. H. Ishibuchi, K. Nozaki, N. Yamamoto, and H. Tanaka. Selecting fuzzy if-then rules for classification problems using genetic algorithms. *IEEE Transactions on Fuzzy Systems*, 3:260–270, 1995.

211. H. Ishigami, T. Fukuda, T. Shibata, and F. Arai. Structure optimization of fuzzy neural network by genetic algorithm. *Fuzzy Sets and Systems*, 71:257–264, 1995.

212. A. K. Jain and R. C. Dubes. *Algorithms for Clustering Data*. Prentice-Hall, Englewood Cliffs, NJ, 1988.

213. A. K. Jain and J. Mao (eds.). Special issue on artificial neural networks and statistical pattern recognition. *IEEE Transactions on Neural Networks*, 8(1), 1997.

214. A. K. Jain, M. N. Murty, and P. J. Flynn. Data clustering: A review. *ACM Computing Surveys*, 31(3):264–323, 1999.

215. J. R. Jang, C. Sun, and E. Mizutani. *Neuro-Fuzzy and Soft Computing: A Computational Approach to Learning and Machine Intelligence*. Pearson Education, 1996.

216. C. J. Janikow. A genetic algorithm method for optimizing the fuzzy component of a fuzzy decision tree. In S. K. Pal and P. P. Wang, editors, *Genetic Algorithms for Pattern Recognition*, pages 253–281. CRC Press, Boca Raton, 1996.

217. A. Jaszkiewicz. Do multiple-objective metaheuristics deliver on their promises? A computational experiment on the set-covering problem. *IEEE Transactions Evolutionary Computation*, 7(2):133–143, 2003.

218. P. Jog, J. Y. Suh, and D. V. Gucht. The effects of population size, heuristic crossover and local improvement on a genetic algorithm for the traveling salesman problem. In J. D. Schaffer, editor, *Proceedings of the 3rd International Conference on Genetic Algorithms*, pages 110–115. Morgan Kaufmann, San Mateo, 1989.

219. G. Jones, P. Willett, and R. C. Glen. Molecular recognition of receptor sites using a genetic algorithm with a description of desolvation. *Journal of Molecular Biology*, 245:43–53, 1995.

220. G. Jones, P. Willett, R. C. Glen, A. R. Leach, and R. Taylor. Further development of a genetic algorithm for ligand docking and its application to screening combinatorial libraries. *American Chemical Society Symposium Series*, 719:271–291, 1999.

221. G. Jones, P. Willett, R. C. Glen, A. R. Leach, and R. J. Taylor. Development and validation of a genetic algorithm for flexible docking. *Journal of Molecular Biology*, 267:727–748, 1997.

222. A. Joshi and R. Krishnapuram. Robust fuzzy clustering methods to support Web mining. In *Proc. Workshop in Data Mining and Knowledge Discovery, SIGMOD*, pages (15)1–(15)8. ACM, USA, 1998.

223. A. Joshi, N. Ramakrishman, E. N. Houstis, and J. R. Rice. On neurobiological, neuro-fuzzy, machine learning, and statistical pattern recognition techniques. *IEEE Transactions on Neural Networks*, 8(1):18–31, 1997.

224. A. Kandel. *Fuzzy Techniques in Pattern Recognition*. Wiley Interscience, New York, 1982.

225. A. Kandel. *Fuzzy Mathematical Techniques with Applications*. Addison-Wesley, New York, 1986.

226. H. Kargupta. The gene expression messy genetic algorithm. In *Proceedings of the IEEE International Conference on Evolutionary Computation*, pages 814–819. IEEE Press, New Jersey, 1996.

227. H. Kargupta. The gene expression messy genetic algorithm. In *Proceedings of the IEEE International Conference on Evolutionary Computation*, pages 631–636, 1996.

228. H. Kargupta and S. Bandyopadhyay. Further experimentations on the scalability of the GEMGA. In *Proceedings of the V Parallel Problem Solving from Nature (PPSN V), Lecture Notes in Computer Science*, volume 1498, pages 315–324. Springer, Berlin Heidelberg New York, 1998.

229. H. Kargupta and S. Bandyopadhyay. A perspective on the foundation and evolution of the linkage learning genetic algorithms. *The Journal of Computer Methods in Applied Mechanics and Engineering, Special issue on Genetic Algorithms*, 186:266–294, 2000.

230. H. Kargupta, K. Deb, and D. E. Goldberg. Ordering genetic algorithms and deception. In R. Manner and B. Manderick, editors, *Proceedings of the Parallel Problem Solving from Nature*, pages 47–56. North-Holland, Amsterdam, 1992.

231. H. Kargupta, B. H. Park, D. Hershberger, and E. Johnson. Collective data mining: A new perspective toward distributed data mining. In *Advances in Distributed and Parallel Knowledge Discovery*. MIT/AAAI Press, 1999.

232. N. K. Kasabov and M. J. Watts. Genetic algorithms for structural optimization, dynamic adaptation and automated design of fuzzy neural networks. In *Proceedings of the IEEE International Conference on Neural Networks*, pages 2546–2549, Houston, 1997.

233. K. Katoh, K. Kuma, and T. Miyata. Genetic algorithm-based maximum-likelihood analysis for molecular phylogeny. *Journal of Molecular Evolution*, 53(4-5):477–484, 2001.

234. L. Kaufman and P. J. Rousseeuw. *Finding Groups in Data: An Introduction to Cluster Analysis*. John Wiley & Sons, NY, US, 1990.

235. A. Kel, A. Ptitsyn, V. Babenko, S. Meier-Ewert, and H. Lehrach. A genetic algorithm for designing gene family-specific oligonucleotide sets used for hybridization: The G protein-coupled receptor protein superfamily. *Bioinformatics*, 14(3):259–270, 1998.

236. O. V. Kel-Margoulis, T. G. Ivanova, E. Wingender, and A. E. Kel. Automatic annotation of genomic regulatory sequences by searching. In *Proceedings of the Pacific Symposium on Biocomputing*, pages 187–198, 2002.

237. J. D. Kelly, Jr. and L. Davis. A hybrid genetic algorithm for classification. In *Proceedings of the 12th International Joint Conference on Artificial Intelligence*, Sydney, 1991.

238. J. D. Kelly, Jr. and L. Davis. Hybridizing the genetic algorithm and the K nearest neighbors classification algorithm. In R. K. Belew and L. B. Booker, editors, *Proceedings of the 4th International Conference on Genetic Algorithms*, pages 377–383. Morgan Kaufmann, San Mateo, 1991.

239. M. Khimasia and P. Coveney. Protein structure prediction as a hard optimization problem: The genetic algorithm approach. *Molecular Simulation*, 19:205–226, 1997.

240. F. Khoshalhan. *A New selection Method in Genetic Algorithms and Its Applications to Production Management Problems*. PhD thesis, Tarbiat Modares University, Tehran, Iran, 2003.

241. S. Kim and B. T. Zhang. Web document retrieval by genetic learning of importance factors for HTML tags. In *Proceedings of the International Workshop on Text and Web Mining*, pages 13–23, 2000.

242. T. H. Kim, L. O. Barrera, M. Zheng, C. Qu, M. A. Singer, T. A. Richmond, Y. Wu, R. D. Green, and B. Ren. A high-resolution map of active promoters in the human genome. *Nature*, 436:876–880, 2005.

243. S. Kirkpatrick, C. D. Gelatt Jr., and M. P. Vechhi. Optimization by simulated annealing. *Science*, 220:671–680, 1983.

244. M. Kleinberg. Authoritative sources in hyperlinked environment. In *Proceedings of the Ninth Annual ACM-SIAM Symposium on Discrete Algorithms*, 1998.

245. L. Knight and S. Sen. PLEASE: A prototype learning system using genetic algorithms. In L. Eshelman, editor, *Proceedings of the Sixth International Conference on Genetic Algorithms*, pages 429–435, San Francisco, CA, 1995. Morgan Kaufmann.

246. J. D. Knowles and D. Corne. Approximating the nondominated front using the Pareto archived evolution strategy. *Evolutionary Computation*, 8(2):149–172, 2000.

247. J. D. Knowles, M. J. Oates, and D. W. Corne. Advanced multiobjective evolutionary algorithms applied to two problems in telecommunications. *BT Technology Journal*, 18(4):51–65, 2000.

248. S. Knudsen. Promoter2.0: For the recognition of PolII promoter sequences. *Bioinformatics*, 15:356–361, 1999.

249. S. Kobayashi, I. Ono, and M. Yamamura. An efficient genetic algorithm for job shop scheduling problems. In L. Eshelman, editor, *Proceedings of the Sixth International Conference on Genetic Algorithms*, pages 506–511, San Francisco, CA, 1995. Morgan Kaufmann.

250. T. Kohonen. *Self-Organization and Associative Memory*. Springer, Berlin Heidelberg New York, 1989.

251. R. Konig and T. Dandekar. Improving genetic algorithms for protein folding simulations by systematic crossover. *BioSystems*, 50:17–25, 1999.

252. R. Kosala and H. Blockeel. Web mining research: A survey. *SIGKDD Explorations*, 2(1):1–15, 2000.

253. R. Kothari and D. Pitts. On finding the number of clusters. *Pattern Recognition Letters*, 20:405–416, 1999.

254. J. R. Koza. Concept formation and decision tree induction using the genetic programming paradigm. In H.-P. Schwefel and R. Manner, editors, *Proceedings of the 1st International Workshop on Parallel Problem Solving From Nature (PPSN-1)*, volume 496, pages 124–128, Dortmund, Germany, 1991. Springer.

255. J. R. Koza. *Genetic Programming: On the Programming of Computers by Means of Natural Selection*. MIT Press, Cambridge, 1992.

256. D. H. Kraft, F. E. Petry, B. P. Buckles, and T. Sadasivan. The use of genetic programming to build queries for information retrieval. In *Proceedings of the IEEE Symposium on Evolutionary Computation*, 1994.

257. N. Krasnogor, W. E. Hart, J. Smith, and D. A. Pelta. Protein structure prediction with evolutionary algorithms. In *Proceedings of the Genetic and Evolutionary Computation Conference*, volume 2, pages 1596–1601, 1999.

258. N. Krasnogor, D. Pelta, P. M. Lopez, P. Mocciola, and E. Canal. Genetic algorithms for the protein folding problem: A critical view. In C. Fyfe and E. Alpaydin, editors, *Proceedings of the Engineering Intelligent Systems*, pages 353–360. ICSC Academic Press, 1998.

259. R. Krishnapuram, A. Joshi, O. Nasraoui, and L. Yi. Low-complexity fuzzy relational clustering algorithms for Web mining. *IEEE-FS*, 9:595–607, Aug. 2001.

260. A. Krone and H. Kiendl. An evolutionary concept for generating relevant fuzzy rules from data. *International Journal of Knowledge-based Intelligent Engineering System*, 1:207–213, 1997.

261. L. I. Kuncheva. Editing the k-nearest neighbor rule by a genetic algorithm. *Pattern Recognition Letters*, 16:809–814, 1995.

262. L. I. Kuncheva. Initializing of an RBF network by a genetic algorithm. *Neurocomputing*, 14:273–288, 1997.

263. T. Kuo and S. Hwang. Using disruptive selection to maintain diversity in genetic algorithms. *Applied Intelligence*, 7:257–267, 1997.

264. F. Kursawe. A variant of evolution strategies for vector optimization. In H.-P. Schwefel and R. Manner, editors, *Parallel Problem Solving from Nature, 1st Workshop, Proceedings, volume 496 of Lecture Notes in Computer Science*, pages 193–197. Springer, Berlin Heidelberg New York, 1991.

265. C. Kwok and D. Weld. Planning to gather information. In *Proceedings of 14th National Conference on AI*, 1996.

266. L. D. Landau and E. M. Lifshitz. *Theory of Elasticity*. Pergamon Press, New York, 1970.

267. M. Laumanns, L. Thiele, K. Deb, and E. Zitzler. Combining convergence and diversity in evolutionary multi-objective optimization. *Evolutionary Computation*, 10(3):263–282, 2002.

268. C. Lawrence, S. Altschul, M. Boguski, J. Liu, A. Neuwald, and J. Wootton. Detecting subtle sequence signals: A Gibbs sampling strategy for multiple alignment. *Science*, 262:208–214, 1993.

269. A. R. Lemmon and M. C. Milinkovitch. The metapopulation genetic algorithm: An efficient solution for the problem of large phylogeny estimation. *Proceedings of the National Academy of Sciences, USA*, 99(16):10516–10521, 2002.

270. G. Leroy, A. M. Lally, and H. Chen. The use of dynamic contexts to improve casual internet searching. *ACM Transactions on Information Systems*, 21(3):229–253, 2003.

271. A. M. Lesk. *Introduction to Bioinformatics*. Oxford University Press, Oxford, 2002.

272. V. G. Levitsky and A. V. Katokhin. Recognition of eukaryotic promoters using a genetic algorithm based on iterative discriminant analysis. *In Silico Biology*, 3: 0008(1-2):81–87, 2003.

273. P. O. Lewis. A genetic algorithm for maximum likelihood phylogeny inference using nucleotide sequence data. *Molecular Biology and Evolution*, 15(3):277–283, 1998.

274. G. E. Liepins and S. Baluja. apGA: An adaptive parallel genetic algorithm. In O. Balci, R. Sharda, and S. A. Zenios, editors, *Computer Science and Operations Research, New Development in Their Interfaces*, pages 399–409. Pergamon Press, 1992.

275. G. E. Liepins, M. R. Hilliard, M. Palmer, and G. Rangarajan. Credit assignment and discovery in classifier systems. *International Journal of Intelligent Systems*, 6:55–69, 1991.

276. G. E. Liepins and M. D. Vose. Deceptiveness and genetic algorithm dynamics. In G. J. E. Rawlins, editor, *Foundations of Genetic Algorithms*, pages 36–50. Morgan Kaufmann, San Mateo, 1991.

277. S. Lin, E. D. Goodman, and W. F. Punch. A genetic algorithm approach to dynamic job shop scheduling problems. In T. Bäck, editor, *Proceedings of the Seventh International Conference on Genetic Algorithms*, pages 481–488, San Mateo, CA, 1997. Morgan Kaufmann.

278. S. Lin, E. D. Goodman, and W. F. Punch III. Investigating parallel genetic algorithms on jop shop scheduling problems. In P. J. Angeline, R. G. Reynolds, J. R. McDonnell, and R. Eberhart, editors, *Evolutionary Programming VI*, pages 383–393, Berlin Heidelberg New York, 1997. Springer. Lecture Notes in Computer Science 1213.

279. R. P. Lippmann. An introduction to computing with neural nets. *IEEE ASSP Magazine*, 4(2):4–22, 1987.

280. J. Lis and A. E. Eiben. A multi-sexual genetic algorithm for multiobjective optimization. In T. Fukuda and T. Furuhashi, editors, *Proceedings of the 1996 International Conference on Evolutionary Computation*, pages 59–64. IEEE Press, 1996.

281. V. Loia and P. Luongo. An evolutionary approach to automatic Web page categorization and updating. In N. Zhong, Y. Yao, J. Liu, and S. Oshuga, editors, *Web Intelligence: Research and Development*, pages 292–302. LNCS, Springer, 2001.

282. C. Lopez-Pujalte, V. P. G. Bote, and F. de Moya Anegon. Order-based fitness functions for genetic algorithms applied to relevance feedback. *Journal of the American Society for Information Science and Technology*, 54(2):152–160, 2003.

283. S. J. Louis. Genetic learning for combinational logic design. *Journal of Soft Computing*, 9(1):38–43, 2004.

284. S. J. Louis and J. McDonnell. Learning with case injected genetic algorithms. *IEEE Transactions on Evolutionary Computation*, 8(4):316–328, 2004.

285. Q. Lu, Z. Peng, F. Chu, and J. Huang. Design of fuzzy controller for smart structures using genetic algorithms. *Smart Mater. Struct.*, 12:979–986, 2003.

286. N. M. Luscombe, D. Greenbaum, and M. Gerstein. What is bioinformatics? A proposed definition and overview of the field. *Yearbook of Medical Informatics*, pages 83–100, 2001.

287. S. K. Madria, S. S. Bhowmick, W. K. Ng, and E. -P. Lim. Research issues in Web data mining. In Mukesh K. Mohania and A. Min Tjoa, editors, *Proceedings of First International Conference on Data Warehousing and Knowledge Discovery DaWaK*, volume 1676 of *Lecture Notes in Computer Science*, pages 303–312. Springer, Berlin Heidelberg New York, 1999.

288. H. Maini, K. Mehrotra, C. Mohan, and S. Ranka. Knowledge-based nonuniform crossover. *Complex Systems*, 8:257–293, 1994.

289. O. L. Mangasarin, R. Setiono, and W. H. Wolberg. Pattern recognition via linear programming: Theory and application to medical diagnosis. In T. F. Coleman and Y. Li, editors, *Large-scale Numerical Optimization*, pages 22–30. SIAM Publications, Philadelphia, 1990.

290. V. Maniezzo. Genetic evolution of the topology and weight distribution of neural networks. *IEEE Transactions on Neural Networks*, 5:39–53, 1994.

291. M. H. Marghny and A. F. Ali. Web mining based on genetic algorithm. In *Proceedings of ICGST International Conference on Artificial Intelligence and Machine Learning (AIML-05)*, December 2005.

292. W. N. Martin, J. Lienig, and J. P. Cohoon. Island (migration) models: evolutionary algorithms based on punctuated equilibria. In T. Bäck, D. Fogel, and Z. Michalewicz, editors, *Evolutionary Computation*, volume 2. Institute of Physics Publishing, Bristol, UK, 2000.

293. M. Martin-Bautista and M.-A. Vila. A survey of genetic feature selection in mining issues. In *Proceedings of the Congress on Evolutionary Computation (CEC 99)*, pages 1314–1321, 1999.

294. T. Maruyama, A. Konagaya, and K. Konishi. An asynchronous fine grained parallel genetic algorithm. In *Proceedings of the Parallel Problem Solving from Nature*, pages 563–572, 1992.

295. B. Masand, M. Spiliopoulou, J. Srivastava, and O. Zaiane. Web mining for usage patterns & profiles. *SIGKDD Explorations Newsletter*, 4(2):125–127, 2002.

296. K. Mathias, D. Whitley, A. Kusuma, and C. Stork. An empirical evaluation of genetic algorithms on noisy objective functions. In S. K. Pal and P. P. Wang, editors, *Genetic Algorithms for Pattern Recognition*, pages 65–86. CRC Press, Boca Raton, 1996.

297. H. Matsuda. Protein phylogenetic inference using maximum likelihood with a genetic algorithm. In *Pacific Symposium on Biocomputing*, pages 512–523. World Scientific, London, 1996.

298. U. Maulik and S. Bandyopadhyay. Genetic algorithm based clustering technique. *Pattern Recognition*, 33:1455–1465, 2000.

299. U. Maulik and S. Bandyopadhyay. Fuzzy partitioning using a real-coded variable-length genetic algorithm for pixel classification. *IEEE Transactins on Geoscience and Remote Sensing*, 41(5):1075– 1081, 2003.

300. U. Maulik and S. Bandyopadhyay. Performance evaluation of some clustering algorithms and validity indices. *IEEE Transactions on Pattern Analysis and Machine Intelligence*, 24(12):1650–1654, 2002.

301. A. C. W. May and M. S. Johnson. Improved genetic algorithm-based protein structure comparisons: Pairwise and multiple superpositions. *Protein Engineering*, 8:873–882, 1995.

302. P. Mazumdar and E. M. Rudnick. *Genetic Algorithms for VLSI Design, Layout and Test Automation*. Prentice-Hall PTR, NJ, 1999.

303. D. B. McGarrah and R. S. Judson. Analysis of the genetic algorithm method of molecular conformation determination. *Journal of Computational Chemistry*, 14(11):1385–1395, 1993.

304. P. R. McMullen. An ant colony optimization approach to addessing a JIT sequencing problem with multiple objectives. *Artificial Intelligence in Engineering*, 15:309–317, 2001.

305. P. Melin and O. Castillo. Adaptive intelligent control of aircraft systems with a hybrid approach combining neural networks, fuzzy logic and fractal theory. *Applied Soft Computing*, 3(4):353–362, 2003.

306. P. Merz and B. Freisleben. Genetic local search for the TSP: New results. In *Proceedings of The IEEE Conference on Evolutionary Computation, IEEE World Congress on Computational Intelligence*, pages 159–164, 1997.

307. J. E. Mezzich. Evaluating clustering methods for psychiatric-diagnosis. *Biological Psychiatry*, 13:265–281, 1978.

308. Z. Michalewicz. *Genetic Algorithms + Data Structures = Evolution Programs*. Springer, Berlin Heidelberg New York, 1992.

309. Z. Michalewicz and C. Z. Janikow. Genetic algorithms for numerical optimization. *Statistics and Computing*, 1:75–91, 1991.

310. G. W. Milligan and C. Cooper. An examination of procedures for determining the number of clusters in a data set. *Psychometrika*, 50(2):159–179, 1985.

311. B. Minaei-Bidgoli, G. Kortemeyer, and W. F. Punch. Optimizing classification ensembles via a genetic algorithm for a Web-based educational system. In *Proceedings Of Joint International Association for Pattern Recognition (IAPR) Workshops on Syntactical and Structural Pattern Recognition (SSPR 2004) and Statistical Pattern Recognition (SPR 2004)*, 2004.

312. M. Minsky and S. Papert. *Perceptrons: An Introduction to Computational Geometry*. MIT Press, Cambridge, 1969.

313. T. M. Mitchell. *Machine learning*. McGraw-Hill, New York, 1997.

314. B. Mobasher, N. Jain, E.H. Han, and J. Srivastava. Web mining: Patterns from WWW transactions. Technical Report TR96050, University of Minnesota, 1997.

315. C. Mohan. Dynamic e-business: Trends in Web services, 2002. Invited talk at the 3rd VLDB Workshop on Technologies for E-Services (TES).

316. D. J. Montana and L. Davis. Training feedforward neural networks using genetic algorithms. In N. S. Sridharan, editor, *Proceedings of the 11th International Joint Conference on Artificial Intelligence*, pages 762–767. Morgan Kaufmann, San Mateo, 1989.

317. G. M. Morris, D. S. Goodsell, R. S. Halliday, R. Huey, W. E. Hart, R. K. Belew, and A. J. Olsoni. Automated docking using a Lamarckian genetic algorithm and an empirical binding free energy function. *Journal of Computational Chemistry*, 19(14):1639–1662, 1998.

318. H. Muhlenbein. Evolution in time and space — the parallel genetic algorithm. In G. J. E. Rawlins, editor, *Foundations of Genetic Algorithms*, pages 316–337. Morgan Kaufmann, San Mateo, 1991.

319. H. Muhlenbein, M. Schomish, and J. Born. The parallel genetic algorithm as function optimizer. In R. K. Belew and L. B. Booker, editors, *Proceedings of the 4th International Conference on Genetic Algorithms*, pages 271–278. Morgan Kaufmann, San Mateo, 1991.

320. H. J. Muller, editor. *Studies in Genetics — selected papers*. Indiana University Press, Bloomington, 1962.

321. M. Murakawa, S. Yoshizawa, I. Kajitani, and T. Higuchi. On-line adaptation of neural networks with evolvable hardware. In T. Bäck, editor, *Proceedings of the Seventh International Conference on Genetic Algorithms*, San Mateo, CA, 1997. Morgan Kaufmann.

322. H. Murao, H. Tamaki, and S. Kitamura. A coevolutionary approach to adapt the genotype-phenotype map in genetic algorithms. In *Proceedings of the Congress on Evolutionary Computation*, volume 2, pages 1612–1617, 2002.

323. T. Murata and H. Ishibuchi. Positive and negative combination effects of crossover and mutation operators in sequencing problems. *Evolutionary Computation*, 20-22:170–175, 1996.

324. C. A. Murthy, D. Bhandari, and S. K. Pal. ϵ-optimal stopping time for genetic algorithms with elitist model. *Fundamenta Informaticae*, 35:91–111, 1998.

325. C. A. Murthy and N. Chowdhury. In search of optimal clusters using genetic algorithms. *Pattern Recognition Letters*, 17:825–832, 1996.

326. S. B. Needleman and C. D. Wunsch. A general method applicable to the search for similarities in the amino acid sequence of two proteins. *Journal of Molecular Biology*, 48:443–453, 1970.

327. R. Ng and J. Han. Efficient and effective clustering method for spatial data mining. In *Proceedings of the 1994 International Conference on Very Large Data Bases*, pages 144–155, Santiago, Chile, 1994.

328. H. D. Nguyen, I. Yoshihara, K. Yamamori, and M. Yasunaga. A parallel hybrid genetic algorithm for multiple protein sequence alignment. In *Proceedings of the Congress on Evolutionary Computation*, volume 1, pages 309–314, 2002.

329. Z. Z. Nick and P. Themis. Web search using a genetic algorithm. *IEEE Internet Comput.*, 5(2):18–26, 2001.

330. C. Notredame and D. G. Higgins. SAGA: Sequence alignment by genetic algorithm. *Nucleic Acids Research*, 24(8):1515–1524, 1996.

331. C. Notredame, E. A. O'Brien, and D. G. Higgins. RAGA: RNA sequence alignment by genetic algorithm. *Nucleic Acids Research*, 25(22):4570–4580, 1997.

332. M. Nunez. The use of background knowledge in decision tree induction. *Machine Learning*, 6:231–250, 1991.

333. I. M. Oliver, D. J. Smith, and J. R. C. Holland. A study of permutation crossover operators on the traveling salesman problem. In *Proceedings of the 2nd International Conference on Genetic Algorithms*, pages 224–230. Lawrence Erlbaum Associates, Hillsdale, 1987.

334. I. Ono, H. Fujiki, M. Ootsuka, N. Nakashima, N. Ono, and S. Tate. Global optimization of protein 3-dimensional structures in NMR by a genetic algorithm. In *Proceedings of the Congress on Evolutionary Computation*, volume 1, pages 303–308, 2002.

335. C. M. Oshiro, I. D. Kuntz, and J. S. Dixon. Flexible ligand docking using a genetic algorithm. *Journal of Computer-Aided Molecular Design*, 9(2):113–130, 1995.

336. A. Osyczka and S. Kundu. A new method to solve generalized multicriteria optimization problems using the simple genetic algorithms. *Structural Optimization*, 10:94–99, 1995.

337. E. Ozcan and C. K. Mohan. Partial shape matching using genetic algorithms. *Pattern Recognition Letters*, 18:987–992, 1997.

338. T. Özyer, R. Alhajj, and K. Barker. A boosting genetic fuzzy classifier for intrusion detection using data mining techniques for rule pre-screening. *Design and Application of Hybrid Intelligent Systems*, pages 983–992, 2003.

339. M. K. Pakhira, S. Bandyopadhyay, and U. Maulik. Validity index for crisp and fuzzy clusters. *Pattern Recognition*, 37(3):487–501, 2004.

340. N. R. Pal and S. Chakraborty. Fuzzy rule extraction from ID3-type decision trees for real data. *IEEE Transactions on Systems, Man and Cybernetics - B*, 31(5):745–753, 2001.

341. N. R. Pal, S. Chakraborty, and A. Bagchi. RID3: An ID3-like algorithm for real data. *Information Sciences*, 96:271–290, 1997.

342. S. K. Pal. Fuzzy set theoretic measures for automatic feature evaluation - II. *Information Sciences*, 64:165–179, 1992.

343. S. K. Pal, S. Bandyopadhyay, and C. A. Murthy. Genetic algorithms for generation of class boundaries. *IEEE Transactions on Systems, Man and Cybernetics*, 28(6):816–828, 1998.

344. S. K. Pal, S. Bandyopadhyay, and C. A. Murthy. Genetic classifiers for remotely sensed images: Comparison with standard methods. *International Journal of Remote Sensing*, 22(13):2545–2569, 2001.

345. S. K. Pal, S. Bandyopadhyay, and S. S. Ray. Evolutionary computation in bioinformatics: A review. *IEEE Transactions on Systems, Man and Cybernetics, Part C*, 36(2), 2006.

346. S. K. Pal and D. Bhandari. Selection of optimal set of weights in a layered network using genetic algorithms. *Information Sciences*, 80:213–234, 1994.

347. S. K. Pal, D. Bhandari, and M. K. Kundu. Genetic algorithms for optimal image enhancement. *Pattern Recognition Letters*, 15:261–271, 1994.

348. S. K. Pal, S. De, and A. Ghosh. Designing Hopfield type networks using genetic algorithms and its comparison with simulated annealing. *International Journal of Pattern Recognition and Artificial Intelligence*, 11:447–461, 1997.

349. S. K. Pal and S. C. K. Liu. *Foundations of Soft Case-Based Reasoning*. Wiley Series on Intelligent Systems, USA, 2004.

350. S. K. Pal and D. Dutta Majumder. Fuzzy sets and decision making approaches in vowel and speaker recognition. *IEEE Transactions on Systems, Man and Cybernetics*, SMC-7:625–629, 1977.

351. S. K. Pal and D. Dutta Majumder. *Fuzzy Mathematical Approach to Pattern Recognition*. John Wiley, New York, 1986.

352. S. K. Pal and D. P. Mandal. Linguistic recognition system based on approximate reasoning. *Information Sciences*, 61:135–161, 1992.

353. S. K. Pal and S. Mitra. *Neuro-Fuzzy Pattern Recognition: Methods in Soft Computing*. John Wiley and Sons, New York, 1999.

354. S. K. Pal, B. L. Narayan, and S. Dutta. A Web surfer model incorporating topic continuity. *IEEE Transactions on Knowledge and Data Engineering*, 17(5):726–729, May 2005.

355. S. K. Pal, V. Talwar, and P. Mitra. Web mining in soft computing framework: Relevance, state of the art and future directions. *IEEE Transactions Neural Networks*, 13(5):1163–1177, 2002.

356. S. K. Pal and P. P. Wang, editors. *Genetic Algorithms for Pattern Recognition*. CRC Press, Boca Raton, 1996.

357. Y. H. Pao. *Adaptive Pattern Recognition and Neural Networks*. Addison-Wesley, New York, 1989.

358. R. V. Parbhane, S. Unniraman, S. S. Tambe, V. Nagaraja, and B. D. Kulkarni. Optimum DNA curvature using a hybrid approach involving an artificial neural network and genetic algorithm. *Journal of Biomol. Struct. Dyn.*, 17(4):665–672, 2000.

359. R. J. Parsons, S. Forrest, and C. Burks. Genetic algorithms for DNA sequence assembly. In *Proceedings of the 1st International Conference on Intelligent Systems in Molecular Biology*, pages 310–318, 1993.

360. R. J. Parsons, S. Forrest, and C. Burks. Genetic algorithms, operators, and DNA fragment assembly. *Machine Learning*, 21(1-2):11–33, 1995.

361. R. J. Parsons and M. E. Johnson. A case study in experimental design applied to genetic algorithms with applications to DNA sequence assembly. *American Journal of Mathematical and Management Sciences,* 17(3-4): 369–396, 1997.

362. R. J. Parsons and M. E. Johnson. DNA fragment assembly and genetic algorithms. New results and puzzling insights. *International Conference on Intelligent Systems in Molecular Biology*, pages 277–284, AAAI Press, Menlo Park, CA, 1995.

363. A. W. P. Patton, III and E. Goldman. A standard GA approach to native protein conformation prediction. In *Proceedings of the International Conference on Genetic Algorithms*, volume Morgan Kaufmann, pages 574–581, 1995.

364. T. Pavlidis. *Structural Pattern Recognition*. Springer, Berlin Heidelberg New York, 1977.

365. Z. Pawlak. *Rough Sets - Theoretical Aspects of Reasoning about Data*. Kluwer Academic Publishers, Dordrecht, 1991.

366. J. T. Pedersen and J. Moult. Protein folding simulations with genetic algorithms and a detailed molecular description. *Journal of Molecular Biology*, 269(2):240–259, 1997.

367. W. Pedrycz. A fuzzy cognitive structure for pattern recognition. *Pattern Recognition Letters*, 9:305–313, 1989.

368. W. Pedrycz. Fuzzy sets in pattern recognition: Methodology and methods. *Pattern Recognition*, 23:121–146, 1990.

369. W. Pedrycz. Genetic algorithms for learning in fuzzy relational structures. *Fuzzy Sets and Systems*, 69:37–52, 1995.

370. W. Pedrycz. *Computational Intelligence: An Introduction*. CRC Press, USA, 1997.

371. W. Pedrycz, editor. *Fuzzy Evolutionary Computation*. Kluwer Academic Publisher, Boston, 1997.

372. W. Pedrycz and M. Reformat. Genetic optimization with fuzzy coding. In F. Herrera and J. L. Verdegay, editors, *Genetic Algorithms and Soft Computing: Studies in Fuzziness and Soft Computing*, volume 8, pages 51–67. Physica-Verlag, 1996.

373. J. Periaux, M. Sefrioui, and B. Mantel. GA multiple objective optimization strategies for electromagnetic backscattering. In D. Quagliarella, J. Periaux, C. Poloni, and G. Winter, editors, *Genetic algorithms and evolution strategies in engineering and computer science. Recent advances and industrial applications*, pages 225–243. John Wiley and Sons, West Sussex, England, 1997.

374. F. E. Petry, B. P. Petry, and D. H. Kraft. The use of genetic programming to build queries for information retrieval. In T. Bäck, D. Fogel, and Z. Michalewicz, editors, *Handbook of Evolutionary Computation*, pages G2.1:1–G2.1:6. IOP Publishing Ltd., Bristol, UK, 1997.

375. F. Picarougne, N. Monmarché, A. Oliver, and G. Venturini. Web mining with a genetic algorithm. In *Eleventh International World Wide Web Conference*, Honolulu, Hawaii, 7-11 May 2002.

376. J. Piper. Genetic algorithm for applying constraints in chromosome classification. *Pattern Recognition Letters*, 16:857–864, 1995.

377. J. Pitkow. In search of reliable usage data on the WWW. In *Proceedings of the Sixth International WWW conference*, pages 451–463, 1997.

378. C. Poloni and V. Pediroda. GA coupled with computationally expensive simulation: Tools to improve efficiency. In D. Quagliarella, J. Periaux, C. Poloni, and G. Winter, editors, *Genetic Algorithms and Evolution Strategies in Engineering and Computer Science*, pages 267–288. John-Wiley and Sons, Sussex, England, 1997.

379. J. Potvin, D. Dube, and C. Robillard. A hybrid approach to vehicle routing using neural networks and genetic algorithms. *Applied Intelligence*, 6:241–252, 1996.

380. M. Prakash and M. N. Murty. A genetic approach for selection of (near-) optimal subsets of principal components for discrimination. *Pattern Recognition Letters*, 16:781–787, 1995.

381. N. Qian and T. J. Sejnowski. Predicting the secondary structure of globular proteins using neural network models. *Journal Molecular Biology*, 202(4):865–884, 1988.

382. J. Quackenbush. Computational analysis of microarray data. *National Review of Genetics*, 2:418–427, 2001.

383. J. R. Quinlan. Induction of decision trees. *Machine Learning*, 3:81–106, 1986.

384. J. R. Quinlan. *C4.5: Programs for machine learning*. Morgan Kaufmann Publishers, San Mateo, California, 1993.

385. J. R. Quinlan. Improved use of continuous attributes in C4.5. *Journal of Artificial Intelligence Research*, 4:77–90, 1996.

386. A. A. Rabow and H. A. Scheraga. Improved genetic algorithm for the protein folding problem by use of a Cartesian combination operator. *Protein Science*, 5:1800–1815, 1996.

387. N. J. Radcliffe. Forma analysis and random respectful recombination. In R. K. Belew and L. B. Booker, editors, *Proceedings of the 4th International Conference on Genetic Algorithms*, pages 222–229. Morgan Kaufmann, San Mateo, 1991.

388. N. J. Radcliffe. Genetic set recombination. In L. D. Whitley, editor, *Foundations of Genetic Algorithms 2*, pages 203–219. Morgan Kaufmann, San Mateo, 1993.

389. A. Rafiee, M. H. Moradi, and M. R. Farzaneh. Novel genetic-neuro-fuzzy filter for speckle reduction from sonography images. *Journal of Digital Imaging*, 17(4):292–300, 2004.

390. E. Ramat, G. Venturini, C. Lente, and M. Slimane. Solving the multiple resource constrained project scheduling problem with hybrid genetic algorithm. In T. Bäck, editor, *Proceedings of the Seventh International Conference on Genetic Algorithms*, San Mateo, CA, 1997. Morgan Kaufmann.

391. C. L. Ramsey and J. J. Grefenstette. Case-based initialization of genetic algorithms. In Stephanie Forrest, editor, *Proceedings of the Fifth Int. Conf. on Genetic Algorithms*, pages 84–91, San Mateo, CA, 1993. Morgan Kaufmann.

392. T. V. Ravi and K. C. Gowda. An ISODATA clustering procedure for symbolic objects using a distributed genetic algorithm. *Pattern Recognition Letters*, 20:659–666, 1999.

393. L. A. Rendell. A doubly layered genetic penetrance learning system. In *Proceedings of the National Conference on Artificial Intelligence*, pages 343–347, Washington, DC, 1983.

394. J. A. Richards. *Remote Sensing Digital Image Analysis: An Introduction.* Springer, Berlin Heidelberg New York, 1993.

395. S. K. Riis and A. Krogh. Improving prediction of protein secondary structure using structured neural networks and multiple sequence alignments. *Journal of Computational Biology*, 3:163–183, 1996.

396. R. Riolo. Modelling simple human category learning with a classifier system. In R. K. Belew and L. B. Booker, editors, *Proceedings of the 4th International Conference on Genetic Algorithms*, pages 324–333. Morgan Kaufmann, San Mateo, 1991.

397. S. G. Romaniuk. Learning to learn with evolutionary growth perceptrons. In S. K. Pal and P. P. Wang, editors, *Genetic Algorithms for Pattern Recognition*, pages 179–211. CRC Press, Boca Raton, 1996.

398. G. H. Rosenfield and K. Fitzpatrik-Lins. Coefficient of agreement as a measure of thematic classification accuracy. *Photogrametic Engineering and Remote Sensing*, 52:223–227, 1986.

399. C. D. Rosin, R. S. Halliday, W. E. Hart, and R. K. Belew. A comparison of global and local search methods in drug docking. In *Proceedings of the International Conference on Genetic Algorithms*, pages 221–228, 1997.

400. U. Rost and P. Oechtering. Knowledge-based genetic learning. In *Proceedings of the Sixth Scandinavian Conference on Artificial Intelligence (SCAI 97)*, pages 107–118, 1997.

401. D. W. Ruck, S. K. Rogers, and M. Kabrisky. Feature selection using a multilayer perceptron. *Journal of Neural Network Computing*, pages 40–48, 1990.

402. D. W. Ruck, S. K. Rogers, M. Kabrisky, M. E. Oxley, and B. W. Suter. The multilayer perceptron as an approximation to a Bayes optimal discriminant function. *IEEE Transactions on Neural Networks*, 1:436–438, 1990.

403. G. Rudolph. Convergence analysis of canonical genetic algorithms. *IEEE Transactions on Neural Networks*, 5:96–101, 1994.

404. X. Rui and D. Wunsch II. Survey of clustering algorithms. *IEEE Transactions on Neural Networks*, 16(3):645– 678, 2005.

405. D. E. Rumelhart, G. E. Hinton, and R. J. Williams. Learning internal representation by error propagation. In D. E. Rumelhart and J. L. McClelland, editors, *Parallel Distributed Processing: Explorations in the Microstructures of Cognition*, volume 1, pages 318–362. MIT Press, Cambridge, 1986.

406. D. E. Rumelhart, J. McClelland, and the PDP Research Group. *Parallel Distributed Processing: Explorations in the Microstructure of Cognition*, volume 1 and 2. MIT Press, Cambridge, 1986.

407. S. Saha and J. P. Christensen. Genetic design of sparse feedforward neural networks. *Information Sciences*, 79:191–200, 1994.

408. A. Salamov and V. Solovyev. Prediction of protein secondary structure by combining nearest-neighbor algorithms and multiple sequence alignments. *Journal of Molecular Biology*, 247:11–15, 1995.

409. S. Salzberg and S. Cost. Predicting protein secondary structure with a nearest-neighbor algorithm. *Journal of Molecular Biology*, 227:371–374, 1992.

410. E. Sanchez, T. Shibata, and L. A. Zadeh, editors. *Genetic Algorithms and Fuzzy Logic Systems: Soft Computing Perspectives*. World Scientific, 1997.

411. M. Sarkar and B. Yegnanarayana. An evolutionary programming-based probabilistic neural networks construction technique. In *Proceedings of the IEEE International Conference on Neural Networks*, pages 456–461, Houston, 1997.

412. M. Sarkar and B. Yegnanarayana. Feedforward neural networks configuration using evolutionary programming. In *Proceedings of the IEEE International Conference on Neural Networks*, pages 438–443, Houston, 1997.

413. M. Sarkar, B. Yegnanarayana, and D. Khemani. A clustering algorithm using an evolutionary programming-based approach. *Pattern Recognition Letters*, 18:975–986, 1997.

414. P. Saxena, I. Whang, Y. Voziyanov, C. Harkey, P. Argos, M. Jayaram, and T. Dandekar. Probing flp: A new approach to analyze the structure of a DNA recognizing protein by combining the genetic algorithm, mutagenesis and non-canonical DNA target sites. *Biochim. Biophys. Acta*, 1340(2):187–204, 1997.

415. J. D. Schaffer. Multiple objective optimization with vector evaluated genetic algorithms. In J. J. Grefenstette, editor, *Genetic Algorithms and Their Applications: Proceedings of the First International Conference on Genetic Algorithms*, pages 93–100. Lawrence Erlbaum, 1985.

416. J. D. Schaffer, R. Caruana, L. J. Eshelman, and R. Das. A study of control parameters affecting online performance of genetic algorithms. In L. Davis, editor, *Genetic Algorithms and Simulated Annealing*, pages 89–103. Pitman, London, 1987.

417. J. D. Schaffer, R. A. Caruana, and L. J. Eshelman. Using genetic search to exploit the emergent behavior of neural networks. *Physica D*, 42:244–248, 1990.

418. J. D. Schaffer and L. J. Eshelman. On crossover as an evolutionarily viable strategy. In R. K. Belew and L. B. Booker, editors, *Proceedings of the 4th International Conference on Genetic Algorithms*, pages 61–68. Morgan Kaufmann, San Mateo, 1991.

419. J. D. Schaffer and J. J. Grefenstette. Multi-objective learning via genetic algorithms. In *Proceedings of the Ninth International Joint Conference on Artificial Intelligence*, pages 593–595. Morgan Kaufmann, 1985.

420. R. Schalkhoff. *Pattern Recognition, Statistical, Structural and Neural Approaches*. John Wiley and Sons, New York, 1992.

421. J. R. Schott. *Fault tolerant design using single and multi-criteria genetic algorithms*. PhD thesis, Dept. of Aeronautics and Astronautics, Massachusetts Institute of Technology, Boston, MA, 1995.

422. S. Schulze-Kremer. Genetic algorithms and protein folding. Methods in molecular biology. *Protein Structure Prediction: Methods and Protocols*, 143:175–222, 2000.

423. H.-P. Schwefel. Collective phenomena in evolutionary systems. In *31st Ann. Meet. Intl. Soc. for General System Research*, pages 1025–1033, Budapest, 1987.

424. G. Seetharaman, A. Narasimahan, and L. Stor. Image segmentation with genetic algorithms: A formulation and implementation. In *Proceedings of the SPIE Conference on Stochastics and Neural Methods in Signal Processing and Computer Vision*, volume 1569, San Diego, 1991.

425. S. Z. Selim and M. A. Ismail. K-means type algorithms: A generalized convergence theorem and characterization of local optimality. *IEEE Transactions on Pattern Analysis and Machine Intelligence*, 6:81–87, 1984.

426. J. Setubal and J. Meidanis. *Introduction to Computational Molecular Biology.* International Thomson Publishing, Boston, MA, 1999.
427. G. Shafer. *A Mathematical Theory of Evidence.* Princeton University Press, Princeton, 1976.
428. B. A. Shapiro and J. Navetta. A massively parallel genetic algorithm for RNA secondary structure prediction. *Journal of Supercomputing,* 8:195–207, 1994.
429. B. A. Shapiro and J. C. Wu. An annealing mutation operator in the genetic algorithms for RNA folding. *Computer Applications in the Biosciences,* 12:171–180, 1996.
430. B. A. Shapiro, J. C. Wu, D. Bengali, and M. J. Potts. The massively parallel genetic algorithm for RNA folding: MIMD implementation and population variation. *Bioinformatics,* 17(2):137–148, 2001.
431. K. Shimojima, T. Fukuda, and Y. Hasegawa. Self-tuning fuzzy modelling with adaptive membership function, rules and hierarchical structure based on genetic algorithm. *Fuzzy Sets and Systems,* 71:295–309, 1995.
432. A. N. Shiryayev. *Optimal Stopping Rules.* Springer, Berlin Heidelberg New York, 1977.
433. M. Shokouhi, P. Chubak, and Z. Raeesy. Enhancing focused crawling with genetic algorithms. In *International Conference on Information Technology: Coding and Computing (ITCC'05),* volume II, pages 503–508, 2005.
434. W. Siedlecki and J. Sklansky. A note on genetic algorithms for large-scale feature selection. *Pattern Recognition Letters,* 10:335–347, 1989.
435. R. Sikora and M. Shaw. A double layered genetic approach to acquiring rules for classification: Integrating genetic algorithms with similarity based learning. *ORSA Journal of Computing,* 6:174–187, 1994.
436. A. Skourikhine. Phylogenetic tree reconstruction using self-adaptive genetic algorithm. In *IEEE International Symposium on Bioinformatics and Biomedical Engineering,* pages 129–134, 2000.
437. R. Smith and C. Bonacina. Mating restriction and niching pressure: Results from agents and implications for general EC. In *Proceedings of the Genetic and Evolutionary Computation Conference,* pages 1382–1393, Berlin Heidelberg New York, 2003. Springer.
438. R. E. Smith, S. Forrest, and A. S. Perelson. Population diversity in an immune system model: Implications for genetic search. In L. D. Whitley, editor, *Foundations of Genetic Algorithms 2,* pages 153–165. Morgan Kaufmann, San Mateo, 1993.
439. S. F. Smith. *A Learning System Based on Genetic Algorithms.* PhD thesis, University of Pittsburg, PA, 1980.
440. T. F. Smith and M. S. Waterman. Identification of common molecular sequences. *Journal of Molecular Biology,* 147:195–197, 1981.
441. H. Spath. *Cluster Analysis Algorithms.* Ellis Horwood, Chichester, UK, 1989.
442. W. M. Spears. Crossover or mutation? In L. D. Whitley, editor, *Foundations of Genetic Algorithms 2,* pages 221–237. Morgan Kaufmann, San Mateo, 1993.
443. W. M. Spears and K. De Jong. An analysis of multi-point crossover. In G. J. E. Rawlins, editor, *Foundations of Genetic Algorithms,* pages 301–315. Morgan Kaufmann, San Mateo, 1991.
444. W. M. Spears and K. De Jong. On the virtues of parameterized uniform crossover. In R. K. Belew and L. B. Booker, editors, *Proceedings of the 4th International Conference on Genetic Algorithms,* pages 230–236. Morgan Kaufmann, San Mateo, 1991.

445. C. M. Sperberg-McQueen. Web services and W3C, 2003.

446. E. Spertus. ParaSite: mining structural information on the web. In *Selected papers from the sixth international conference on World Wide Web*, pages 1205–1215, Essex, UK, 1997. Elsevier Science Publishers Ltd.

447. A. Spink, D. Wolfram, M. B. J. Jansen, and T. Saracevic. Searching the Web: The public and their queries. *Journal of the American Society for Information Science and Technology*, 52(3):226–234, 2001.

448. R. Srikanth, R. George, N. Warsi, D. Prabhu, F. E. Petry, and B. P. Buckles. A variable-length genetic algorithm for clustering and classification. *Pattern Recognition Letters*, 16:789–800, 1995.

449. M. Srinivas and L. M. Patnaik. Binomially distributed populations for modelling genetic algorithms. In *Proceedings of the 5th International Conference on Genetic Algorithms*, pages 138–145. Morgan Kaufmann, San Mateo, 1993.

450. M. Srinivas and L. M. Patnaik. Adaptive probabilities of crossover and mutation in genetic algorithm. *IEEE Transactions on Systems, Man and Cybernetics*, 24:656–667, 1994.

451. Z. Sun, X. Xia, Q. Guo, and D. Xu. Protein structure prediction in a 210-type lattice model: Parameter optimization in the genetic algorithm using orthogonal array. *Journal of Protein Chemistry*, 18(1):39–46, 1999.

452. J. Suzuki. A Markov chain analysis on simple genetic algorithms. *IEEE Transactions on Systems, Man and Cybernetics*, 25:655–659, 1995.

453. G. Syswerda. Uniform crossover in genetic algorithms. In J. D. Schaffer, editor, *Proceedings of the 3rd International Conference on Genetic Algorithms*, pages 2–9. Morgan Kaufmann, San Mateo, 1989.

454. J. D. Szustakowski and Z. Weng. Protein structure alignment using a genetic algorithm. *Proteins*, 38(4):428–440, 2000.

455. K. Tai and S. Akhtar. Structural topology optimization using a genetic algorithm with a morphological geometric representation scheme. *Structural and Multidisciplinary Optimization*, 30(2):113–127, 2005.

456. K. C. Tan, E. F. Khor, T. H. Lee, and Y. J. Yang. A tabu-based exploratory evolutionary algorithm for multiobjective optimization. *Artif. Intell. Rev.*, 19(3):231–260, 2003.

457. A. Tettamanzi. Evolutionary algorithms and fuzzy logic: a two-way integration. In *2nd Joint Conference on Information Sciences*, pages 464–467, Wrightsville Beach, NC, 1995.

458. S. Theodoridis and K. Koutroumbas. *Pattern Recognition*. Academic Press, 1999.

459. J. D. Thompson, D. G. Higgins, and T. J. Gibson. CLUSTAL W: Improving the sensitivity of progressive multiple sequence alignment through sequence weighting, position-specific gap penalties and weight matrix choice. *Nucleic Acids Research*, 22:4673–4680, 1994.

460. D. Tominaga, M. Okamoto, Y. Maki, S. Watanabe, and Y. Eguchi. Nonlinear numerical optimization technique based on a genetic algorithm for inverse problems: Towards the inference of genetic networks. In *Proceedings of the German Conference on Bioinformatics (Computer Science and Biology)*, pages 127–140, 1999.

461. J. T. Tou and R. C. Gonzalez. *Pattern Recognition Principles*. Addison-Wesley, Reading, 1974.

462. H. K. Tsai, J. M. Yang, and C. Y. Kao. Applying genetic algorithms to finding the optimal order in displaying the microarray data. In *Proceedings of the Genetic and Evolutionary Computation Conference (GECCO)*, pages 610–617, 2002.

463. H. K. Tsai, J. M. Yang, Y. F. Tsai, and C. Y. Kao. An evolutionary approach for gene expression patterns. *IEEE Transactions on Information Technology in Biomedicine*, 8(2):69–78, 2004.

464. W. H. Tsai and S. S. Yu. Attributed string matching with merging for shape recognition. *IEEE Transactions on Pattern Analysis and Machine Intelligence*, 7:453–462, 1985.

465. L. Tseng and S. Yang. Genetic algorithms for clustering, feature selection, and classification. In *Proceedings of the IEEE International Conference on Neural Networks*, pages 1612–1616, Houston, 1997.

466. P. D. Turney. Cost sensitive classification: Empirical evaluation of a hybrid genetic decision tree induction algorithm. *Journal of Artificial Intelligence Research*, 2:369–409, 1995.

467. R. Unger and J. Moult. Genetic algorithms for protein folding simulations. *Journal of Molecular Biology*, 231(1):75–81, 1993.

468. R. Unger and J. Moult. A genetic algorithms for three dimensional protein folding simulations. In *Proceedings of the International Conference on Genetic Algorithms*, pages 581–588. Morgan Kaufmann, 1993.

469. R. Unger and J. Moult. On the applicability of genetic algorithms to protein folding. In *Proceedings of the Hawaii International Conference on System Sciences*, volume 1, pages 715–725, 1993.

470. M. Valenzuela-Rendon and E. Uresti-Charre. A non-generational genetic algorithm for multiobjective optimization. In T. Bäck, editor, *Proceedings of the Seventh International Conference on Genetic Algorithms*, pages 658–665. Morgan Kaufmann, San Mateo, California, 1997.

471. P. J. M. van Laarhoven and E. H. L. Aarts. *Simulated Annealing: Theory and Applications*. Kluwer Academic Publishers, MA, USA, 1987.

472. D. V. Veldhuizen. *Multiobjective Evolutionary Algorithms: Classification, Analyses, and New Innovations*. PhD thesis, Air Force Institute of Technology, Dayton, OH, 1999.

473. V. Vemuri and W. Cedeno. Industrial applications of genetic algorithms. *International Journal of Knowledge-based Intelligent Engineering Systems*, 1:1–12, 1997.

474. V. Venkatasubramanian, K. Chan, and J. Caruthers. Computer aided molecular design using genetic algorithms. *Comp. and Chemical Eng.*, 18(9):833–844, 1994.

475. M. D. Vose. *The Simple Genetic Algorithm: Foundations and Theory*. The MIT Press, Cambridge, MA, 1999.

476. M. D. Vose and G. E. Liepins. Punctuated equilibria in genetic search. *Complex Systems*, 5:31–44, 1991.

477. M. D. Vose and A. H. Wright. The Walsh transform and the theory of simple genetic algorithm. In S. K. Pal and P. P. Wang, editors, *Genetic Algorithms for Pattern Recognition*, pages 25–44. CRC Press, Boca Raton, 1996.

478. D. Vrajitoru. Crossover improvement for the genetic algorithm in information retrieval. *Information Processing and Management*, 34(4):405–415, 1998.

479. D. Vrajitoru. Simulating gender separation with genetic algorithms. In *Proceedings of the Genetic and Evolutionary Computation Conference (GECCO)*, 2002.

480. B. W. Wah, A. Ieumwananonthachai, and Y. Li. Generalization of heuristics learned in genetics-based learning. In S. K. Pal and P. P. Wang, editors, *Genetic Algorithms for Pattern Recognition*, pages 87–126. CRC Press, Boca Raton, 1996.

481. L. H. Wang, C. Kao, M. Ouh-Young, and W. Chen. Molecular binding: A case study of the population-based annealing genetic algorithms. *IEEE International Conference on Evolutionary Computation*, pages 50–55, 1995.

482. L. H. Wang, C. Kao, M. Ouh-Young, and W. C. Cheu. Using an annealing genetic algorithm to solve global energy minimization problem in molecular binding. In *Proceedings of the Sixth International Conference on Tools with Artificial Intelligence*, pages 404–410, 1994.

483. S. Y. Wang and K. Tai. Structural topology design optimization using genetic algorithms with a bit-array representation. *Computer Methods in Applied Mechanics and Engineering*, 194:3749–3770, 2005.

484. S. Y. Wang, K. Tai, and M. Y. Wang. An enhanced genetic algorithm for structural topology optimization. *International Journal for Numerical Methods in Engineering*, 65:18–44, 2006.

485. Y. Wang, K. Fan, and J. Horng. Genetic-based search for error-correcting graph isomorphism. *IEEE Transactions on Systems, Man and Cybernetics*, 27:588–597, 1997.

486. M. Waterman. RNA structure prediction. In *Methods in Enzymology*, volume 164. Academic Press, USA, 1988.

487. R. A. Watson and J. B. Pollack. Incremental commitment in genetic algorithms. In Wolfgang Banzhaf, Jason Daida, Agoston E. Eiben, Max H. Garzon, Vasant Honavar, Mark Jakiela, and Robert E. Smith, editors, *Proceedings of the Genetic and Evolutionary Computation Conference*, volume 1, pages 710–717, Orlando, Florida, USA, 13-17 1999. Morgan Kaufmann.

488. A. Webb. *Statistical Pattern Recognition*. John Wiley & Sons, New York, second edition, 2002.

489. D. R. Westhead, D. E. Clark, D. Frenkel, J. Li, C. W. Murray, B. Robson, and B. Waszkowycz. PRO-LIGAND: An approach to de novo molecular design. 3. A genetic algorithm for structure refinement. *Journal of Computer Aided Molecule Design*, 9(2):139–148, 1995.

490. D. Whitley and T. Starkweather. GENITOR-II: A distributed genetic algorithm. *Journal of Experimental and Theoretical Artificial Intelligence*, 2:189–214, 1990.

491. D. Whitley, T. Starkweather, and C. Bogart. Genetic algorithms and neural networks: Optimizing connections and connectivity. *Parallel Computing*, 14:347–361, 1990.

492. L. D. Whitley. Fundamental principles of deception in genetic search. In G. J. E. Rawlins, editor, *Foundations of Genetic Algorithms*, pages 221–241. Morgan Kaufmann, San Mateo, 1991.

493. L. D. Whitley, K. Mathias, and P. Fitzhorn. Delta coding: An iterative strategy for genetic algorithms. In *Proceedings of the 4th International Conference on Genetic Algorithms*, pages 77–84. Morgan Kaufmann, San Mateo, 1991.

494. K. C. Wiese and E. Glen. A permutation-based genetic algorithm for the RNA folding problem: A critical look at selection strategies, crossover operators, and representation issues. *Biosystems*, 72(1-2):29–41, 2003.

495. P. Willet. Genetic algorithms in molecular recognition and design. *Trends in Biotechnology*, 13(12):516–521, 1995.

496. G. Winter, J. Periaux, M. Galan, and P. Cuesta, editors. *Genetic Algorithms in Engineering and Computer Science*. John Wiley and Sons, Chichester, 1995.

497. A. S. Wu and I. Garibay. The proportional genetic algorithm: Gene expression in a genetic algorithm. *Genetic Programming and Evolvable Hardware*, 3(2):157–192, 2002.

498. A. S. Wu and I. Garibay. Intelligent automated control of life support systems using proportional representations. *IEEE Transactions on Systems, Man and Cybernetics Part B*, 34(3):1423–1434, 2003.

499. A. S. Wu and R. K. Lindsay. A survey of intron research in genetics. In *Proceedings of the 4th Conference on Parallel Problem Solving from Nature*, Lecture Notes in Computer Science, pages 101–110, Berlin Heidelberg New York, 1996. Springer.

500. K. Wu and M. Yang. A cluster validity index for fuzzy clustering. *Pattern Recogn. Lett.*, 26(9):1275–1291, 2005.

501. Y. Xiao and D. Williams. Genetic algorithms for docking of Actinomycin D and Deoxyguanosine molecules with comparison to the crystal structure of Actinomycin D-Deoxyguanosine complex. *Journal of Physical Chemistry*, 98:7191–7200, 1994.

502. H. Xue, M. Jamshidi, and E. Tunstel. Genetic algorithms for optimization of fuzzy systems in prediction and control. *International Journal of Knowledge-based Intelligent Engineering System*, 1:13–21, 1997.

503. S. M. Yamany, K. J. Khiani, and A. A. Farag. Application of neural networks and genetic algorithms in the classification of endothelial cells. *Pattern Recognition Letters*, 18(11-13):1205–1210, 1997.

504. J. Yang and R. Korfhage. Query modification using genetic algorithms in vector space models. Technical Report TR LIS045/I592001, University of Pittsburgh, 1992.

505. J. Yang and R. Korfhage. Query optimization in information retrieval using genetic algorithms. In *Proceedings of the 5th International Conference on Genetic Algorithms*, pages 603–613, San Francisco, CA, USA, 1993. Morgan Kaufmann Publishers Inc.

506. J. M. Yang and C. Y. Kao. A family competition evolutionary algorithm for automated docking of flexible ligands to proteins. *IEEE Transactions on Information Technology in Biomedicine*, 4(3):225–237, 2000.

507. T. Yokoyama, T. Watanabe, A. Taneda, and T. Shimizu. A Web server for multiple sequence alignment using genetic algorithm. *Genome Informatics*, 12:382–383, 2001.

508. Y. Yuan and H. Zhuang. A genetic algorithm for generating fuzzy classification rules. *Fuzzy Sets and Systems*, 84(1):1–19, 1996.

509. L. A. Zadeh. Fuzzy sets. *Information and Control*, 8:338–353, 1965.

510. O. Zaïane, J. Srivastava, M. Spiliopoulou, and B. Masand, editors. *WEBKDD 2002 – Mining Web Data for Discovering Usage Patterns and Profiles: 4th International Workshop*, volume 2703 of *Lecture Notes in Artificial Intelligence*. Springer, Berlin Heidelberg New York, 2003.

511. C. Zhang. A genetic algorithm for molecular sequence comparison. *IEEE International Conference on Systems, Man, and Cybernetics*, 2:1926–1931, 1994.

512. C. Zhang and A. K. C. Wong. A genetic algorithm for multiple molecular sequence alignment. *Bioinformatics*, 13:565–581, 1997.

513. C. Zhang and A. K. C. Wong. Toward efficient multiple molecular sequence alignment: A system of genetic algorithm and dynamic programming. *IEEE Transactions on Systems, Man, and Cybernetics: B*, 27(6):918–932, 1997.

514. C. Zhang and A. K. C. Wong. A technique of genetic algorithm and sequence synthesis for multiple molecular sequence alignment. *IEEE International Conference on Systems, Man, and Cybernetics*, 3:2442–2447, 1998.

515. T. Zhang, R. Ramakrishnan, and M. Livny. Birch: An efficient data clustering method for very large databases. In *Proceedings of the 1996 ACM SIGMOD International Conference on Management of Data*, pages 103–114. ACM Press, 1996.

516. Q. Zhao and T. Higuchi. Efficient learning of NN-MLP based on individual evolutionary algorithm. *Neurocomputing*, 13:201–215, 1996.

517. Q. Zhao and T. Higuchi. Minimization of nearest neighbor classifiers based on individual evolutionary algorithm. *Pattern Recognition Letters*, 17:125–131, 1996.

518. Z. Y. Zhu and K. S. Leung. Asynchronous self-adjustable island genetic algorithm for multi-objective optimization problems. In *Proceedings of the Congress on Evolutionary Computation*, pages 837–842. IEEE Press, 2002.

519. E. Zitzler. *Evolutionary algorithms for multiobjective optimization: Methods and Applications*. PhD thesis, Swiss Federal Institute of Technology (ETH), Zurich, Switzerland, 1999.

520. E. Zitzler, K. Deb, and L. Thiele. Comparison of multiobjective evolutionary algorithms: Empirical results. *Evolutionary Computation*, 8(2):173–195, 2000.

521. E. Zitzler, M. Laumanns, and L. Thiele. SPEA2: Improving the Strength Pareto Evolutionary Algorithm. Technical Report 103, Gloriastrasse 35, CH-8092 Zurich, Switzerland, 2001.

522. E. Zitzler and L. Thiele. Multiobjective evolutionary algorithms: A comparative case study and the strength Pareto approach. *IEEE Transactions on Evolutionary Computation*, 3(4):257–271, November 1999.

523. M. Zuker and P. Stiegler. Optimal computer folding of large RNA sequences using thermo-dynamics and auxiliary information. *Nucleic Acids Research*, 9:133–148, 1981.

Index

Natural Computing Series

W.M. Spears: **Evolutionary Algorithms. The Role of Mutation and Recombination.** XIV, 222 pages, 55 figs., 23 tables. 2000

H.-G. Beyer: **The Theory of Evolution Strategies.** XIX, 380 pages, 52 figs., 9 tables. 2001

L. Kallel, B. Naudts, A. Rogers (Eds.): **Theoretical Aspects of Evolutionary Computing.** X, 497 pages. 2001

G. Păun: **Membrane Computing. An Introduction.** XI, 429 pages, 37 figs., 5 tables. 2002

A.A. Freitas: **Data Mining and Knowledge Discovery with Evolutionary Algorithms.** XIV, 264 pages, 74 figs., 10 tables. 2002

H.-P. Schwefel, I. Wegener, K. Weinert (Eds.): **Advances in Computational Intelligence. Theory and Practice.** VIII, 325 pages. 2003

A. Ghosh, S. Tsutsui (Eds.): **Advances in Evolutionary Computing. Theory and Applications.** XVI, 1006 pages. 2003

L.F. Landweber, E. Winfree (Eds.): **Evolution as Computation.** DIMACS Workshop, Princeton, January 1999. XV, 332 pages. 2002

M. Hirvensalo: **Quantum Computing.** 2nd ed., XI, 214 pages. 2004 (first edition published in the series)

A.E. Eiben, J.E. Smith: **Introduction to Evolutionary Computing.** XV, 299 pages. 2003

A. Ehrenfeucht, T. Harju, I. Petre, D.M. Prescott, G. Rozenberg: **Computation in Living Cells. Gene Assembly in Ciliates.** XIV, 202 pages. 2004

L. Sekanina: **Evolvable Components. From Theory to Hardware Implementations.** XVI, 194 pages. 2004

G. Ciobanu, G. Rozenberg (Eds.): **Modelling in Molecular Biology.** X, 310 pages. 2004

R.W. Morrison: **Designing Evolutionary Algorithms for Dynamic Environments.** XII, 148 pages, 78 figs. 2004

R. Paton[†], H. Bolouri, M. Holcombe, J.H. Parish, R. Tateson (Eds.): **Computation in Cells and Tissues. Perspectives and Tools of Thought.** XIV, 358 pages, 134 figs. 2004

M. Amos: **Theoretical and Experimental DNA Computation.** XIV, 170 pages, 78 figs. 2005

M. Tomassini: **Spatially Structured Evolutionary Algorithms.** XIV, 192 pages, 91 figs., 21 tables. 2005

G. Ciobanu, G. Păun, M.J. Pérez-Jiménez (Eds.): **Applications of Membrane Computing.** X, 441 pages, 99 figs., 24 tables. 2006

K.V. Price, R.M. Storn, J.A. Lampinen: **Differential Evolution.** XX, 538 pages, 292 figs., 48 tables and CD-ROM. 2006

J. Chen, N. Jonoska, G. Rozenberg: **Nanotechnology: Science and Computation.** XII, 385 pages, 126 figs., 10 tables. 2006

A. Brabazon, M. O'Neill: **Biologically Inspired Algorithms for Financial Modelling.** XVI, 275 pages, 92 figs., 39 tables. 2006

T. Bartz-Beielstein: **Experimental Research in Evolutionary Computation.** XIV, 214 pages, 66 figs., 36 tables. 2006

S. Bandyopadhyay, S.K. Pal: **Classification and Learning Using Genetic Algorithms.** 328 pages, 87 figs., 43 tables. 2007